Matrices and Tensors in Physics

By the same author

Elements of Group Theory for Physicists

MATRICES and TENSORS in Physics

A. W. JOSHI
Meerut University Institute of Advanced Studies

A HALSTED PRESS BOOK

JOHN WILEY & SONS
New York London Sydney Toronto

Reprinted 1977

Published in the U.S.A., Canada, Latin America
and the Middle East by Halsted Press,
a Division of John Wiley & Sons, Inc., New York.

Library of Congress Cataloging in Publication Data

Joshi, A W
Matrices and tensors in physics.
"A Halsted Press book."
Includes bibliographies.

1. Matrices. 2. Calculus of tensors. 3. Mathematical physics. I. Title.
QC20.7.M3J67 1975 530.1′22 75-26772

ISBN 0-470-45086-X

Printed in India at

Raj Bandhu Industrial Co., New Delhi-64.

To my grandfather

from whom

I have inherited the most

Preface

"Through understanding of what, pray, does all this world become understood, Sir?"
Muṇḍaka Upaniṣad, 1-1-3

The present century has seen the development of modern physics along two major directions—the theories of relativity and quantum mechanics. For a proper understanding of the basic concepts of quantum mechanics, a sound foundation in matrix algebra and vector spaces is essential. Similarly, tensor analysis and differential geometry provide the mathematical foundation for the theories of relativity. It is from this point of view that matrix algebra and tensor analysis have been selected for presentation in this book.

Matrices and tensors are generally taught to B.Sc. and M.Sc. (physics) students in Indian universities in a course on mathematics or mathematical methods in physics. Matrices are generally taken up directly without recourse to vector spaces and transformations. However, the connecting link between matrices and quantum mechanics is vector spaces, and it is often found that students fail to grasp the fundamental principles of quantum mechanics because of a lack of understanding of vector spaces. The first chapter of this book starts by introducing vector spaces and transformations, and matrices arising from it. This also gives the proper historical perspective because the study of matrices arose in connection with transformations in vector spaces. Emphasis is given to important concepts needed in quantum mechanics, such as linear combinations, linear dependence and independence of vectors and matrices, number of independent parameters of various special matrices, the most general matrix of a given type, etc. Section 9 contains an improved proof of the theorem that two matrices can be simultaneously diagonalized if and only if they commute with each other. Topics such as nondiagonalizable matrices and functions of a matrix are also included. Infinite (discrete and continuous) matrices are also discussed briefly at the end of chapter I.

The second chapter deals with the algebra and the calculus of general tensors in an *N*-dimensional Riemannian space. Many physical examples are given showing the necessity of the use of second and higher rank tensors. These include the conductivity tensor, the effective mass tensor, stress and strain tensors, tensors of elastic stiffness and compliance constants, piezoelectric strain coefficient tenser, dielectric susceptibility tensor, moment of inertia tensor, curvature tensor, etc. In this chapter, emphasis is given on the proper use of free and dummy indices in tensor equations. Einstein's summation convention is explained in detail and common errors arising in its use are pointed out. F ules are given to check the correctness of indices in a tensor equation.

Although it is desirable to have an elementary knowledge of vector spaces for a better understanding of the use of matrices and of the eigenvalue problem, the matter is so arranged that a reader wishing to proceed straight to matrices without going through vector spaces may skip Section 1 without much loss of continuity. However, it is felt that this is not the correct approach.

There are two appendices in the book. Appendix I discusses the logic of the phrases such as "if", "only if", and "if and only if". Appendix II gives a proof of the result that a finite nonsingular matrix (that is, one possessing both left and right inverses) must be square.

Altogether about 100 solved examples are included, and about 200 exercises are given in the book. Many of the exercises anticipate what is to come in the succeeding sections. Answers to selected exercises are given at the end of the book and such exercises are marked with asterisks.

The book has developed from courses given to graduate and postgraduate students during the past six years. Students' problems and queries have helped a lot in the preparation and presentation of the material. I am also thankful to my colleagues and senior students for their help during the preparation of the manuscript. Suggestions and comments from readers will be welcome.

A. W. JOSHI

June, 1975

Contents

Set theoretical notation

$\in$: belongs to, belonging to
$\ni$: such that
$\exists$: there exists
$\forall$: for every, for all
$\subset$: is contained in
$\supset$: contains
$\Rightarrow$: implies, only if
$\Leftarrow$: is implied by, if, follows from
$\Leftrightarrow$: implies and is impli d by, if and only if
$\not\Rightarrow$: does not imply
$\not\Leftarrow$: is not implied by, does not follow from

EXAMPLE: An expression such as

$$\exists\, e \in G \ni xe = ex = x \;\forall\, x \in G$$

means that "there exists an element e belonging to G such that $xe = ex = x$ for all x belonging to G".

I Matrix Algebra

Matrices occur in physics mainly in two ways: first, in the solution of systems of linear equations, and second, in the solution of eigenvalue problems in classical and quantum mechanics. Both these types of problems arise, in turn, from transformations of vectors in vector spaces and the operation of linear operators on vector spaces. The methods of matrix algebra are not only useful but essential in handling such problems. In this chapter, we shall discuss various operations with matrices and different situations in which they can be applied.

1. Vector spaces and transformations

We shall begin by discussing how matrices arise in connection with vector spaces. For the sake of completeness, it will be convenient to define a *group* and a *field* before coming to vector spaces.

Group

Consider the set $R=\{x: x=\text{a real number}\}$, the set of all real numbers. The elements of R, endowed with a law of addition, satisfy four properties: (a) for any two real numbers x and y belonging to R, their sum $x+y$ belongs to R; (b) there exists an element 0 in R, called *zero*, such that $x+0=0+x=x$ for every $x\in R$; (c) for every $x\in R$, there exists a unique element y in R such that $x+y=y+x=0$; (d) for any three elements x, y and z of R, the associative law is satisfied, that is, $x+(y+z)=(x+y)+z$.

Similarly, consider another set, $U(1)=\{z: |z|=1\}$, the set of all complex numbers of unit magnitude. The elements of $U(1)$, endowed with a law of multiplication, satisfy four properties: (a) for any two complex numbers of unit magnitude x and y, their product is also a complex numbers of unit magnitude and hence belongs to $U(1)$; (b) there exists an element $1+0i$, where $i=\sqrt{-1}$, or 1 for short, in $U(1)$ such that $x1=1x=x$ for every element x of $U(1)$; (c) for every $x\in U(1)$, there exists a unique element y in $U(1)$ such that $xy=yx=1$; (d) for any three elements x, y and z of $U(1)$, the associative law is satisfied, that is, $x(yz)=(xy)z$.

We notice that the four properties satisfied by the two sets are similar in nature. In fact, both the sets considered above are examples of a group.

In general, a *group* is defined as a set $G=\{x, y, z\dots\}$ endowed with a binary law of composition for its elements such that it satisfies the four properties listed below by using the set theoretical notation:

(a) Closure: $xy\in G\,\forall x, y\in G$;
(b) Existence of Identity: $\exists e\in G \ni xe=ex=x\,\forall x\in G$;
(c) Existence of Inverse: $\forall x\in G\,\exists y\in G\ni xy=yx=e$; (1)
(d) Associative Property: $x(yz)=(xy)z\,\forall x, y, z\in G$.

These are known as the *group axioms*. The element e satisfying property (b) is known as the *identity element* of the group, and if $xy=yx=e$ [property (c)], y is known as the *inverse* of x and vice versa.

It is not necessary that the law of composition of group elements be commutative. In fact, in general, it is not, so that $xy \neq yx \forall x, y \in G$, where G is a group. If, in particular, $xy = yx \forall x, y \in G$, G is said to be an *abelian group*. Both the examples discussed above are abelian groups.

Field

A *field* $F = \{a, b, c, d, \ldots\}$ is a set of elements, endowed with two binary laws of composition for its elements, one denoted by $+$ called *addition*, and the other denoted by $\circ$ called *multiplication*, such that the following two conditions are satisfied:

(a) F is an abelian group under addition with the identity element denoted by 0 and called *zero*; and

(b) the set of the nonzero elements of F is an abelian group under multiplication with the identity element denoted by 1 and called *unity*.

0 is called the *additive identity* while 1 the *multiplicative identity* of the field.

Examples of a field are:

1. The set R of all real numbers with the additive identity 0 and the multiplicative identity 1.
2. The set C of all complex numbers with the additive identity $0+0i$ and the multiplicative identity $1+0i$.
3. The set $\{0, 1, 2, \ldots, p-1\}$ of p integers, where p is a prime number greater than 1, with the two binary operations of addition modulo p and multiplication modulo p; a finite field is called a *Galois field*.

The elements of a field are called *scalars*.

Vector space

A set $L = \{\mathbf{u}, \mathbf{v}, \mathbf{w}, \ldots\}$ is said to be a *vector space over a field* F if the following two conditions are satisfied:

(a) An operation of addition denoted by $+$ is defined in L such that L is an abelian group under addition. The identity element of this group will be denoted by $\mathbf{0}$.

(b) A scalar of the field F and an element of the set L can be combined by an operation called *scalar multiplication* to give an element of L such that for every $\mathbf{u}, \mathbf{v} \in L$ and $a, b \in F$, we have

$$a(\mathbf{u}+\mathbf{v})=a\mathbf{u}+a\mathbf{v}\in L; \quad (a+b)\,\mathbf{u}=a\mathbf{u}+b\mathbf{u}\in L,$$
$$a(b\mathbf{u})=(a \circ b)\,\mathbf{u}, \quad 1\mathbf{u}=\mathbf{u},\ 0\mathbf{u}=\mathbf{0}. \tag{2}$$

The elements of a vector space are called *vectors.* Also note that 0 is an element (the additive identity) of F, whereas $\mathbf{0}$ is the *null vector* of L. The phrases *linear vector space* and *linear space* are also used for a vector space. Henceforth, we shall drop the multiplication symbol for scalars and write, for example, $a \circ b$ simply as ab.

The familiar three-dimensional space of position vectors is an example of a vector space over the field of real numbers. First, it is evident that the set of all position vectors is an abelian group. For, if $\mathbf{u}$ and $\mathbf{v}$ are any two vectors of this space, $\mathbf{u}+\mathbf{v}=\mathbf{v}+\mathbf{u}$ is also a vector of this space. The identity element is the null vector $\mathbf{0}$. The 'inverse' of a vector $\mathbf{u}$ is the vector $-\mathbf{u}$, because $\mathbf{u}+(-\mathbf{u})=\mathbf{0}$. Moreover, the law of vector addition is associative. Second, the position vectors satisfy the properties listed in Eqs. (2) for all a, b belonging to the field R of real numbers.

Linear independence of vectors

Two nonzero vectors $\mathbf{u}$ and $\mathbf{v}$ of a vector space are said to be *linearly dependent* if one is a multiple of the other, i.e., if $\mathbf{u}=c\mathbf{v}$, where c is some scalar. In other words, $\mathbf{u}$ and $\mathbf{v}$ are linearly dependent if it is possible to find scalars a and b different from zero such that $a\mathbf{u}+b\mathbf{v}=\mathbf{0}$. Note that this is equivalent to saying that $\mathbf{u}$ is a multiple of $\mathbf{v}$ and vice versa.

Conversely, two vectors $\mathbf{u}$ and $\mathbf{v}$ are said to be *linearly independent* of each other if one is not a multiple of the other. In this case, it is impossible to satisfy the equation $a\mathbf{u}+b\mathbf{v}=\mathbf{0}$ except when $a=b=0$.

The concepts of linear dependence and independence of vectors can be extended to more than two vectors. Consider a set of n vectors $\mathbf{x}_1, \mathbf{x}_2,\ldots,\mathbf{x}_n$, none of which is a null vector. The vectors of this set are said to be *linearly dependent* if it is possible to find scalars $a_1, a_2,\ldots,a_n$, at least two of which are nonzero, such that

$$a_1\mathbf{x}_1+a_2\mathbf{x}_2+\ldots+a_n\mathbf{x}_n=\mathbf{0}. \tag{3}$$

Suppose the coefficient a_i is different from zero. Then dividing Eq. (3) by a_i, we have

$$\mathbf{x}_i=b_1\mathbf{x}_1+\ldots+b_{i-1}\mathbf{x}_{i-1}+b_{i+1}\mathbf{x}_{i+1}+\ldots+b_n\mathbf{x}_n, \tag{4}$$

with $b_j=-a_j/a_i$. Thus we can say that *the n vectors of a set are linearly dependent if at least one of them can be expressed as a linear combination of the remaining $n-1$ vectors.*

Conversely, the n vectors of a set are *linearly independent* if the only solution of Eq. (3) is $a_i=0$ for $1 \leqslant i \leqslant n$. In other words, the vectors of a set are linearly independent if it is impossible to construct the null vector from a linear combination of the vectors except when all the coefficients vanish.

As an example, in the space of three-dimensional position vectors, it is possible to find a set of three, but not more than three, linearly independent vectors. If we choose the three linearly independent vectors to be $\mathbf{x}_1$, $\mathbf{x}_2$, $\mathbf{x}_3$, any other vector of the space can be expressed as a linear combination of these three vectors in the form

$$\mathbf{u}=a_1\mathbf{x}_1+a_2\mathbf{x}_2+a_3\mathbf{x}_3. \tag{5}$$

The *dimension* of a space is defined as the maximum number of linearly independent vectors in the space. A set of n linearly independent vectors in an n-dimensional vector space is called a *basis* for the vector space. Clearly, the basis is not unique and we may choose the basis in an infinite number of ways.

Vector space of n-tuplets

The vector space of n-tuplets is the most general vector space in the sense that *every vector space is a set of n-tuplets*. An n-tuplet is an ordered set of n numbers, real or complex, such as $\mathbf{u}=(u_1, u_2, \ldots, u_n)$, where u_i are any numbers, all of the same type (that is, real or complex). If we consider the set of all such elements $(\mathbf{u}, \mathbf{v}, \ldots)$ obtained by giving all possible values to the components u_i, we have the vector space of n-tuplets; u_i are said to be the *components* of the vector $\mathbf{u}$. A vector all of whose components are real numbers is said to be a real vector. A vector space whose vectors are real is called a *real vector space*. We shall in general consider the vector space of complex n-tuplets.

A vector all of whose components are zero is called the *null vector*: $\mathbf{0}=(0, 0, \ldots, 0)$. Two vectors of a vector space are said to be equal to each other if, and only if, their respective components are equal to each other. Thus,

$$\mathbf{u}=\mathbf{v} \Leftrightarrow u_i=v_i \text{ for } 1 \leqslant i \leqslant n. \tag{6}$$

The addition of two vectors and the scalar multiplication are defined as follows:

$$\mathbf{u}+\mathbf{v}=\mathbf{v}+\mathbf{u}=(u_1+v_1, u_2+v_2, \ldots, u_n+v_n), \tag{7a}$$

$$a\mathbf{u}=\mathbf{u}a=(au_1, au_2, \ldots, au_n), \tag{7b}$$

both of which can be seen to be elements of the vector space.

Inner product space

A vector space L defined over a field F, where F is the field of real or complex numbers, is called an *inner product space* if with every pair of elements $\mathbf{u}, \mathbf{v} \in L$, there is associated a unique number belonging to the field F—denoted by $(\mathbf{u}, \mathbf{v})$ and called the *inner product* or the *scalar product* of $\mathbf{u}$ and $\mathbf{v}$—for which the following properties hold:

$$
\begin{aligned}
(\mathbf{u}, \mathbf{v}) &= (\mathbf{v}, \mathbf{u})^*, \\
(a\mathbf{u}, b\mathbf{v}) &= a^*b\,(\mathbf{u}, \mathbf{v}), \\
(\mathbf{w}, a\mathbf{u}+b\mathbf{v}) &= a(\mathbf{w}, \mathbf{u})+b(\mathbf{w}, \mathbf{v}),
\end{aligned}
\tag{8}
$$

where the asterisk denotes the complex conjugate.

The vector space of n-tuplets of real or complex numbers can be made an inner product space if we define the inner product of two vectors by

$$(\mathbf{u}, \mathbf{v}) = \sum_{i=1}^{n} u_i^* v_i. \tag{9}$$

The ordinary three-dimensional space of position vectors is also an inner product space with the familiar rule for taking the scalar product of two vectors.

EXAMPLE 1: Determine whether the four vectors

$$
\begin{aligned}
\mathbf{u} &= (1, 2, 3), \quad \mathbf{v} = (2, 0, -1), \\
\mathbf{w} &= (1, -1, 1), \quad \mathbf{x} = (2, 1, 0)
\end{aligned}
\tag{10}
$$

are linearly dependent or independent.

Solution: We solve the equation

$$a\mathbf{u}+b\mathbf{v}+c\mathbf{w}+d\mathbf{x}=\mathbf{0} \tag{11}$$

for the unknown coefficients a, b, c, d. Substituting for the vectors from Eqs. (10) and using the rules of vector addition and scalar multiplication given in Eqs. (7), Eq. (11) becomes

$$(a+2b+c+2d,\ 2a-c+d,\ 3a-b+c)=(0, 0, 0). \tag{12}$$

This gives us the three simultaneous equations

$$a+2b+c+2d=0,\ 2a-c+d=0,\ 3a-b+c=0, \tag{13}$$

which have the solution

$$b=12a/5,\ c=-3a/5,\ d=-13a/5,\ a \text{ arbitrary}. \tag{14}$$

Thus Eq. (11) can be satisfied without all the coefficients being zero. We can write Eq. (11), using Eq. (14) and taking $a=5$, as

$$5\mathbf{u}+12\mathbf{v}-3\mathbf{w}-13\mathbf{x}=\mathbf{0}. \tag{15}$$

The given vectors are therefore linearly dependent.

EXAMPLE 2: It can be shown that we can find a maximum of n linearly independent vectors in the vector space of n-tuplets. We can conveniently choose these n independent vectors or 'coordinate axes' as

$$\begin{aligned} \mathbf{e}_1 &= (1, 0, 0, \ldots, 0), \\ &\vdots \\ \mathbf{e}_i &= (0, 0, \ldots, 1_i, \ldots, 0), \\ &\vdots \\ \mathbf{e}_n &= (0, 0, 0, \ldots, 1), \end{aligned} \tag{16}$$

where 1_i means that unity occurs in the i-th position. We shall now prove the following two results.

(a) *The n vectors* $(\mathbf{e}_1, \mathbf{e}_2, \ldots, \mathbf{e}_n)$ *are linearly independent.*

Solution: We consider the equation

$$a_1\mathbf{e}_1 + a_2\mathbf{e}_2 + \ldots + a_n\mathbf{e}_n = \mathbf{0}. \tag{17}$$

Substituting for the vectors from Eqs. (16), we see that Eq. (17) becomes

$$(a_1, a_2, \ldots, a_n) = (0, 0, \ldots, 0). \tag{18}$$

The only solution of Eq. (17) is therefore $a_i = 0$ for $1 \leqslant i \leqslant n$, showing that the n vectors are linearly independent.

(b) *Any vector of the space* (that is, *any n-tuplet*) *can be uniquely expressed as a linear combination of the n vectors* $\mathbf{e}_i$.

Solution: Consider a vector $\mathbf{u} = (u_1, u_2, \ldots, u_n)$. Let us try to express it in the form

$$\mathbf{u} = b_1\mathbf{e}_1 + b_2\mathbf{e}_2 + \ldots + b_n\mathbf{e}_n. \tag{19}$$

This equation has the solution $b_i = u_i$ for $1 \leqslant i \leqslant n$, so that Eq. (19) becomes

$$\mathbf{u} = \sum_{i=1}^{n} u_i \mathbf{e}_i. \tag{20}$$

If we take the scalar product of a vector with itself, we find from Eq. (9) that

$$(\mathbf{u}, \mathbf{u}) = \sum_{i=1}^{n} | u_i |^2. \tag{21}$$

The positive square root of this quantity is defined as the *norm* of the vector $\mathbf{u}$ and is denoted by

$$\| \mathbf{u} \| = (\mathbf{u}, \mathbf{u})^{1/2} = \left[\sum_{i=1}^{n} | u_i |^2 \right]^{1/2}. \tag{22}$$

It is easy to recognize that this is the length of a vector in familiar language. If the norm of a vector is unity, it is called a *unit vector* or a *normalized vector.* It can be seen that all the vectors $\mathbf{e}_1, \mathbf{e}_2, \ldots, \mathbf{e}_n$ defined in Eqs. (16) are unit vectors.

If the scalar product of two vectors is zero, the vectors are said to be *orthogonal.* Thus two vectors $\mathbf{u}$ and $\mathbf{v}$ are orthogonal if, and only if,

$$(\mathbf{u}, \mathbf{v})=0. \tag{23}$$

Once again it can be verified that any two vectors of the set $\{\mathbf{e}_i\}$ are orthogonal. Combining this with the earlier result, we have

$$(\mathbf{e}_i, \mathbf{e}_j)=\delta_{ij}. \tag{24}$$

A set of vectors each of which is orthogonal to all the remaining vectors of the set is called an *orthogonal set.* If each of the vectors is further normalized to unity, it is called an *orthonormal set.* A vector can be normalized by dividing it by its norm. Thus, if $\mathbf{u}$ is any vector, $\mathbf{u}/\|\mathbf{u}\|$ is the normalized vector parallel to $\mathbf{u}$.

Schmidt's orthogonalization method

It is possible to obtain a set of orthogonal, or in fact orthonormal, vectors starting from a set of nonorthogonal but linearly independent vectors. This can be done by a procedure known as the *Schmidt's orthogonalization method* which is discussed below.

Let $\mathbf{u}_1, \mathbf{u}_2, \ldots, \mathbf{u}_n$ be a set of linearly independent vectors which are not necessarily orthogonal. It is required to obtain a set of orthogonal vectors $\mathbf{v}_1, \mathbf{v}_2, \ldots, \mathbf{v}_n$ starting from the original set of vectors. We proceed along the following steps:

1. Take $\mathbf{v}_1=\mathbf{u}_1$.
2. Let

$$\mathbf{v}_2=\mathbf{u}_2+a_{21}\,\mathbf{v}_1, \tag{25}$$

where a_{21} is a constant to be determined from the condition that $\mathbf{v}_2$ be orthogonal to $\mathbf{v}_1$, i.e., $(\mathbf{v}_1, \mathbf{v}_2)=0$. Taking the scalar product of $\mathbf{v}_1$ with $\mathbf{v}_2$ of Eq. (25) and equating it to zero, we get

$$(\mathbf{v}_1, \mathbf{u}_2)+a_{21}\,(\mathbf{v}_1, \mathbf{v}_1)=0 \Rightarrow a_{21}=-(\mathbf{v}_1, \mathbf{u}_2)/(\mathbf{v}_1, \mathbf{v}_1). \tag{26}$$

Thus we have two orthogonal vectors, $\mathbf{v}_1$ and $\mathbf{v}_2$.

3. Let

$$\mathbf{v}_3=\mathbf{u}_3+a_{32}\,\mathbf{v}_2+a_{31}\,\mathbf{v}_1, \tag{27}$$

where a_{31} and a_{32} are constants to be determined from the conditions that $\mathbf{v}_3$ be orthogonal to $\mathbf{v}_1$ and $\mathbf{v}_2$. This gives

$$(\mathbf{v}_1, \mathbf{v}_3)\equiv 0=(\mathbf{v}_1, \mathbf{u}_3)+a_{31}\,(\mathbf{v}_1, \mathbf{v}_1) \Rightarrow a_{31}=-(\mathbf{v}_1, \mathbf{u}_3)/(\mathbf{v}_1, \mathbf{v}_1);$$
$$(\mathbf{v}_2, \mathbf{v}_3)\equiv 0=(\mathbf{v}_2, \mathbf{u}_3)+a_{32}\,(\mathbf{v}_2, \mathbf{v}_2) \Rightarrow a_{32}=-(\mathbf{v}_2, \mathbf{u}_3)/(\mathbf{v}_2, \mathbf{v}_2). \tag{28}$$

Now we have three mutually orthogonal vectors, $\mathbf{v}_1$, $\mathbf{v}_2$ and $\mathbf{v}_3$.

The same procedure can be continued. In the i-th step, we take the vector

$$\mathbf{v}_i=\mathbf{u}_i+a_{i1}\mathbf{v}_1+a_{i2}\mathbf{v}_2+\ldots+a_{i,\,i-1}\,\mathbf{v}_{i-1} \tag{29a}$$

and determine the coefficients a_{ij} $(1\leqslant j\leqslant i-1)$ from the conditions that $\mathbf{v}_i$ be orthogonal to all the previously obtained vectors $\mathbf{v}_1$, $\mathbf{v}_2$. ..., $\mathbf{v}_{i-1}$. We would obtain

$$a_{ij}=-(\mathbf{v}_j,\ \mathbf{u}_i)/(\mathbf{v}_j,\ \mathbf{v}_j). \tag{29b}$$

Finally, all the vectors can be normalized to obtain an orthonormal set $\{\mathbf{x}_i\}$, where

$$\mathbf{x}_i=\mathbf{v}_i/\|\ \mathbf{v}_i\ \|. \tag{30}$$

EXAMPLE 3: Obtain a set of four orthonormal vectors by the Schmidt's method from the vectors

$$\mathbf{u}_1=(1,1,0,1),\qquad \mathbf{u}_2=(2,0,0,1),$$
$$\mathbf{u}_3=(0,\ 2,\ 3,\ -2),\ \mathbf{u}_4=(1,\ 1,\ 1,\ -5). \tag{31}$$

Solution: Let $\mathbf{v}_1=\mathbf{u}_1$. Let also $\mathbf{v}_2=\mathbf{u}_2+a_{21}\mathbf{v}_1$ with a given by Eq. (26). We have

$$(\mathbf{v}_1,\mathbf{u}_2)=1\times2+1\times0+0\times0+1\times1=3, \tag{32a}$$
$$(\mathbf{v}_1,\mathbf{v}_1)=1\times1+1\times1+0\times0+1\times1=3, \tag{32b}$$

so that $a_{21}=-1$. Hence

$$\mathbf{v}_2=\mathbf{u}_2-\mathbf{v}_1=(2,\ 0,\ 0,\ 1)-(1,1,0,1)=(1,\ -1,\ 0,\ 0). \tag{33}$$

Now take $\mathbf{v}_3=\mathbf{u}_3+a_{31}\mathbf{v}_1+a_{32}\mathbf{v}_2$ with coefficients given by Eq. (28). We have

$$(\mathbf{v}_1,\ \mathbf{u}_3)=0\Rightarrow a_{31}=0;$$
$$(\mathbf{v}_2,\ \mathbf{u}_3)=-2,\ (\mathbf{v}_2,\ \mathbf{v}_2)=2\Rightarrow a_{32}=1. \tag{34}$$

Hence

$$\mathbf{v}_3=\mathbf{u}_3+\mathbf{v}_2=(1,\ 1,\ 3,\ -2). \tag{35}$$

Finally, we take $\mathbf{v}_4=\mathbf{u}_4+a_{41}\mathbf{v}_1+a_{42}\mathbf{v}_2+a_{43}\mathbf{v}_3$. We have

$$a_{41}=-(\mathbf{v}_1,\ \mathbf{u}_4)/\|\ \mathbf{v}_1\ \|=\ \ 1,$$
$$a_{42}=-(\mathbf{v}_2,\ \mathbf{u}_4)/\|\ \mathbf{v}_2\ \|=\ \ 0, \tag{36}$$
$$a_{43}=-(\mathbf{v}_3,\ \mathbf{u}_4)/\|\ \mathbf{v}_3\ \|=-1.$$

Hence

$$\mathbf{v}_4=\mathbf{u}_4+\mathbf{v}_1-\mathbf{v}_3=(1,\ 1,\ -2,\ -2). \tag{37}$$

The desired orthogonal set is therefore

$$\mathbf{v}_1=(1,\ 1,\ 0,\ 1),\quad \mathbf{v}_2=(1,\ -1,\ 0,\ 0),$$
$$\mathbf{v}_3=(1,\ 1,\ 3,\ -2),\ \mathbf{v}_4=(1,\ 1,\ -2,\ -2). \tag{38a}$$

The corresponding orthonormal set is

$$\mathbf{x}_1=\frac{1}{\sqrt{3}}(1, 1, 0, 1), \quad \mathbf{x}_2=\frac{1}{\sqrt{2}}(1, -1, 0, 0),$$
$$\mathbf{x}_3=\frac{1}{\sqrt{15}}(1, 1, 3, -2), \quad \mathbf{x}_4=\frac{1}{\sqrt{10}}(1, 1, -2, -2). \tag{38b}$$

Linear transformations

We are often required to consider the transformation of the components of a vector from one coordinate system to another. When the transformation is such that the origin of coordinates remains unaltered, that is, no translation of the coordinate system is involved, it is said to be a *homogeneous transformation*. When the transformation contains a translation of the coordinate system, it is called an *inhomogeneous transformation*. However, since the parallel displacement of a coordinate system and a vector relative to each other does not change the components of the vector in a Euclidean space, we may disregard any displacement of the origin and restrict ourselves to homogeneous transformations only.

The simplest of such transformations are the *linear transformations* in which the new components of a vector are linear functions of the old ones. Thus, if x_i are the components of a vector in one coordinate system and y_i those in another coordinate system after a linear transformation, we have

$$y_i=b_{i1}x_1+b_{i2}x_2+\ldots+b_{in}x_n=\sum_{k=1}^{n} b_{ik}x_k, \ 1\leqslant i\leqslant n. \tag{39}$$

Now consider a further linear transformation of the coordinate system in which the same vector has components z_i, which are linearly related to the components y_i by

$$z_i=\sum_{k=1}^{n} a_{ik}y_k, \ 1\leqslant i\leqslant n. \tag{40}$$

It is possible to eliminate the intermediate coordinate system and obtain a transformation directly from the components x_i to the components z_i. Substituting Eq. (39) in Eq. (40), we have

$$z_i=\sum_{j=1}^{n} \sum_{k=1}^{n} a_{ik}b_{kj}x_j, \ 1\leqslant i\leqslant n. \tag{41}$$

This could be written in the form of a linear homogeneous transformation as

$$z_i=\sum_{j=1}^{n} c_{ij}x_j,\ 1\leqslant i\leqslant n, \tag{42}$$

where

$$c_{ij}=\sum_{k=1}^{n} a_{ik}b_{kj},\ 1\leqslant i, j\leqslant n. \tag{43}$$

In dealing with such transformations it is convenient to introduce the concept of *matrices.* In fact, it is in the study of such successive linear transformations that the branch of *matrix algebra* historically developed. The study of the basic rules of matrix algebra, the various types of special matrices and how they can be used to handle linear transformations in vector spaces will engage us in the next few sections.

Exercise 1

1. In each of the following cases, find the norm of the vector and write it as a normalized vector:

*(a) (2, 1, 3, −1); (b) (1, 0, −2, 3, 5); *(c) (1, i);
(d) (1+i, −2+3i, 3, 4i); (e) (2+i, 1−i, 0, 3+3i).

2. Determine whether the following sets of vectors are linearly dependent or independent:

*(a) (1, 2, −1), (0, 5, −3) and (−4, 1, 3);
(b) (1, −2, 3, 1), (2, 0, −2, 2), (6, 2, 1, −5) and (−3, 2, 0, 1);
*(c) (1, 2, 0, −2, −1), (3, −1, 4, 1, −2), (0, 2, −3, 4, 1), (3, 3, 3, 4, 1) and (−1, 0, 2, 1, 3).

3. Show that the three vectors (1, 1, 1), (1, 0, −1) and (1, −2, 1) are a set of linearly independent vectors. Verify that they are also orthogonal vectors. Express each of the vectors (i) (1, 0, 0), (ii) (2, −1, 3) as a linear combination of the three given vectors.

4. If $\mathbf{x}_i$ $(1\leqslant i\leqslant n)$ is a set of linearly independent vectors and if

$$\sum_{i=1}^{n} a_i\mathbf{x}_i=\sum_{i=1}^{n} b_i\mathbf{x}_i,$$

where a_i and b_i are scalars, show that $a_i=b_i$ for $1\leqslant i\leqslant n$.

5. In the following cases, find the values of x for which the given sets of vectors are linearly dependent:

*(a) (1, 2, 3), (4, 5, 6), (x, 8, 9);
(b) (1, 1, 0), (x, 2, −1), (0, 0, 1);
*(c) (0, −1, 2), (0, 1, 6), (1, 2, x);
(d) (2, 3, 1, 4), (1, 1, 1, 1), (0, 1, 2, x), (1, −1, 1, −1).

*Exercises whose answers are given at the end of the book have been marked with asterisks.

6. Express each of the vectors of Exercises 1(a), 1(d) and 1(e) as a linear combination of the four vectors of Exercise 2(b).

*7. State with reason whether the following statements are true or false:

(a) If the vectors of a set are pairwise orthogonal, it necessarily follows that it is an orthogonal set.

(b) If the vectors of a set are pairwise linearly independent, it necessarily follows that it is a set of linearly independent vectors.

8. If $\mathbf{u}_i$, $1 \leqslant i \leqslant n$, is a set of n vectors, their *Gram-Schmidt determinant*, or simply the *Gram determinant*, is defined as an $n \times n$ determinant whose elements are the scalar products of the vectors with each other, i.e.,

$$D = \begin{vmatrix} (\mathbf{u}_1, \mathbf{u}_1) & (\mathbf{u}_1, \mathbf{u}_2) \ldots (\mathbf{u}_1, \mathbf{u}_n) \\ (\mathbf{u}_2, \mathbf{u}_1) & (\mathbf{u}_2\ \mathbf{u}_2) \ldots (\mathbf{u}_2, \mathbf{u}_n) \\ \vdots & \\ (\mathbf{u}_n, \mathbf{u}_1) & (\mathbf{u}_n, \mathbf{u}_2) \ldots (\mathbf{u}_n, \mathbf{u}_n) \end{vmatrix}.$$

Prove that the set of vectors is linearly dependent *if, and only if*, $D=0$.

*9. Show that the vectors of the set

$$(1, 2, 3, 4), (1, 0, 2, 1), (2, -1, 4, 1), (-1, 3, 0, 4)$$

are linearly dependent. Find the condition for which a general vector (x, y, z, w) can be expressed as a linear combination of the first three of the above vectors.

*10. A transformation $\mathbf{A}$ on vectors $\mathbf{r}$ has the effect $\mathbf{Ar} = \mathbf{a} \times \mathbf{r}$, where $\mathbf{a}$ is a given vector and $\times$ denotes the vector product of two vectors. Show that $\mathbf{A}^3 + a^2\mathbf{A} = \mathbf{0}$, the null operator, where $a = \|\,\mathbf{a}\,\|$.

11. Let $P(\boldsymbol{\theta})$ denote an operation of rotation through an angle θ about an axis parallel to the vector $\boldsymbol{\theta}$ and passing through the origin of coordinates. Here θ is the magnitude of the vector $\boldsymbol{\theta}$. For infinitesimal values of θ, show that the effect of the transformation $P(\boldsymbol{\theta})$ is $P(\boldsymbol{\theta})\ \mathbf{r} = \mathbf{r} + \boldsymbol{\theta} \times \mathbf{r}$.

2. The algebra of matrices

In this section, we shall define a matrix and discuss some of the simple operations by which two or more matrices can be combined.

Definition and notation

A rectangular array of numbers (real or complex) is called a *matrix*. The array is usually enclosed within square or curved brackets. Thus, the rectangular arrays

$$\begin{bmatrix} 2 & -3 \\ 5 & 7 \end{bmatrix}, \begin{bmatrix} 3 & -5 & 9 \\ 8 & 6 & -2 \end{bmatrix}, \begin{bmatrix} a & b & c \\ j & k & l \\ r & s & t \\ x & y & z \end{bmatrix}, \begin{bmatrix} 1+2i & 3-4i \\ 5+id & x+iy \\ s+4it & u+7i \end{bmatrix} \tag{1}$$

are examples of a matrix. The individual members of the array are called the *elements* of the matrix. Although we have defined a matrix here with reference to numbers, we can easily extend the definition to a matrix whose elements are functions. An example of such a matrix is

$$\begin{bmatrix} f_1(x) & f_2(x) & f_3(x) \\ f_4(x) & f_5(x) & f_6(x) \end{bmatrix}, \tag{2}$$

where $f_i(x)$ are functions of x.

It is convenient to think of every element of a matrix as belonging to a certain *row* and a certain *column* of the matrix. Thus, referring to the third matrix in (1), we would say that $(a \;\; b \;\; c)$ is the first row of the matrix, $(j \;\; k \;\; l)$ the second row, $(r \;\; s \;\; t)$ the third row and $(x \;\; y \;\; z)$ the fourth row of the matrix. Similarly, the matrix has three columns which are

$$\begin{bmatrix} a \\ j \\ r \\ x \end{bmatrix}, \begin{bmatrix} b \\ k \\ s \\ y \end{bmatrix} \text{ and } \begin{bmatrix} c \\ l \\ t \\ z \end{bmatrix}. \tag{3}$$

It is then obvious that every element of a matrix can be uniquely characterized by a row index and a column index. The element s, for example, belongs to the third row and the second column, the element z belongs to the fourth row and the third column, etc.

If a matrix has m rows and n columns, we shall say that the matrix is of order $m \times n$ (called 'm by n'). A general matrix of order $m \times n$ can be conveniently written as

$$\mathbf{A}=\begin{bmatrix} a_{11} & a_{12} & a_{13} & \dots & a_{1n} \\ a_{21} & a_{22} & a_{23} & \dots & a_{2n} \\ \vdots & & & & \\ a_{m1} & a_{m2} & a_{m3} & \dots & a_{mn} \end{bmatrix}, \tag{4}$$

where the elements may be real or complex numbers or functions. This may be condensed to the shorthand notation

$$\mathbf{A}\equiv[a_{ij}]_{m\times n}, \tag{5a}$$

which means that $\mathbf{A}$ is a matrix of order $m\times n$ whose ij-th element[1] is a_{ij}. It should be clear that $1\leqslant i\leqslant m$ and $1\leqslant j\leqslant n$. If the order of the matrix $\mathbf{A}$ need not be written explicitly, we may also express this as

$$(\mathbf{A})_{ij}=a_{ij}, \tag{5b}$$

which merely states that the ij-th element of $\mathbf{A}$ is a_{ij}

The zero matrix

A matrix $\mathbf{A}$ of arbitrary order is said to be a *zero matrix* if, and only if, every element of $\mathbf{A}$ equals zero. We shall denote a zero matrix by $\mathbf{0}$ (not to be confused with the null element of a vector space, which is also denoted by the same symbol). Thus, if $\mathbf{A}$ is of order $m\times n$, then

$$\mathbf{A}=\mathbf{0}\Leftrightarrow(\mathbf{A})_{ij}=0 \text{ for } 1\leqslant i\leqslant m,\ 1\leqslant j\leqslant n. \tag{6}$$

If it is necessary to specify the order of a zero matrix, we may write it as $\mathbf{0}_{m\times n}$. It is evident that for any arbitrary matrix $\mathbf{A}$,

$$\mathbf{A}+(-\mathbf{A})=\mathbf{A}-\mathbf{A}=\mathbf{0}. \tag{7}$$

Combination of matrices

Before we proceed to discuss the combination of matrices and the various identities and relations among matrices, the first logical step would be to define the *equality of two matrices.*

In order that two matrices $\mathbf{A}$ and $\mathbf{B}$ be equal to each other, it is necessary, though not sufficient, that they be of the same order. Granted that $\mathbf{A}$ and $\mathbf{B}$ are of the same order, say $m\times n$, we say that $\mathbf{A}$ is equal to $\mathbf{B}$ *if, and only if,* every element of $\mathbf{A}$ is equal to the corresponding element of $\mathbf{B}$. Thus, if $\mathbf{A}\equiv[a_{ij}]_{m\times n}$ and $\mathbf{B}\equiv[b_{ij}]_{m\times n}$, then

$$\mathbf{A}=\mathbf{B}\Leftrightarrow a_{ij}=b_{ij}, \text{ for } 1\leqslant i\leqslant m,\ 1\leqslant j\leqslant n. \tag{8}$$

[1]An element in the i-th row and the j-th column of a matrix will be called the ij-th element of the matrix.

This set of mn conditions extends the concept of equality of numbers to that of equality of matrices.

Matrix addition

We may now define the addition and subtraction of two matrices for which, again, both the matrices must be of the same order. If **A** is the matrix defined earlier and $\mathbf{C} \equiv [c_{ij}]_{m\times n}$, then the sum of **A** and **C** is defined as a matrix of the same order and whose ij-th element equals the sum of the ij-th elements of **A** and **C**. Thus,

$$\mathbf{A}+\mathbf{C} \equiv [a_{ij}+c_{ij}]_{m\times n}, \quad \text{or} \quad (\mathbf{A}+\mathbf{C})_{ij}=a_{ij}+c_{ij}. \tag{9}$$

Similarly, we define the difference of the two matrices **A** and **C** as a matrix of the same order and whose elements are given by

$$\mathbf{A}-\mathbf{C} \equiv [a_{ij}-c_{ij}]_{m\times n} = -(\mathbf{C}-\mathbf{A}), \tag{10a}$$

or

$$(\mathbf{A}-\mathbf{C})_{ij} = -(\mathbf{C}-\mathbf{A})_{ij} = a_{ij}-c_{ij}. \tag{10b}$$

Because of the fact that matrix addition is a simple extension of the concept of addition of numbers, *the law of matrix addition is commutative.* This means that just as any two scalars satisfy the property $a+c=c+a$, for any two matrices whose sum can be defined (that is, which are of the same order), we have

$$\mathbf{A}+\mathbf{C}=\mathbf{C}+\mathbf{A}. \tag{11}$$

EXAMPLE 1: Find the sum of the two matrices

$$\mathbf{A}=\begin{bmatrix} 2 & 5 & 0 & 7 \\ -1 & 6 & 2 & 4 \\ 3 & -4 & 8 & -2 \end{bmatrix}, \quad \mathbf{B}=\begin{bmatrix} -1 & 0 & 3 & -3 \\ 2 & 2 & -2 & 5 \\ -3 & 5 & 6 & 4 \end{bmatrix}.$$

Solution: Both the matrices are of order 3×4, so that their sum is defined and is given by

$$\mathbf{A}+\mathbf{C}=\begin{bmatrix} 2-1 & 5+0 & 0+3 & 7-3 \\ -1+2 & 6+2 & 2-2 & 4+5 \\ 3-3 & -4+5 & 8+6 & -2+4 \end{bmatrix}=\begin{bmatrix} 1 & 5 & 3 & 4 \\ 1 & 8 & 0 & 9 \\ 0 & 1 & 14 & 2 \end{bmatrix}.$$

The concept can obviously be extended to more than two matrices. Thus, for any matrices **A**, **B**, **C**, **D**,..., all of the same order, we have

$$(\mathbf{A}+\mathbf{B}+\mathbf{C}+\mathbf{D}+\dots)_{ij}=(\mathbf{A})_{ij}+(\mathbf{B})_{ij}+(\mathbf{C})_{ij}+(\mathbf{D})_{ij}+\dots. \tag{12}$$

Moreover, it is easy to see that in the triple sum $\mathbf{A}+\mathbf{B}+\mathbf{C}$, we may either first add **A** to **B** and then add the resulting matrix to **C**, or first add **B** to **C** and then add the resulting matrix to **A**. In mathematical notation, we have

$$(\mathbf{A}+\mathbf{B})+\mathbf{C}=\mathbf{A}+(\mathbf{B}+\mathbf{C}). \tag{13}$$

We say that *the law of matrix addition is associative.*

This may seem to be a little trivial, but this is an important property satisfied only by certain laws of higher algebra. To give a simple example, it is easy to see that *the law of matrix subtraction is not associative*; thus,

$$(\mathbf{A}-\mathbf{B})-\mathbf{C}\neq\mathbf{A}-(\mathbf{B}-\mathbf{C}). \tag{14}$$

Matrix multiplication

The *multiplication*, or *product*, of two matrices is not such a simple extension of the concept of multiplication of numbers. The product of two matrices, as we shall define it here, is prompted by the theory of successive linear transformations discussed briefly at the end of the previous section.

The necessary condition for the product of two matrices, say **A** and **B**, to be defined is that the number of columns of **A** be equal to the number of rows of **B**. It is then immediately clear that the matrix product depends on the order in which the matrices occur in the product. For example, if **A** is of order $m\times n$ and **B** is of order $n\times p$, then the matrix product **AB** is defined but the product of **B** with **A** in that order is not, unless $m=p$. The elements of the product matrix **AB** are obtained in the following way. Let $\mathbf{A}\equiv[a_{ij}]_{m\times n}$ and $\mathbf{B}\equiv[b_{ij}]_{n\times p}$. Then *the ij-th element of the product* **AB** *is the sum of the products of the corresponding elements of the i-th row of* **A** *and the j-th column of* **B**. To develop the mathematical notation, we write down below the i-th row of **A** and the j-th column of **B**:

$$i\text{-th row of } \mathbf{A}=[a_{i1}\; a_{i2}\ldots a_{in}]; \tag{15a}$$

$$j\text{-th column of } \mathbf{B}=\begin{bmatrix} b_{1j} \\ b_{2j} \\ \vdots \\ b_{nj} \end{bmatrix}. \tag{15b}$$

Taking the sum of the products of the corresponding elements in accordance with the above rule, we find that

$$(\mathbf{AB})_{ij}=a_{i1}b_{1j}+a_{i2}b_{2j}+\ldots+a_{in}b_{nj}$$

$$=\sum_{k=1}^{n} a_{ik}b_{kj}, \text{ for } 1\leqslant i\leqslant m,\ 1\leqslant j\leqslant p. \tag{16}$$

The product matrix **AB** clearly has m rows and p columns. Note that this is similar to Eq. (1.43) and, in fact, becomes identical to it when $m=p=n$.

EXAMPLE 2: Find the matrix products **AB** and **BA**, if defined, where

$$\mathbf{A}=\begin{bmatrix} -2 & 3 & 5 \\ 0 & -1 & 6 \\ 8 & -6 & 1 \\ 5 & 4 & -3 \end{bmatrix}, \qquad \mathbf{B}=\begin{bmatrix} 2 & 3 \\ 5 & -7 \\ -1 & 3 \end{bmatrix}. \tag{17}$$

Solution: **A** is of order 4×3 and **B** of order 3×2, so that the matrix product **BA** is not possible. The product **AB** is equal to

$$\mathbf{AB}=\begin{bmatrix} -2\times 2+3\times 5+5(-1) & -2\times 3+3(-7)+5\times 3 \\ 0\times 2+(-1)5+6(-1) & 0\times 3+(-1)(-7)+6\times 3 \\ 8\times 2+(-6)5+1(-1) & 8\times 3+(-6)(-7)+1\times 3 \\ 5\times 2+4\times 5+(-3)(-1) & 5\times 3+4(-7)+(-3)3 \end{bmatrix}$$

$$=\begin{bmatrix} 6 & -12 \\ -11 & 25 \\ -15 & 69 \\ 33 & -22 \end{bmatrix}, \tag{18}$$

which is a matrix of order 4×2.

Column and row vectors

We can develop a more sophisticated notation for writing down the matrix product by introducing what are known as *row vectors* and *column vectors*. Consider the i-th row of **A** written in Eq. (15a). This itself can be treated as a matrix of order $1\times n$, for it has just one row and n columns. Such a matrix is called a *row vector* of dimension n. Let us denote this particular row vector by

$$\mathbf{a}_i'=[a_{i1}\ a_{i2}\ldots a_{in}]. \tag{19}$$

Similarly, the j-th column of the matrix **B** [see Eq. (15b)] has n rows and one column. A matrix having n rows and only one column is called a *column vector* of dimension n. Let the particular column vector of Eq. (15b) be denoted by[2]

$$\mathbf{b}_j=\begin{bmatrix} b_{1j} \\ b_{2j} \\ \vdots \\ b_{nj} \end{bmatrix}. \tag{20a}$$

[2]Column vectors and row vectors will be denoted by small boldface letters.

To save space, we shall frequently write a column vector by using curled brackets as

$$\mathbf{b}_j = \{b_{1j}\ b_{2j} \ldots b_{nj}\}. \tag{20b}$$

Now the two matrices $\mathbf{a}_i'$ and $\mathbf{b}_j$ of Eqs. (19) and (20) satisfy the requirement for matrix product, i.e., the number of columns of $\mathbf{a}_i'$ equals the number of rows of $\mathbf{b}_j$. Their matrix product $\mathbf{a}_i'\mathbf{b}_j$ will then be a matrix of order 1×1, which is just a scalar, given by

$$\mathbf{a}_i'\mathbf{b}_j = a_{i1}b_{1j} + a_{i2}b_{2j} + \ldots + a_{in}b_{nj}, \tag{21}$$

which is the same as the ij-th element of **AB** given in Eq. (16). Thus, we can write the matrix product **AB** as

$$\mathbf{AB} = \begin{bmatrix} \mathbf{a}_1'\mathbf{b}_1 & \mathbf{a}_1'\mathbf{b}_2 \ldots \mathbf{a}_1'\mathbf{b}_p \\ \mathbf{a}_2'\mathbf{b}_1 & \mathbf{a}_2'\mathbf{b}_2 \ldots \mathbf{a}_2'\mathbf{b}_p \\ \vdots & \\ \mathbf{a}_m'\mathbf{b}_1 & \mathbf{a}_m'\mathbf{b}_2 \ldots \mathbf{a}_m'\mathbf{b}_p \end{bmatrix}. \tag{22}$$

We have remarked earlier that if the matrix product **AB** is defined, the product of **B** with **A** is not necessarily defined. Given two matrices **A** and **B**, both the matrix products **AB** and **BA** will be possible if, for example, **A** is of order $m \times n$ and **B** is of order $n \times m$. **AB** will then be of order $m \times m$ and **BA** of order $n \times n$. We still cannot define the equality of **AB** and **BA** since they are of different orders. It is clear that both **AB** and **BA** will be of the same order if and only if $m = n$, in which case **A**, **B**, **AB** and **BA** are all of order $n \times n$. Even in this case, **AB** *is not necessarily equal* to **BA**, as will be clear from the following example.

EXAMPLE 3: Show that $\mathbf{AB} \neq \mathbf{BA}$, where

$$\mathbf{A} = \begin{bmatrix} 3 & 7 & -9 \\ 1 & -2 & 3 \\ 5 & 0 & -6 \end{bmatrix}, \quad \mathbf{B} = \begin{bmatrix} 6 & 9 & -3 \\ 2 & 0 & 5 \\ -8 & 1 & 7 \end{bmatrix}. \tag{23}$$

Solution: We have

$$\mathbf{AB} = \begin{bmatrix} 3 & 7 & -9 \\ 1 & -2 & 3 \\ 5 & 0 & -6 \end{bmatrix} \begin{bmatrix} 6 & 9 & -3 \\ 2 & 0 & 5 \\ -8 & 1 & 7 \end{bmatrix} = \begin{bmatrix} 104 & 18 & -37 \\ -22 & 12 & 8 \\ 78 & 39 & -57 \end{bmatrix},$$

while

$$\mathbf{BA} = \begin{bmatrix} 6 & 9 & -3 \\ 2 & 0 & 5 \\ -8 & 1 & 7 \end{bmatrix} \begin{bmatrix} 3 & 7 & -9 \\ 1 & -2 & 3 \\ 5 & 0 & -6 \end{bmatrix} = \begin{bmatrix} 12 & 24 & -9 \\ 31 & 14 & -48 \\ 12 & -58 & 33 \end{bmatrix}.$$

Evidently, $\mathbf{AB} \neq \mathbf{BA}$.

The commutator

The difference between the two matrices **AB** and **BA** is known as the *commutator* of **A** and **B** and is denoted by

$$[\mathbf{A}, \mathbf{B}]=\mathbf{AB}-\mathbf{BA}. \tag{24a}$$

It should be clear that

$$[\mathbf{B}, \mathbf{A}]=-[\mathbf{A}, \mathbf{B}]. \tag{24b}$$

If, in particular, **AB** is equal to **BA**, the two matrices **A** and **B** are said to *commute* with each other. The two matrices **A** and **B** of Eqs. (23) do not commute with each other, but the two matrices

$$\begin{bmatrix} 13 & -2 & -5 \\ -2 & 10 & -2 \\ -5 & -2 & 13 \end{bmatrix} \text{ and } \begin{bmatrix} 0 & 1 & 2 \\ 1 & 1 & 1 \\ 2 & 1 & 0 \end{bmatrix}$$

do, as can be readily verified.

If **A** commutes with **B** and **B** commutes with **C**, *it does not necessarily follow* that **A** commutes with **C** (see Exercise 6).

Thus, we have seen that *the law of matrix multiplication is not commutative*. However, like the law of matrix addition, *it is associative*, which means that if **A**, **B** and **C** are any three matrices such that the matrix products **AB** and **BC** are defined, they satisfy the property

$$(\mathbf{AB})\mathbf{C}=\mathbf{A}(\mathbf{BC}). \tag{25}$$

On the other hand, division, like subtraction, is *not* associative. If a, b, c are any three nonzero numbers, we have

$$(a/b)/c\neq a/(b/c). \tag{26}$$

The validity of Eq. (25) can be readily verified by showing that the elements of the matrices on both sides of Eq. (25) are respectively equal. To this end, let us suppose that

$$\mathbf{A}\equiv[a_{ij}]_{m\times n},\ \mathbf{B}\equiv[b_{ij}]_{n\times p},\ \mathbf{C}\equiv[c_{ij}]_{p\times q}. \tag{27}$$

The ij-th element of the left hand side of Eq. (25) is then

$$\begin{aligned}[(\mathbf{AB})\mathbf{C}]_{ij}&=\sum_{k=1}^{p}(\mathbf{AB})_{ik}\,(\mathbf{C})_{kj}\\ &=\sum_{k=1}^{p}\sum_{l=1}^{n}(\mathbf{A})_{il}\,(\mathbf{B})_{lk}\,(\mathbf{C})_{kj}\\ &=\sum_{k=1}^{p}\sum_{l=1}^{n}a_{il}\,b_{lk}\,c_{kj},\end{aligned} \tag{28}$$

while the ij-th element of the right hand side of Eq. (25) is

$$[\mathbf{A}(\mathbf{BC})]_{ij}=\sum_{l=1}^{n}(\mathbf{A})_{il}\,(\mathbf{BC})_{lj}$$
$$=\sum_{l=1}^{n}(\mathbf{A})_{il}\sum_{k=1}^{p}(\mathbf{B})_{lk}\,(\mathbf{C})_{kj}$$
$$=\sum_{l=1}^{n}\sum_{k=1}^{p}a_{il}\,b_{lk}\,c_{kj}. \tag{29}$$

From Eqs. (28) and (29), it is evident that

$$[(\mathbf{AB})\mathbf{C}]_{ij}=[\mathbf{A}(\mathbf{BC})]_{ij},\ 1\leqslant i\leqslant m,\ 1\leqslant j\leqslant q. \tag{30}$$

From the definition of the equality of two matrices, therefore, Eq. (25) immediately follows.

The law of matrix multiplication satisfies one more important property, which is discussed in the following example.

EXAMPLE 4: If **A**, **B** and **C** are any matrices such that the addition **B**+**C** and the products **AB** and **AC** are defined, show that

$$\mathbf{A}(\mathbf{B}+\mathbf{C})=\mathbf{AB}+\mathbf{AC}. \tag{31}$$

This property is expressed by saying that *matrix multiplication is distributive with respect to matrix addition.*

Solution: To prove this, let

$$\mathbf{A}\equiv[a_{ij}]_{m\times n},\ \mathbf{B}\equiv[b_{ij}]_{n\times p},\ \mathbf{C}\equiv[c_{ij}]_{n\times p}, \tag{32}$$

so that, as desired, the addition **B**+**C** and the products **AB** and **AC** are defined. Consider now the *ij*-th element of the left hand side of Eq. (31), which is given by

$$[\mathbf{A}(\mathbf{B}+\mathbf{C})]_{ij}=\sum_{k=1}^{n}(\mathbf{A})_{ik}\,(\mathbf{B}+\mathbf{C})_{kj}\quad\text{[from Eq. (16)]}$$
$$=\sum_{k=1}^{n}(\mathbf{A})_{ik}\,[(\mathbf{B})_{kj}+(\mathbf{C})_{kj}]\quad\text{[from Eq. (9)]}$$
$$=\sum_{k=1}^{n}a_{ik}\,(b_{kj}+c_{kj}). \tag{33}$$

The *ij*-th element of the right hand side of Eq. (31) is

$$[\mathbf{AB}+\mathbf{AC}]_{ij}=(\mathbf{AB})_{ij}+(\mathbf{AC})_{ij}\quad\text{[from Eq. (9)]}$$
$$=\sum_{k=1}^{n}(\mathbf{A})_{ik}\,(\mathbf{B})_{kj}+\sum_{k=1}^{n}(\mathbf{A})_{ik}\,(\mathbf{C})_{kj}\quad\text{[from Eq. (16)]}$$
$$=\sum_{k=1}^{n}a_{ik}\,(b_{kj}+c_{kj}). \tag{34}$$

Since the element in Eq. (33) is the same as that in Eq. (34), the validity of Eq. (31) follows.

Scalar multiplication

It is possible to combine a matrix of arbitrary order and a scalar by the law of *scalar multiplication*. If **A** is a matrix of order $m \times n$ and c is a scalar, we define $c\mathbf{A}$ to be a matrix of the same order as **A**. Moreover, every element of $c\mathbf{A}$ is c times the corresponding element of **A**. Thus, if $(\mathbf{A})_{ij}=a_{ij}$, we have

$$(c\mathbf{A})_{ij}=c(\mathbf{A})_{ij}=ca_{ij}. \tag{35}$$

The scalar can be commuted with the matrix, i.e.,

$$c\mathbf{A}=\mathbf{A}c. \tag{36}$$

EXAMPLE 5: Find $4\mathbf{A}$ and $-3\mathbf{A}$, where

$$\mathbf{A}=\begin{bmatrix} 3 & 2 & 1 \\ 5 & -6 & 3 \end{bmatrix}.$$

Solution: The desired matrices are

$$4\mathbf{A}=\begin{bmatrix} 12 & 8 & 4 \\ 20 & -24 & 12 \end{bmatrix}; \quad -3\mathbf{A}=\begin{bmatrix} -9 & -6 & -3 \\ -15 & 18 & -9 \end{bmatrix}.$$

We shall now define three elementary operations which can be carried out on matrices to give new matrices from a given matrix. These are known as *derived matrices*.

Complex conjugation

If $\mathbf{A}\equiv[a_{ij}]_{m\times n}$ is an arbitrary matrix whose elements may be complex numbers, the *complex conjugate matrix* denoted by $\mathbf{A}^*$ is also a matrix of order $m \times n$ every element of which is the complex conjugate of the corresponding element of **A**, i.e.,

$$(\mathbf{A}^*)_{ij}=(\mathbf{A})_{ij}^*=a_{ij}^*. \tag{37}$$

It is clear that if c is any scalar, then

$$(c\mathbf{A})^*=c^*\mathbf{A}^*. \tag{38}$$

Transposition

If the rows and the columns of a matrix are interchanged, the resulting matrix is called the *transposed matrix*. We shall denote the transposed matrix by $\tilde{\mathbf{A}}$ (called **A** tilde) or $\mathbf{A}^T$.[3] Thus, if **A** is the $m \times n$ matrix of Eq. (4), its transposed matrix is

[3]For convenience, we shall denote the transpose of a single matrix by using the tilde and the transpose of the product of a number of matrices by using the superscript T.

$$\widetilde{\mathbf{A}}=\begin{bmatrix} a_{11} & a_{21}\ldots a_{m1} \\ a_{12} & a_{22}\ldots a_{m2} \\ \vdots & \\ a_{1n} & a_{2n}\ldots a_{mn} \end{bmatrix}, \tag{39}$$

which is clearly a matrix of order $n\times m$. In terms of the elements, we have

$$(\widetilde{\mathbf{A}})_{ij}=(\mathbf{A})_{ji} \text{ for } 1\leqslant i\leqslant n,\ 1\leqslant j\leqslant m. \tag{40}$$

It should be evident that the transposed matrix of a column vector would be a row vector and vice versa.

Hermitian conjugation

When the two operations of complex conjugation and transposition are carried out one after another on a matrix, the resulting matrix is called[4] the *hermitian conjugate* of the original matrix and will be denoted by $\mathbf{A}^\dagger$ (called A dagger). The order of the two operations is immaterial, as is shown below:

$$\begin{array}{ccc} \mathbf{A}=\begin{bmatrix} a_{11} & a_{12}\ldots a_{1n} \\ a_{21} & a_{22}\ldots a_{2n} \\ \vdots & \\ a_{m1} & a_{m2}\ldots a_{mn} \end{bmatrix} & \rightarrow & \mathbf{A}^*=\begin{bmatrix} a_{11}^* & a_{12}^*\ldots a_{1n}^* \\ a_{21}^* & a_{22}^*\ldots a_{2n}^* \\ \vdots & \\ a_{m1}^* & a_{m2}^*\ldots a_{mn}^* \end{bmatrix} \\ \downarrow & & \downarrow \\ \widetilde{\mathbf{A}}=\begin{bmatrix} a_{11} & a_{21}\ldots a_{m1} \\ a_{12} & a_{22}\ldots a_{m2} \\ \vdots & \\ a_{1n} & a_{2n}\ldots a_{mn} \end{bmatrix} & \rightarrow & \mathbf{A}^\dagger=\begin{bmatrix} a_{11}^* & a_{21}^*\ldots a_{m1}^* \\ a_{12}^* & a_{22}^*\ldots a_{m2} \\ \vdots & \\ a_{1n}^* & a_{2n}^*\ldots a_{mn}^* \end{bmatrix}. \end{array} \tag{41}$$

From this, it is clear that

$$\mathbf{A}^\dagger=(\mathbf{A}^*)^T=(\widetilde{\mathbf{A}})^*. \tag{42}$$

In terms of the elements, we have

$$(\mathbf{A}^\dagger)_{ij}=a_{ji}^*,\ 1\leqslant i\leqslant n,\ 1\leqslant j\leqslant m. \tag{43}$$

$\mathbf{A}^\dagger$, like $\widetilde{\mathbf{A}}$, is seen to be a matrix of order $n\times m$ when $\mathbf{A}$ is of order $m\times n$.

[4]Some authors, particularly mathematicians, denote the complex conjugate (our $\mathbf{A}^*$) by $\bar{\mathbf{A}}$ and the hermitian conjugate (our $\mathbf{A}^\dagger$) by $\mathbf{A}^*$. Mathematicians also refer to $\mathbf{A}^\dagger$ as the *adjoint* matrix. However, what we mean by the adjoint is defined under Inverse Matrix (Section 5).

EXAMPLE 6: For the given matrix $\mathbf{A}$, write down $\mathbf{A}^*$, $\widetilde{\mathbf{A}}$ and $\mathbf{A}^\dagger$:

$$\mathbf{A}=\begin{bmatrix} 2+3i & 1-i & 5i & -3 \\ 1+i & 6-i & 1+3i & -1-2i \\ 5-6i & 3 & 0 & -4 \end{bmatrix}. \tag{44a}$$

Solution: We have

$$\mathbf{A}^*=\begin{bmatrix} 2-3i & 1+i & -5i & -3 \\ 1-i & 6+i & 1-3i & -1+2i \\ 5+6i & 3 & 0 & -4 \end{bmatrix}, \tag{44b}$$

$$\widetilde{\mathbf{A}}=\begin{bmatrix} 2+3i & 1+i & 5-6i \\ 1-i & 6-i & 3 \\ & 1+3i & 0 \\ -3 & -1-2i & -4 \end{bmatrix}, \tag{44c}$$

and

$$\mathbf{A}^\dagger=\begin{bmatrix} 2-3i & 1-i & 5+6i \\ 1+i & 6+i & 3 \\ -5i & 1-3i & 0 \\ -3 & -1+2i & -4 \end{bmatrix}. \tag{44d}$$

EXAMPLE 7: A result of considerable importance in matrix algebra is[5]

$$(\mathbf{AB})^T=\widetilde{\mathbf{B}}\,\widetilde{\mathbf{A}}, \tag{45}$$

which, when expressed in words, states that *the transpose of the product of two matrices equals the product of the transposed matrices taken in reverse order*.

Solution: To prove this *ab initio*, as usual, we must show that every element of the left hand side is equal to the corresponding element of the right hand side. The *ij*-th element of the left hand side of Eq. (45) is given by

$$\begin{aligned} [(\mathbf{AB})^T]_{ij} &= (\mathbf{AB})_{ji} \quad \text{[by Eq. (40)]} \\ &= \sum_k (\mathbf{A})_{jk}\,(\mathbf{B})_{ki} \quad \text{[by Eq. (16)]}. \end{aligned} \tag{46}$$

[5]Hereafter, whenever we write $\mathbf{A}+\mathbf{B}$ or $\mathbf{AB}$, it will be implicitly assumed that the orders of the matrices are such that their sum or product is defined. Thus, we shall not use such explicit phrases as 'let $\mathbf{A}$ and $\mathbf{B}$ be two matrices whose sum (or product) is defined'. Note that if the matrix product $\mathbf{AB}$ is defined, then the product $\widetilde{\mathbf{B}}\,\widetilde{\mathbf{A}}$ also is defined. Moreover the matrices $(\mathbf{AB})^T$ and $\widetilde{\mathbf{B}}\,\widetilde{\mathbf{A}}$ are of the same order, so that their equality is defined. Try to prove this as a simple exercise.

The ij-th element of the right hand side of Eq. (45) is

$$(\tilde{\mathbf{B}}\,\tilde{\mathbf{A}})_{ij}=\sum_k (\tilde{\mathbf{B}})_{ik}\,(\tilde{\mathbf{A}})_{kj} \quad \text{[by Eq. 16)]}$$

$$=\sum_k (\mathbf{B})_{ki}\,(\mathbf{A})_{jk} \quad \text{[by Eq. (40)]}$$

$$=\sum_k (\mathbf{A})_{jk}\,(\mathbf{B})_{ki}, \tag{47}$$

where, in the last step, we have commuted $(\mathbf{A})_{jk}$ with $(\mathbf{B})_{ki}$ because they are just numbers. The equality of the two elements in Eqs. (46) and (47) immediately leads to the result of Eq. (45).

Linear transformation and matrices

A linear transformation can be concisely expressed in the matrix notation. Thus, the transformation of Eq. (1.39) can be written as

$$\mathbf{y}=\mathbf{Bx}, \tag{48}$$

where $\mathbf{y}$ and $\mathbf{x}$ are the column vectors

$$\mathbf{y}=\{y_1, y_2, \ldots, y_n\}, \quad \mathbf{x}=\{x_1, x_2, \ldots, x_n\}, \tag{49}$$

and $\mathbf{B}\equiv[b_{ij}]$ is the $n\times n$ matrix of coefficients known as the *matrix of transformation.* Similarly, Eq. (1.40) can be condensed to

$$\mathbf{z}=\mathbf{Ay}, \tag{50}$$

where $\mathbf{z}\equiv\{z_i\}$ is an $n\times 1$ column vector and $\mathbf{A}\equiv[a_{ij}]$ is the $n\times n$ matrix of coefficients. Substituting for $\mathbf{y}$ from Eq. (48) into Eq. (50) we get

$$\mathbf{z}=\mathbf{ABx}=\mathbf{Cx}, \tag{51}$$

where $\mathbf{C}=\mathbf{AB}$ is the product matrix whose elements are given by Eq. (1.43).

We can think of a linear transformation from one set of variables to another where the two sets contain different numbers of elements. Thus, let $\mathbf{x}=\{x_1, x_2, \ldots, x_p\}$ and $\mathbf{y}=\{y_1, y_2, \ldots, y_n\}$ be two column vectors whose components are the p and n variables x_i and y_i, respectively. A linear transformation from one set to another can be written as

$$y_i=\sum_{k=1}^{p} b_{ik}\,x_k, \quad 1\leqslant i\leqslant n, \tag{52}$$

or as $\mathbf{y}=\mathbf{Bx}$, where $\mathbf{B}\equiv[b_{ik}]$ is now an $n\times p$ matrix. If the variables y_i are further related to the m variables z_i by a linear transformation of the form

$$z_i = \sum_{k=1}^{n} a_{ik}\, y_k, \quad 1 \leqslant i \leqslant m, \tag{53}$$

which can be written in the matrix notation as $\mathbf{z}=\mathbf{A}\mathbf{y}$, where $\mathbf{z}\equiv\{z_i\}$ is a column vector of order $m\times 1$ and $\mathbf{A}\equiv[a_{ij}]$ is a matrix of order $m\times n$, then the combined transformation can again be written as in Eq. (51) with the difference that all the matrices are now rectangular. The product matrix $\mathbf{C}=\mathbf{AB}$ will be of order $m\times p$ whose elements are given by Eq. (16). This is the reason why we defined matrix multiplication in the particular way we did.

When we regard vectors as column matrices, their inner product defined in Eq. (1.9) can also be written in a concise way in the matrix notation. If $\mathbf{u}=\{u_1, u_2, \ldots, u_n\}$ and $\mathbf{v}=\{v_1, v_2, \ldots, v_n\}$ are two column vectors, $\mathbf{u}^\dagger=[u_1^*, u_2^*, \ldots, u_n^*]$ and

$$\mathbf{u}^\dagger\mathbf{v}=[u_1^*, u_2^*, \ldots, u_n^*]\,\{v_1, v_2, \ldots, v_n\}=\sum_{i=1}^{n} u_i^*\, v_i, \tag{54}$$

which is the same expression as in Eq. (1.9). Hence the inner product can be expressed as[6]

$$(\mathbf{u}, \mathbf{v})=\mathbf{u}^\dagger\,\mathbf{v}. \tag{55}$$

The orthogonality condition of Eq. (1.23) for two vectors in this notation becomes

$$\mathbf{u}^\dagger\,\mathbf{v}=0. \tag{56}$$

If the components of the vectors are real, the inner product reduces to

$$\tilde{\mathbf{u}}\,\mathbf{v}=\sum_{i=1}^{n} u_i v_i, \tag{57}$$

and the orthogonality condition to

$$\tilde{\mathbf{u}}\,\mathbf{v}=0. \tag{58}$$

Exercise 2

1. Obtain $\mathbf{A}+\mathbf{B}$ and $\mathbf{A}-\mathbf{B}$ in each of the following cases:

(a) $\mathbf{A}=\begin{bmatrix} 3 & -9 \\ 2 & -7 \\ 5 & -6 \end{bmatrix}$, $\mathbf{B}=\begin{bmatrix} 1 & 6 \\ -3 & 0 \\ 8 & -11 \end{bmatrix}$;

[6]Other notations such as $\mathbf{u}.\mathbf{v}$ and $\langle \mathbf{u}|\mathbf{v}\rangle$ are also used for the inner product. Since we are mainly concerned with matrices in this chapter, we shall use the matrix notation $\mathbf{u}^\dagger\mathbf{v}$ for the inner product of two vectors in the remaining part of this chapter.

(b) $\mathbf{A}=\begin{bmatrix} a^2 & b^2 \\ 2a & ac \end{bmatrix}$, $\mathbf{B}=\begin{bmatrix} b^2 & bc \\ ac & c^2 \end{bmatrix}$;

*(c) $\mathbf{A}=\begin{bmatrix} 1/3 & 1/5 & 3/5 & 2/7 \\ 3/8 & 1/6 & 5/6 & 9/4 \\ 1/11 & 2/7 & 3/11 & 1/4 \end{bmatrix}$, $\mathbf{B}=\begin{bmatrix} 2 & 1/3 & 9 & 5 \\ -3 & 1/6 & -1/6 & 11 \\ 0 & 3/7 & 6/11 & -3/4 \end{bmatrix}$.

2. In each of the following cases, find the product **AB** if it is defined:

(a) $\mathbf{A}=\begin{bmatrix} 3 & 9 \\ -1 & 7 \end{bmatrix}$, $\mathbf{B}=\begin{bmatrix} -2 & 7 \\ 1 & 5 \end{bmatrix}$;

*(b) $\mathbf{A}=\begin{bmatrix} 3 & 5 & 9 \\ 1 & -2 & 3 \\ -5 & 0 & 2 \end{bmatrix}$, $\mathbf{B}=\begin{bmatrix} 7 \\ 1 \\ -3 \end{bmatrix}$;

(c) $\mathbf{A}=[1 \quad 3 \quad 2]$, $\mathbf{B}=\begin{bmatrix} 1 & 7 & -6 \\ 2 & 0 & 3 \\ 5 & -2 & 6 \end{bmatrix}$;

(d) $\mathbf{A}=[3 \quad -7 \quad 2]$, $\mathbf{B}=\begin{bmatrix} 1 \\ 5 \\ -7 \end{bmatrix}$;

*(e) $\mathbf{A}=\begin{bmatrix} 2 \\ -1 \\ 4 \end{bmatrix}$, $\mathbf{B}=[-3 \quad 7 \quad 4]$;

(f) The matrices of Exercise 1 (a);

(g) The matrices of Exercise 1 (b).

3. Given the two matrices

$$\mathbf{A}=\begin{bmatrix} 2 & 3 & -4 \\ 6 & 1 & 2 \\ 8 & 4 & -2 \end{bmatrix}, \quad \mathbf{B}=\begin{bmatrix} -5 & 5 & 10 \\ 14 & -14 & -28 \\ 8 & -8 & -16 \end{bmatrix},$$

show that $\mathbf{AB}=\mathbf{0}$, but that **BA** is not equal to the zero matrix. (This is a peculiarity of matrix multiplication that *the product of two matrices can be a zero matrix without any one of them being a zero matrix.*)

4. In each of the following cases, find **AB** and **BA** and determine whether **AB** equals **BA**:

*(a) $\mathbf{A}=\begin{bmatrix} 5 & 6 \\ -3 & 2 \end{bmatrix}$, $\mathbf{B}=\begin{bmatrix} -4 & 3 \\ 2 & 1 \end{bmatrix}$;

(b) $\mathbf{A}=\begin{bmatrix} 3 & -1 & 2 \\ 5 & 9 & 0 \\ -3 & 6 & 1 \end{bmatrix}$, $\mathbf{B}=\begin{bmatrix} 2 & 0 & -3 \\ 7 & -1 & 3 \\ -4 & 6 & 2 \end{bmatrix}$;

(c) The matrices of Exercise 2 (a);

*(d) The matrices of Exercise 2 (d);

(e) The matrices of Exercise 2 (e).

*5. If

$$\mathbf{A}=\begin{bmatrix} 1 & 3 & 5 \\ 2 & -4 & 6 \\ -1 & 0 & 1 \end{bmatrix},$$

find $\mathbf{A}^2=\mathbf{AA}$, $\mathbf{A}^3=\mathbf{AAA}$, and $\mathbf{A}^4=\mathbf{AAAA}$.

6. Let

$$\mathbf{A}=\begin{bmatrix} 0 & 1 & 0 \\ 1 & 0 & 1 \\ 0 & 1 & 0 \end{bmatrix}, \quad \mathbf{B}=\begin{bmatrix} 1 & 0 & 0 \\ 0 & 1 & 0 \\ 0 & 0 & 1 \end{bmatrix}, \quad \mathbf{C}=\begin{bmatrix} 1 & 0 & 0 \\ 0 & 0 & 0 \\ 0 & 0 & -1 \end{bmatrix}.$$

Show that **A** commutes with **B** and **B** commutes with **C**, but that **A** does not commute with **C**.

7. Show that

(a) $[\mathbf{A}, [\mathbf{B}, \mathbf{C}]]+[\mathbf{B}, [\mathbf{C}, \mathbf{A}]]+[\mathbf{C}, [\mathbf{A}, \mathbf{B}]]=0$;

(b) $[\mathbf{A}, \{\mathbf{B}, \mathbf{C}\}]=\{[\mathbf{A}, \mathbf{B}], \mathbf{C}\}+\{[\mathbf{A}, \mathbf{C}], \mathbf{B}\}$,

where $\{\mathbf{A}, \mathbf{B}\}=\mathbf{AB}+\mathbf{BA}$ is called the *anticommutator* of **A** and **B**.

*8. Let p_i be a set of n numbers $(1\leqslant i\leqslant n)$, not all zero, and let **A** be a matrix of order $n\times n$ with elements $(\mathbf{A})_{ij}=p_i p_j$. Find the condition on the elements p_i if $\mathbf{A}^2=\mathbf{A}$.

9. Show that (a) $(\mathbf{AB})^*=\mathbf{A}^*\mathbf{B}^*$, (b) $(\mathbf{A}+\mathbf{B})^*=\mathbf{A}^*+\mathbf{B}^*$.

10. Show that (a) $(\mathbf{A}+\mathbf{B})^T=\widetilde{\mathbf{A}}+\widetilde{\mathbf{B}}$, (b) $(\mathbf{A}+\mathbf{B})^\dagger=\mathbf{A}^\dagger+\mathbf{B}^\dagger$.

11. Show that $(\mathbf{AB})^\dagger=\mathbf{B}^\dagger\mathbf{A}^\dagger$.

12. Show that (a) $(\mathbf{ABC})^T=\widetilde{\mathbf{C}}\,\widetilde{\mathbf{B}}\,\widetilde{\mathbf{A}}$, (b) $(\mathbf{ABC})^\dagger=\mathbf{C}^\dagger\mathbf{B}^\dagger\mathbf{A}^\dagger$.

*13. Extend the results of Exercise 12 to the product of an arbitrary but finite number of matrices.

14. Show that $(c\mathbf{A})^\dagger=c^*\mathbf{A}^\dagger$, where c is any scalar and **A** is any matrix.

15. Let **A** and **B** be the matrices of Eq. (17). Obtain $\widetilde{\mathbf{A}}$, $\widetilde{\mathbf{B}}$, $(\mathbf{AB})^T$ and $\widetilde{\mathbf{B}}\,\widetilde{\mathbf{A}}$. Hence show that these two matrices satisfy Eq. (45).

*16. Let

$$\mathbf{A}=\begin{bmatrix} 1+i & 2i & 3 \\ 3-2i & -1+i & -2-4i \\ -5i & 0 & 6-i \end{bmatrix}, \quad \mathbf{B}=\begin{bmatrix} 2 & -1+6i & 0 \\ 3+i & i & 5 \\ 7+2i & 6+i & -3 \end{bmatrix}.$$

Obtain $\mathbf{AB}$, $\mathbf{A}^\dagger$, $\mathbf{B}^\dagger$, $(\mathbf{AB})^\dagger$ and $\mathbf{B}^\dagger\mathbf{A}^\dagger$. Hence show that these matrices satisfy the relation given in Exercise 11.

17. If **A** is an $m\times n$ matrix, show that $\mathbf{A}^\dagger\mathbf{A}$ is a diagonal matrix *if, and only if,* the columns of **A** are orthogonal to each other, while $\mathbf{AA}^\dagger$ is diagonal *if, and only if,* the rows of **A** are orthogonal to each other.

18. Let **P** be a hermitian matrix with the property $\mathbf{P}^2=\mathbf{P}$. Show that for any vector **x**, the vectors $\mathbf{Px}$ and $(\mathbf{I}-\mathbf{P})\mathbf{x}$ are orthogonal.

3. Special matrices, I

In this section, we shall define a number of special matrices which receive their names from the properties their elements satisfy.

Square, diagonal and constant matrices

If the number of rows of a matrix equals the number of its columns (that is, $m=n$), the matrix is said to be a *square matrix*. Such matrices give rise to various types of special matrices which frequently occur in physics. If **A** is a square matrix of order $n\times n$, we shall simply say that **A** is of order n.

Let $\mathbf{A}\equiv[a_{ij}]_n$ be a square matrix of order n with $1\leqslant i, j\leqslant n$. The diagonal going from the top left corner to the bottom right corner of the square matrix, which consists of the elements $a_{11}, a_{22},\ldots,a_{nn}$, is called the *principal diagonal* of the matrix. The elements a_{ii} $(1\leqslant i\leqslant n)$ are called the *diagonal elements* of the square matrix **A**. All the remaining elements a_{ij} for $i\neq j$ are called the *off-diagonal* or *nondiagonal elements.*

A square matrix which has nonvanishing elements only along the principal diagonal and zeros elsewhere is called a *diagonal matrix.* Thus, a diagonal matrix would have the from

$$\mathbf{A}=\begin{bmatrix} a_{11} & 0 & 0 & \ldots & 0 \\ 0 & a_{22} & 0 & \ldots & 0 \\ \vdots & & & & \\ 0 & 0 & 0 & \ldots & a_{nn} \end{bmatrix}, \tag{1a}$$

from which we have

$$(\mathbf{A})_{ij}=a_{ii}\,\delta_{ij}. \tag{1b}$$

If all the nonvanishing elements of a diagonal matrix happen to be equal to each other, it is said to be a *constant matrix*. The elements of a constant matrix are thus given by

$$(\mathbf{A})_{ij}=a\delta_{ij}, \tag{2}$$

where a is any scalar.

If we go one step further and take every nonvanishing element of a constant matrix equal to unity [i.e., $a=1$ in Eq. (2)], then we have a *unit matrix*. A unit matrix is generally denoted by the symbol **I**. Whenever we wish to be specific about the order of the unit matrix, we shall write $\mathbf{I}_n$ for the unit matrix of order n. Thus,

$$\mathbf{I}=\begin{bmatrix} 1 & 0 & 0 & \dots & 0 \\ 0 & 1 & 0 & \dots & 0 \\ \vdots & & & & \\ 0 & 0 & 0 & \dots & 1 \end{bmatrix}. \tag{3}$$

If **A** is a constant matrix defined in Eq. (2), we have

$$\mathbf{A}=\begin{bmatrix} a & 0 & 0 & \dots & 0 \\ 0 & a & 0 & \dots & 0 \\ \vdots & & & & \\ 0 & 0 & 0 & \dots & a \end{bmatrix}=a\begin{bmatrix} 1 & 0 & 0 & \dots & 0 \\ 0 & 1 & 0 & \dots & 0 \\ \vdots & & & & \\ 0 & 0 & 0 & \dots & 1 \end{bmatrix}, \tag{4a}$$

or

$$\mathbf{A}=a\mathbf{I}. \tag{4b}$$

This shows that *a constant matrix is a constant multiple of the unit matrix*. It is for this reason that a matrix of the form of Eq. (4a) is called a constant matrix.

There are a couple of interesting results regarding the commutability of diagonal and constant matrices with other matrices, which we shall discuss below in the form of examples.

EXAMPLE 1: If a matrix **B** commutes with a diagonal matrix no two diagonal elements of which are equal to each other, show that **B** must be a diagonal matrix.

Solution: Let **A** be a diagonal matrix of order n whose elements are

$$(\mathbf{A})_{ij}=a_i\,\delta_{ij}, \tag{5}$$

where a_i are scalars such that $a_i \neq a_j$ if $i \neq j$, and $1 \leqslant i, j \leqslant n$. Let the ij-th element of the matrix **B** be b_{ij}. We are given that

$$\mathbf{AB}=\mathbf{BA}. \tag{6}$$

Taking the ij-th element of both sides of Eq. (6), we have

$$\sum_{p=1}^{n} (\mathbf{A})_{ip}\,(\mathbf{B})_{pj}=\sum_{p=1}^{n} (\mathbf{B})_{ip}\,(\mathbf{A})_{pj}. \tag{7}$$

On using Eq. (5), this becomes

$$\sum_{p=1}^{n} a_i\,\delta_{ip}\,b_{pj}=\sum_{p=1}^{n} b_{ip}\,a_j\,\delta_{pj}$$

or

$$a_i\,b_{ij}=b_{ij}\,a_j \Rightarrow (a_i-a_j)\,b_{ij}=0. \tag{8}$$

This shows that if $i \neq j$, then $b_{ij}=0$. The only elements of **B** which are likely to be different from zero are the diagonal elements b_{ii} for $1 \leqslant i \leqslant n$, proving that **B** is a diagonal matrix.

EXAMPLE 2: Let $\mathbf{E}_{kl}$ $(1 \leqslant k, l \leqslant n)$ be a square matrix of order n

whose only nonvanishing element is the kl-th element which is equal to unity, that is,

$$(\mathbf{E}_{kl})_{ij} = \delta_{ik}\ \delta_{jl}. \tag{9}$$

(There are clearly n^2 such matrices.) If a matrix of order n commutes with the n matrices $\mathbf{E}_{kl}$ for $1 \leqslant k \leqslant n$ and fixed l, then show that it must be a constant matrix.

Solution: Let $\mathbf{B} \equiv [b_{ij}]$ be the matrix which commutes with $\mathbf{E}_{kl}$, that is

$$\mathbf{BE}_{kl} = \mathbf{E}_{kl}\ \mathbf{B},\ 1 \leqslant k \leqslant n, \text{ fixed } l. \tag{10}$$

Equating the ij-th element of both sides, we have

$$\sum_{p=1}^{n} b_{ip}\,(\mathbf{E}_{kl})_{pj} = \sum_{p=1}^{n} (\mathbf{E}_{kl})_{ip}\, b_{pj},$$

or

$$\sum_{p=1}^{n} b_{ip}\,\delta_{kp}\,\delta_{lj} = \sum_{p=1}^{n} \delta_{ik}\,\delta_{lp}\, b_{pj},$$

or

$$b_{ik}\,\delta_{lj} = \delta_{ik}\, b_{lj}. \tag{11}$$

If, in particular, we take $j = l$, we get

$$b_{ik} = b_{ll}\,\delta_{ik},\ 1 \leqslant i,\ k \leqslant n. \tag{12}$$

The above equation shows two things. First, all the nondiagonal elements of $\mathbf{B}$ for which $i \neq k$ vanish. Second, if $i = k$, the diagonal element $b_{ii} = b_{ll}$ for $1 \leqslant i \leqslant n$ and fixed l. This proves that $\mathbf{B}$ is a constant matrix.

Real, symmetric and hermitian matrices

In the previous section, we have defined three operations on matrices which, starting from a matrix $\mathbf{A}$, give us the matrices $\mathbf{A}^*$, $\widetilde{\mathbf{A}}$ and $\mathbf{A}^\dagger$. Of these, $\mathbf{A}$ and $\mathbf{A}^*$ are of the same order, say $m \times n$, while $\widetilde{\mathbf{A}}$ and $\mathbf{A}^\dagger$ would then be of order $n \times m$. We get the following six special matrices by imposing various relations between any two of these four matrices.

If

$$\mathbf{A}^* = \mathbf{A}, \tag{13a}$$

it is clear, on equating the ij-th element of both sides, that

$$a_{ij}^* = a_{ij},\ 1 \leqslant i \leqslant m,\ 1 \leqslant j \leqslant n. \tag{13b}$$

This shows that each element of the matrix $\mathbf{A}$ is real. A matrix satisfying Eq. (13a) or Eq. (13b) is called a *real matrix*.

If

$$\mathbf{A}^* = -\mathbf{A} \Rightarrow a_{ij}^* = -a_{ij},\ 1 \leqslant i \leqslant m,\ 1 \leqslant j \leqslant n, \tag{14}$$

it is evident that every element of **A** must be purely imaginary or zero. A matrix obeying Eq. (14) is called a *purely imaginary matrix*.

Let us now consider the equality of **A** with $\widetilde{\mathbf{A}}$ or $\mathbf{A}^\dagger$. The equality of **A** and $\widetilde{\mathbf{A}}$ or of **A** and $\mathbf{A}^\dagger$ will be defined only if **A** is a square matrix with $m=n$. We then get the following four special matrices.

A matrix is called *symmetric* if

$$\widetilde{\mathbf{A}}=\mathbf{A}\Rightarrow a_{ij}=a_{ji},\ 1\leqslant i, j\leqslant n. \tag{15}$$

A matrix is called *antisymmetric* (or *skew-symmetric*) if

$$\widetilde{\mathbf{A}}=-\mathbf{A}\Rightarrow a_{ij}=-a_{ji},\ 1\leqslant i,\ j\leqslant n. \tag{16}$$

A matrix is called *hermitian* if

$$\mathbf{A}^\dagger=\mathbf{A}\Rightarrow a_{ij}{}^*=a_{ji},\ 1\leqslant i,\ j\leqslant n. \tag{17}$$

Finally, a matrix is called *antihermitian* (or *skew-hermitian*) if

$$\mathbf{A}^\dagger=-\mathbf{A}\Rightarrow a_{ij}{}^*=-a_{ji},\ 1\leqslant i,\ j\leqslant n. \tag{18}$$

In any of the above six cases, note that one may think of the particular condition being imposed on the matrix or on its elements. Both the views are, of course, equivalent.

EXAMPLE 3: Show that every square matrix can be uniquely written as the sum of a hermitian and an antihermitian matrix.

Solution: We shall prove this by two methods.

First proof. Let $\mathbf{A}\equiv[a_{ij}]$ be a square matrix of order n. Suppose **A** is expressible as

$$\mathbf{A}=\mathbf{B}+\mathbf{C}, \tag{19}$$

where **B** is a hermitian matrix and **C** an antihermitian matrix, both of order n. Now, if we are able to determine the elements of **B** and **C** in terms of those of **A**, the proof will be complete.

Let $(\mathbf{B})_{ij}=b_{ij}$ and $(\mathbf{C})_{ij}=c_{ij}$. Then, equating the ij-th element of both sides of Eq. (19), we get

$$a_{ij}=b_{ij}+c_{ij}. \tag{20}$$

Taking the hermitian conjugate of Eq. (19) and using Eqs. (17), (18) and the result of Exercise 2.10(b), we have

$$\mathbf{A}^\dagger=(\mathbf{B}+\mathbf{C})^\dagger=\mathbf{B}^\dagger+\mathbf{C}^\dagger=\mathbf{B}-\mathbf{C}. \tag{21}$$

Equating the ij-th element of both sides of the above equation gives

$$a_{ji}{}^*=b_{ij}-c_{ij}. \tag{22}$$

Finally, adding and subtracting Eqs. (20) and (22), we find

$$b_{ij}=(a_{ij}+a_{ji}{}^*)/2,\ c_{ij}=(a_{ij}-a_{ji}{}^*)/2. \tag{23}$$

This uniquely determines the elements of **B** and **C** in terms of those of **A**. As a check, we can verify that **B** is hermitian and **C** is antihermitian. Thus, from Eqs. (23), we have

$$b_{ij}^*=(a_{ij}^*+a_{ji})/2=b_{ji}, \tag{24a}$$

$$c_{ij}^*=(a_{ij}^*-a_{ji})/2=-(a_{ji}-a_{ij}^*)/2=-c_{ji}. \tag{24b}$$

Second proof. As before, we suppose that **A** is expressible as in Eq. (19). Taking the hermitian conjugate of both sides of Eq. (19), we get

$$\mathbf{A}^\dagger=\mathbf{B}-\mathbf{C}. \tag{25}$$

Adding and subtracting Eqs. (19) and (25), we find

$$\mathbf{B}=(\mathbf{A}+\mathbf{A}^\dagger)/2,\ \mathbf{C}=(\mathbf{A}-\mathbf{A}^\dagger)/2. \tag{26}$$

Taking the hermitian conjugate of Eqs. (26), we get

$$\mathbf{B}^\dagger=(\mathbf{A}^\dagger+\mathbf{A})/2=\mathbf{B},\ \mathbf{C}^\dagger=(\mathbf{A}^\dagger-\mathbf{A})/2=-\mathbf{C}. \tag{27}$$

Comparing this with Eqs. (17) and (18), it is clear that **B** is hermitian and **C** is antihermitian.

We notice from the above example that there are in general two methods for solving problems involving identities in matrix algebra. In the first method, one may proceed by taking the elements of both sides of a matrix equation and use the basic definition of the equality of matrices, that two matrices are equal if and only if their corresponding elements are equal. In the second method, one uses the various matrix operations and obtains the same result without going to the elements. The second method is obviously shorter and more elegant. One more example will perhaps be useful.

EXAMPLE 4. If **A** is an antisymmetric matrix of order n and **x** a column vector of dimension n, show that $\tilde{\mathbf{x}}\mathbf{A}\mathbf{x}=0$.

Solution: Again, we shall prove this by two methods.

First proof. Let $\mathbf{A}\equiv[a_{ij}]$ be an antisymmetric matrix so that $a_{ji}=-a_{ij}$ and let $\mathbf{x}=\{x_1\ x_2\ldots x_n\}$ be a column vector. It is clear that the triple matrix product $\tilde{\mathbf{x}}\mathbf{A}\mathbf{x}$ is a matrix of order 1×1, i.e., a number. By the definition of matrix multiplication, we have

$$\tilde{\mathbf{x}}\mathbf{A}\mathbf{x}=[\tilde{\mathbf{x}}\mathbf{A}\mathbf{x}]_{11}=\sum_{i,j=1}^{n}[\tilde{\mathbf{x}}]_{1i}\ [\mathbf{A}]_{ij}\ [\mathbf{x}]_{j1}. \tag{28}$$

Now, $(\mathbf{x})_{j1}=x_j$, so that $(\tilde{\mathbf{x}})_{1i}=x_i$, and the above equation becomes

$$(\tilde{\mathbf{x}}\,\mathbf{A}\mathbf{x})_{11}=\sum_{i,j=1}^{n} a_{ij}\, x_i\, x_j. \tag{29}$$

Both i and j are dummy summation indices; interchanging i and j in the above equation, we get

$$(\tilde{\mathbf{x}}\,\mathbf{A}\mathbf{x})_{11}=\sum_{i,j=1}^{n} a_{ji}\, x_j\, x_i=-\sum_{i,j=1}^{n} a_{ij}\, x_i\, x_j=-(\tilde{\mathbf{x}}\,\mathbf{A}\mathbf{x})_{11}, \tag{30}$$

which shows that

$$(\tilde{\mathbf{x}}\,\mathbf{A}\mathbf{x})_{11}\equiv\tilde{\mathbf{x}}\,\mathbf{A}\mathbf{x}=0. \tag{31}$$

Second proof. Since $\mathbf{A}$ is antisymmetric, we have $\tilde{\mathbf{A}}=-\mathbf{A}$. Taking the transpose of the matrix $\tilde{\mathbf{x}}\,\mathbf{A}\mathbf{x}$, we find

$$(\tilde{\mathbf{x}}\,\mathbf{A}\mathbf{x})^T=\tilde{\mathbf{x}}\,\tilde{\mathbf{A}}\,\mathbf{x}=-\tilde{\mathbf{x}}\,\mathbf{A}\mathbf{x}. \tag{32}$$

But the matrix $\tilde{\mathbf{x}}\,\mathbf{A}\mathbf{x}$ has just one row and one column, hence its transpose must be identical to itself, i.e.,

$$(\tilde{\mathbf{x}}\,\mathbf{A}\mathbf{x})^T=\tilde{\mathbf{x}}\,\mathbf{A}\mathbf{x}. \tag{33}$$

Comparing Eqs. (32) and (33), we have

$$\tilde{\mathbf{x}}\,\mathbf{A}\mathbf{x}=-\tilde{\mathbf{x}}\,\mathbf{A}\mathbf{x}\Rightarrow\tilde{\mathbf{x}}\,\mathbf{A}\mathbf{x}=0. \tag{34}$$

Triangular matrix

A square matrix is said to be *triangular* if all the elements on one side of the principal diagonal vanish, i.e., $(\mathbf{A})_{ij}=0$ for $i<j$ or $i>j$. Thus,

$$\mathbf{A}=\begin{bmatrix} 3 & 4 & -1 \\ 0 & 2 & 5 \\ 0 & 0 & -6 \end{bmatrix} \text{ and } \mathbf{B}=\begin{bmatrix} 1 & 0 & 0 & 0 \\ 2 & -2 & 0 & 0 \\ 3 & 5 & 3 & 0 \\ 4 & 6 & -4 & 5 \end{bmatrix}$$

are examples of a triangular matrix. It is clear that there are two types of a triangular matrix. A matrix for which $(\mathbf{A})_{ij}=0$ for $i>j$ is called a *right-triangular* matrix (matrix $\mathbf{A}$ above), whereas a matrix with $(\mathbf{A})_{ij}=0$ for $i<j$ is called a *left-triangular* matrix (matrix B above).

Number of independent parameters

A general matrix of order $m\times n$ whose elements may be complex clearly has $2mn$ independent *real* parameters. A general square matrix of order n has $2n^2$ independent real parameters.

If the matrix is, however, one of the special types, its elements satisfy certain relations among themselves, and this reduces the number of independent parameters required to characterize the matrix completely. For example, if a matrix of order $m \times n$ is known to be real or purely imaginary, it can be completely specified only by mn independent real parameters. The reason is that the elements of these matrices satisfy mn relations among themselves [Eq. (13) or Eq. (14)] which reduce the number of independent parameters. It is fairly easy to see that, in general, the number of independent parameters required to characterize a matrix completely equals $2mn$ minus the number of independent conditions satisfied by the matrix elements.

Let us consider the various special matrices discussed earlier in this section. Take for example a general symmetric matrix whose elements satisfy Eq. (15). We shall determine the number of independent parameters of a general symmetric matrix in two ways; first directly, and then by determining the number of conditions satisfied by the elements and subtracting this from $2n^2$ (because $m=n$ for a symmetric matrix).

In a symmetric matrix $\mathbf{A} \equiv [a_{ij}]_n$, we need to know only the elements on one side of the diagonal (including the diagonal elements), that is, those elements for which, say, $i \leqslant j$. The other elements for which $i > j$ are related to these elements. The number of elements for which $i \leqslant j$ is $1+2+3+\ldots+n=n(n+1)/2$. The total number of independent real parameters is therefore $n(n+1)$, for each element may be complex.

To determine the number of independent conditions, we note that we have $a_{ij}=a_{ji}$ for all i and j between 1 and n. The independent equations among these are obtained only for $1 \leqslant i < j \leqslant n$, which are $1+2+\ldots+(n-1)=n(n-1)/2$ in number. Since each complex equation is equivalent to two real conditions, we have a set of $n(n-1)$ real conditions on the elements of a symmetric matrix. The number of independent real parameters is therefore $2n^2-n(n-1)=n(n+1)$, which agrees with the earlier calculation.

An antisymmetric matrix of order n would have $n(n-1)$ real independent parameters. The number of real independent parameters of a hermitian or an antihermitian matrix of order n turns out to be n^2, a verification of which is left to the reader.

EXAMPLE 5: Write down the most general symmetric and the most general hermitian matrix of order 2.

Solution: The most general symmetric matrix of order 2 is

$$A_S=\begin{bmatrix} a+ib & e+if \\ e+if & c+id \end{bmatrix}, \tag{35}$$

which has six real independent parameters. The most general hermitian matrix of order 2 can be written in terms of four real independent parameters in the form

$$A_H=\begin{bmatrix} a & c+id \\ c-id & b \end{bmatrix}. \tag{36}$$

Independent matrices and linear combinations

If two matrices A_1 and A_2, both of the same order $m\times n$ and none of which is a zero matrix, are not multiples of each other, we shall say that A_1 and A_2 are *linearly independent* of each other.[1] In other words, A_1 and A_2 are linearly independent matrices if and only if there exists no constant c such that $A_1=cA_2$.

We can extend this concept to more than two matrices. A set of matrices $A_1, A_2,\ldots, A_p$, all of the same order $m\times n$ and none of which is a zero matrix, is said to be *linearly independent* if the equation

$$\sum_{i=1}^{p} c_i\, A_i=0, \tag{37}$$

where c_i are scalars, implies that

$$c_i=0,\ 1\leqslant i\leqslant p. \tag{38}$$

On the other hand, the matrices of the set are *linearly dependent* if it is possible to find scalars c_i, at least some of which are nonzero, such that Eq. (37) is satisfied. Suppose a particular coefficient c_k is different from zero. Transferring this term to the other side of the equation and dividing by c_k, we have

$$A_k=-\sum_{i\neq k} (c_i/c_k)\, A_i, \tag{39}$$

showing that the matrix A_k can be expressed as a *linear combination* of the remaining $p-1$ matrices. *It is in this sense that the matrices of the set are called linearly dependent*. Contrary to this, *the matrices*

[1]Compare this with the linear dependence and independence of vectors; Section 1.

of a set are linearly independent if none of them can be expressed as a linear combination of the others.

For $p=2$, Eq. (37) reduces to $c_1 \mathbf{A}_1 + c_2 \mathbf{A}_2 = \mathbf{0}$, so that if c_1 and c_2 are nonzero, we have

$$\mathbf{A}_1 = -(c_2/c_1)\,\mathbf{A}_2, \tag{40}$$

showing that $\mathbf{A}_1$ is a constant multiple of $\mathbf{A}_2$, as said earlier.

The concepts developed here will be found to be related to those of vectors treated in Section 1. The maximum number of linearly independent matrices of order $m \times n$ is equal to mn (assuming that we allow complex coefficients in taking linear combinations). Any other matrix (of the same order) can be expressed as a linear combination of a set of mn linearly independent matrices with complex coefficients. If we are dealing with a set of special matrices, the maximum number of linearly independent matrices (of the kind under consideration) equals the number of independent parameters of the matrix when only real coefficients are allowed and half the number of independent parameters when complex coefficients are allowed. Thus, the maximum number of *real* independent matrices of order $m \times n$ is mn, the maximum number of linearly independent symmetric matrices of order n is $n(n+1)/2$, the maximum number of linearly independent hermitian or antihermitian matrices of order n is n^2, etc.

EXAMPLE 6: (a) Construct a set of mn linearly independent matrices of order $m \times n$, and (b) show that any matrix of order $m \times n$ can be expressed as a linear combination of these mn matrices.

Solution: (a) Let us define a matrix $\mathbf{E}_{kl}$ of order $m \times n$ which has unity in the kl-th position and zeros elsewhere. This is a generalization of the matrices defined in Example 2. Evidently, we have mn such matrices here, and we shall show that they are linearly independent.

Let a_{kl} be a set of mn constants (which may be complex in general) and consider the matrix equation

$$\sum_{k=1}^{m} \sum_{l=1}^{n} a_{kl}\, \mathbf{E}_{kl} = \mathbf{0}. \tag{41a}$$

Consider the ij-th element of the matrix on the left hand side of Eq. (41a). Since only the matrix $\mathbf{E}_{ij}$ has unity in the ij-th position, it is obvious that this element equals a_{ij}. In fact, writing out the left hand side of Eq. (41a) explicitly, we have

$$a_{11}\begin{bmatrix} 1 & 0 & \dots & 0 \\ 0 & 0 & \dots & 0 \\ \vdots \\ 0 & 0 & \dots & 0 \end{bmatrix} + a_{12}\begin{bmatrix} 0 & 1 & 0 & \dots & 0 \\ 0 & 0 & 0 & \dots & 0 \\ \vdots \\ 0 & 0 & 0 & \dots & 0 \end{bmatrix} + \dots$$

$$+ a_{mn}\begin{bmatrix} 0 & 0 & \dots & 0 \\ 0 & 0 & \dots & 0 \\ \vdots \\ 0 & 0 & \dots & 1 \end{bmatrix} = \begin{bmatrix} a_{11} & a_{12} & \dots & a_{1n} \\ a_{21} & a_{22} & \dots & a_{2n} \\ \vdots \\ a_{m1} & a_{m2} & \dots & a_{mn} \end{bmatrix}. \quad (41b)$$

Comparing this with the right hand side of Eq. (41a), we see that for this equation to be valid, we must have

$$a_{ij}=0, \ 1\leqslant i\leqslant m, \ 1\leqslant j\leqslant n. \quad (42)$$

Since this is the only solution of Eq. (41a), it follows that the matrices $\mathbf{E}_{kl}$ are linearly independent.

This set of linearly independent matrices is, of course, not unique, and such a set may be chosen in an infinite number of ways.

(b) Let $\mathbf{B}\equiv[b_{ij}]_{m\times n}$ be any matrix whose elements may be complex. It is then clear that we can write it in the form

$$\mathbf{B}=\sum_{k=1}^{m}\sum_{l=1}^{n} b_{kl}\,\mathbf{E}_{kl}, \quad (43)$$

which proves that any matrix of order $m\times n$ can be expressed as a linear combination of the mn matrices $\mathbf{E}_{kl}$; mn is thus the maximum number of linearly independent matrices, that is, there is no other matrix of order $m\times n$ which is linearly independent of the above set.

EXAMPLE 7: Find a set of four linearly independent hermitian matrices of order 2.

Solution: Let us try the set of four hermitian matrices

$$\mathbf{A}_1=\begin{bmatrix} 1 & 0 \\ 0 & 0 \end{bmatrix}, \mathbf{A}_2=\begin{bmatrix} 0 & 0 \\ 0 & 1 \end{bmatrix}, \mathbf{A}_3=\begin{bmatrix} 0 & 1 \\ 1 & 0 \end{bmatrix}, \mathbf{A}_4=\begin{bmatrix} 0 & -i \\ i & 0 \end{bmatrix}. \quad (44)$$

To show that these are linearly independent, once again we solve the equation

$$\sum_{i=1}^{4} a_i\mathbf{A}_i=\mathbf{0} \quad (45)$$

for the coefficients a_i. Substituting from Eqs. (44), Eq. (45) becomes

$$\begin{bmatrix} a_1 & a_3-ia_4 \\ a_3+ia_4 & a_2 \end{bmatrix}=\begin{bmatrix} 0 & 0 \\ 0 & 0 \end{bmatrix}. \quad (46)$$

Equating the corresponding elements in the above equation, we find that the only solution is $a_i=0$ for $1\leqslant i\leqslant 4$, showing that the four matrices $\mathbf{A}_1$, $\mathbf{A}_2$, $\mathbf{A}_3$ and $\mathbf{A}_4$ are linearly independent.

Once again, this set is not unique, and any four independent linear combinations can be chosen. The conventional choice is to take the sum and the difference of $\mathbf{A}_1$ and $\mathbf{A}_2$, giving the four matrices

$$\mathbf{I}=\begin{bmatrix} 1 & 0 \\ 0 & 1 \end{bmatrix}, \sigma_z=\begin{bmatrix} 1 & 0 \\ 0 & -1 \end{bmatrix}, \sigma_x=\begin{bmatrix} 0 & 1 \\ 1 & 0 \end{bmatrix}, \sigma_y=\begin{bmatrix} 0 & -i \\ i & 0 \end{bmatrix}. \quad (47)$$

The first of the above four matrices is the unit matrix of order 2 and the last three matrices are known as the *Pauli spin matrices* for a spin-$\frac{1}{2}$ particle in quantum mechanics.

Matrices with polynomial elements

Consider a matrix of order $m\times n$ given by

$$\mathbf{A}=\begin{bmatrix} f_{11}(x) & f_{12}(x) & \dots & f_{1n}(x) \\ f_{21}(x) & f_{22}(x) & \dots & f_{2n}(x) \\ \vdots & & & \\ f_{m1}(x) & f_{m2}(x) & \dots & f_{mn}(x) \end{bmatrix}, \quad (48)$$

whose elements $f_{ij}(x)$ are polynomials in x of degree not more than p. Then it can be shown that this can be written uniquely in the form

$$\mathbf{A}=\mathbf{B}_0+\mathbf{B}_1x+\mathbf{B}_2x^2+\dots+\mathbf{B}_px^p, \quad (49)$$

where $\mathbf{B}_k$ $(0\leqslant k\leqslant p)$ are matrices of order $m\times n$ whose elements are independent of x.

For this, let the function $f_{ij}(x)$ be a polynomial of the form

$$f_{ij}(x)=\sum_{k=0}^{p} a_{ij}(k)x^k, \quad (50)$$

where the coefficients $a_{ij}(k)$ are independent of x. Then we assert that the matrices $\mathbf{B}_k$ have the elements

$$(\mathbf{B}_k)_{ij}=a_{ij}(k). \quad (51)$$

To verify the truth of this assertion, consider the ij-th element of the right hand side of Eq. (49), which, on using Eq. (51), is found to be

$$\begin{aligned}(\mathbf{B}_0+\mathbf{B}_1x+\mathbf{B}_2x^2+\dots+\mathbf{B}_px^p)_{ij} \\ =(\mathbf{B}_0)_{ij}+(\mathbf{B}_1)_{ij}x+\dots+(\mathbf{B}_p)_{ij}x^p \\ =a_{ij}(0)+a_{ij}(1)x+\dots+a_{ij}(p)x^p,\end{aligned} \quad (52)$$

which is just the ij-th element of $\mathbf{A}$. Since the polynomial coefficients $a_{ij}(k)$ are unique and independent of x, the matrices $\mathbf{B}_k$ are also unique and independent of x.

EXAMPLE 8: Express the matrix

$$A=\begin{bmatrix} 3x^3+2x-1 & 5x^2-3x+2 & -2x^3+3x^2-5x+1 \\ 9x^3-3 & 6x+4 & 7x^2-4x \\ 2x^2-3x-10 & x^3+6x^2+x & 6x^3-4x^2+8 \end{bmatrix}$$

as a polynomial in x with matrix coefficients.

Solution : It is easily seen that

$$A=\begin{bmatrix} -1 & 2 & 1 \\ -3 & 4 & 0 \\ -10 & 0 & 8 \end{bmatrix}+\begin{bmatrix} 2 & -3 & -5 \\ 0 & 6 & -4 \\ -3 & 1 & 0 \end{bmatrix}x$$
$$+\begin{bmatrix} 0 & 5 & 3 \\ 0 & 0 & 7 \\ 2 & 6 & -4 \end{bmatrix}x^2+\begin{bmatrix} 3 & 0 & -2 \\ 9 & 0 & 0 \\ 0 & 1 & 6 \end{bmatrix}x^3. \quad (53)$$

Trace of a matrix

The trace of a square matrix $A\equiv[a_{ij}]$ is defined as *the sum of its diagonal elements.* It is also called the *spur* or the *diagonal sum* and is denoted by Tr A or Sp A. Thus,

$$\text{Tr } A=\sum_{i=1}^{n} a_{ii}. \quad (54)$$

EXAMPLE 9 : Show that the trace of the product of a finite number of matrices is invariant under any cyclic permutation of the matrices.

Solution : We shall first prove the theorem for two matrices. Let $A\equiv[a_{ij}]_{m\times n}$ and $B\equiv[b_{ij}]_{n\times m}$. Then

$$\text{Tr }(AB)=\sum_{i=1}^{m}(AB)_{ii}=\sum_{i=1}^{m}\sum_{j=1}^{n} a_{ij}b_{ji}. \quad (55a)$$

Also

$$\text{Tr }(BA)=\sum_{j=1}^{n}(BA)_{jj}=\sum_{j=1}^{n}\sum_{i=1}^{m} b_{ji}a_{ij}. \quad (55b)$$

Since a_{ij} and b_{ij} are scalars, it is evident that

$$\text{Tr }(AB)=\text{Tr}(BA). \quad (56)$$

Now, let $A_1, A_2, \ldots, A_k$ be k matrices such that the products $A_1A_2, A_2A_3, \ldots, A_kA_1$ are defined. Let

$$C=A_1A_2\ldots A_r,\ D=A_{r+1}\ldots A_k,\ 1\leqslant r\leqslant k-1. \quad (57)$$

Then, from Eq. (56), we have Tr (CD)=Tr (DC). Substituting for C and D from Eq. (57), we have

$$\mathrm{Tr}\,(\mathbf{A}_1\mathbf{A}_2\ldots\mathbf{A}_k)=\mathrm{Tr}(\mathbf{A}_{r+1}\ldots\mathbf{A}_k\mathbf{A}_1\ldots\mathbf{A}_r),\ 1\leqslant r\leqslant k-1, \tag{58}$$

which proves the result.

It is important to note that the trace of the product of a number of matrices is not invariant under *any* permutation, but only under a cyclic permutation of the matrices.

Exercise 3

1. Write the matrix

$$\begin{bmatrix} 3+2i & 1-i & 5 \\ -2+7i & 6i & -1+4i \\ 4-3i & 2+2i & 6+i \end{bmatrix}$$

as *(a) the sum of a real and a purely imaginary matrix, *(b) the sum of a symmetric and an antisymmetric matrix, and (c) the sum of a hermitian and an antihermitian matrix.

2. (a) Show that every matrix can be uniquely expressed as the sum of a real and a purely imaginary matrix. (b) Show that every square matrix can be uniquely expressed as the sum of a symmetric and an antisymmetric matrix.

3. If **A** and **B** are both symmetric or both antisymmetric, show that **BA** is the transpose of **AB**.

*4. (a) Show that the sum of two hermitian matrices is a hermitian matrix. (b) Show that the product of two hermitian matrices is hermitian *if and only if* the two matrices commute.

5. (a) If **A** and **B** are hermitian matrices, find the condition for which any linear combination of the form $a\mathbf{A}+b\mathbf{B}$, where a and b are arbitrary scalars, is hermitian. (b) If **A** and **B** are antihermitian matrices, find the condition for which any linear combination $a\mathbf{A}+b\mathbf{B}$ with scalar coefficients is antihermitian.

6. If $\mathbf{H}_1$ is hermitian and $\mathbf{H}_2$ antihermitian, show that both $\mathbf{H}_1+i\mathbf{H}_2$ and $\mathbf{H}_1-i\mathbf{H}_2$ are hermitian.

*7. Find scalars a and b such that $\mathbf{I}+a\mathbf{A}+b\mathbf{A}^2=\mathbf{0}$, where $\mathbf{A}=\begin{bmatrix} 2 & -5 \\ 3 & 1 \end{bmatrix}$.

8. If $\mathbf{E}_{kl}$ are the n^2 matrices defined in Eq. (9), show that (a) $\mathbf{E}_{kl}\mathbf{E}_{pq}=\mathbf{E}_{kq}\delta_{lp}$, (b) $\mathbf{E}_{kl}{}^2=\mathbf{E}_{kl}\,\delta_{kl}$.

9. Given

$$\mathbf{I}=\begin{bmatrix} 1 & 0 \\ 0 & 1 \end{bmatrix},\ \mathbf{A}=\begin{bmatrix} 1 & 0 \\ 0 & -1 \end{bmatrix},\ \mathbf{B}=\tfrac{1}{2}\begin{bmatrix} -1 & \sqrt{3} \\ \sqrt{3} & 1 \end{bmatrix},$$

$$\mathbf{C}=\tfrac{1}{2}\begin{bmatrix} -1 & -\sqrt{3} \\ -\sqrt{3} & 1 \end{bmatrix},\ \mathbf{D}=\tfrac{1}{2}\begin{bmatrix} -1 & \sqrt{3} \\ -\sqrt{3} & -1 \end{bmatrix},\ \mathbf{F}=\tfrac{1}{2}\begin{bmatrix} -1 & -\sqrt{3} \\ \sqrt{3} & -1 \end{bmatrix},$$

show that (a) $\mathbf{A}^2=\mathbf{B}^2=\mathbf{C}^2=\mathbf{I}$, (b) $\mathbf{AB}=\mathbf{D}$, (c) $\mathbf{AC}=\mathbf{BA}=\mathbf{F}$.

10. Let

$$\mathbf{M}_x=\frac{1}{\sqrt{2}}\begin{bmatrix} 0 & 1 & 0 \\ 1 & 0 & 1 \\ 0 & 1 & 0 \end{bmatrix},\ \mathbf{M}_y=\frac{1}{\sqrt{2}}\begin{bmatrix} 0 & -i & 0 \\ i & 0 & -i \\ 0 & i & 0 \end{bmatrix},\ \mathbf{M}_z=\begin{bmatrix} 1 & 0 & 0 \\ 0 & 0 & 0 \\ 0 & 0 & -1 \end{bmatrix}.$$

[These are the representations of a spin-1 particle in quantum mechanics.] Find the matrices $\mathbf{M}^{\pm} = \mathbf{M}_x \pm i\mathbf{M}_y$ and show that

(a) $[\mathbf{M}_x, \mathbf{M}_y] = i\mathbf{M}_z$, $[\mathbf{M}_y, \mathbf{M}_z] = i\mathbf{M}_x$, $[\mathbf{M}_z, \mathbf{M}_x] = i\mathbf{M}_y$;

(b) $\mathbf{M}^2 \equiv \mathbf{M}_x^2 + \mathbf{M}_y^2 + \mathbf{M}_z^2 = 2\mathbf{I}$;

(c) $[\mathbf{M}^2, \mathbf{M}_i] = \mathbf{0}$, $i = x, y, z$;

(d) $[\mathbf{M}_z, \mathbf{M}^{\pm}] = \pm \mathbf{M}^{\pm}$;

(e) $[\mathbf{M}^{+}, \mathbf{M}^{-}] = 2\mathbf{M}_z$;

(f) $(\mathbf{M}^{+})^{\dagger} = \mathbf{M}^{-}$, $(\mathbf{M}^{-})^{\dagger} = \mathbf{M}^{+}$;

(g) $\mathbf{M}_x$, $\mathbf{M}_y$, $\mathbf{M}_z$ are hermitian matrices.

11. Show that every even power of the matrix

$$\begin{bmatrix} 1 & 0 & 0 \\ 0 & 1 & 0 \\ a & b & -1 \end{bmatrix}$$

where a and b are arbitrary scalars, equals the unit matrix and every odd power equals itself.

12. Show that the product of two triangular matrices of the same type (that is, both left- or both right-triangular) is a triangular matrix of the same type. Hence show that all the powers of a triangular matrix are also triangular matrices of the same type.

13. Show that any hermitian matrix of order two can be expressed uniquely as a linear combination of the four matrices of Eq. (47).

14. (a) Show that the maximum number of linearly independent hermitian matrices of order 3 is 9. (b) Construct a set of 9 linearly independent hermitian matrices of order 3. (c) Write down the most general hermitian matrix of order 3 and show that it involves 9 real independent parameters. (d) Show that the most general matrix of (c) can be expressed uniquely as a linear combination of the 9 matrices constructed in (b).

*15. Find all the matrices whose square equals the matrix

$$\begin{bmatrix} 2 & 1 \\ 0 & 2 \end{bmatrix}.$$

*16. What is the maximum number of linearly independent antisymmetric matrices of order n? Write down a set of independent antisymmetric matrices of order 3. Write the most general antisymmetric matrix of order 3 and express it as a linear combination of the matrices constructed above (with complex coefficients).

17. Show that the trace of the sum of any number of matrices equals the sum of the traces, i.e.,

$$\mathrm{Tr}(\mathbf{A}+\mathbf{B}+\mathbf{C}+\ldots) = \mathrm{Tr}\ \mathbf{A} + \mathrm{Tr}\ \mathbf{B} + \mathrm{Tr}\ \mathbf{C} + \ldots.$$

18. If $\mathbf{T}$ is a triangular matrix whose diagonal elements are t_{ii}, show that the diagonal elements of $\mathbf{T}^k$, where k is a positive integer, are $t_{ii}{}^k$.

19. Show that $\mathrm{Tr}[(\mathbf{AB})^k] = \mathrm{Tr}\ [(\mathbf{BA})^k]$.

4. Determinants

The reader is expected to possess an elementary knowledge of determinants, their evaluation and various properties. Yet, for the sake of completeness we have devoted this section to some basic definitions and important properties of determinants which will be needed in the rest of this book.

If **A** is a square matrix of order n, we denote its determinant by det (**A**) or det **A** or $|\mathbf{A}|$; it is an entity of the same kind as the elements of **A**, that is, a real or a complex number or a function.

Let $\mathbf{A} \equiv [a_{ij}]$ be a square matrix of order n and consider the square matrix of order $n-1$ obtained from **A** by deleting the i-th row and the j-th column of **A**. The determinant of this matrix of order $n-1$ is called the *minor* of the element a_{ij} in det **A**. The *cofactor* of a_{ij}, denoted by A^{ij} is defined as $(-1)^{i+j}$ times the minor of a_{ij}. The determinant of **A** then equals

$$\det \mathbf{A} = \sum_{k=1}^{n} a_{ik} A^{ik}, \ \det \mathbf{A} = \sum_{j=1}^{n} a_{ji} A^{ji}, \ 1 \leqslant i \leqslant n. \tag{1}$$

The first of Eqs. (1) is known as the *expansion of the determinant in the i-th row* and the second as the *expansion in the i-th column*. The determinant of a square matrix has a unique value.

It can be seen that when the expression for the determinant is written out explicitly, it contains $n!$ terms. Each term contains n factors in such a way that there is one and only one element from each row and each column. In fact, in a shorthand notation, det **A** can be written as

$$\det \mathbf{A} = \sum_{P(i_1 i_2 \ldots i_n)} (-1)^k a_{1i_1} a_{2i_2} \ldots a_{ni_n}, \tag{2}$$

where each index $i_1, i_2, \ldots, i_n$ takes values form 1 to n such that no two indices have the same value, the summation is over the $n!$ permutations, and k equals $+1$ or -1 according as whether $(i_1 i_2 \ldots i_n)$ is an even or an odd permutation of $(1\,2 \ldots n)$.

Consider the expressions defined by

$$S = \sum_{k=1}^{n} a_{ik} A^{jk}, \quad P = \sum_{k=1}^{n} a_{ki} A^{kj}. \tag{3}$$

It can be seen that the expression S (P) denotes the determinant of a matrix obtained from **A** by replacing its j-th row (column) by the

i-th row (column). Since the determinant of a matrix with two identical rows or columns identically vanishes, we have

$$\sum_{k=1}^{n} a_{ik} A^{jk} = \sum_{k=1}^{n} a_{ki} A^{kj} = 0, \; i \neq j. \tag{4}$$

Combining this with Eqs. (1), we have

$$\sum_{k=1}^{n} a_{ik} A^{jk} = \sum_{k=1}^{n} a_{ki} A^{kj} = (\det \mathbf{A}) \, \delta_{ij}, \; 1 \leqslant i, j \leqslant n. \tag{5}$$

Occasionally, we require an expression for the derivative of a determinant. Suppose that the elements a_{ij} of the matrix **A** are functions of a parameter s. What is $d\,|\mathbf{A}|/ds$?

From Eq. (2), it is clear that when det **A** is developed explicitly, each element, say a_{ij}, will appear exactly in $(n-1)!$ terms on the right hand side of Eq. (2). In fact, the coefficient of a_{ij} in det **A** is just the cofactor of a_{ij}. When the expression is differentiated with respect to the parameter s, it will therefore lead to the formula

$$d(\det \mathbf{A})/ds = \sum_{i,j=1}^{n} \text{cofactor of } a_{ij} \, (da_{ij}/ds). \tag{6}$$

Exercise 4

1. Evaluate the following determinants:

*(a) $\begin{vmatrix} b^2+c^2 & a^2 & a^2 \\ b^2 & c^2+a^2 & b^2 \\ c^2 & c^2 & a^2+b^2 \end{vmatrix}$; (b) $\begin{vmatrix} 1 & a & a^2 \\ 1 & b & b^2 \\ 1 & c & c^2 \end{vmatrix}$;

*(c) $\begin{vmatrix} 1 & 1 & 1 \\ a & b & c \\ a^3 & b^3 & c^3 \end{vmatrix}$; (d) $\begin{vmatrix} a-b-c & 2a & 2a \\ 2b & b-c-a & 2b \\ 2c & 2c & c-a-b \end{vmatrix}$;

(e) $\begin{vmatrix} 0 & a & b & c \\ a & 0 & c & b \\ b & c & 0 & a \\ c & b & a & 0 \end{vmatrix}$.

2. Show that the determinant of the unit matrix of any order is unity.

3. If **A** is a square matrix and c is a scalar, show that $\det(c\mathbf{A}) = c^n \det(\mathbf{A})$, where n is the order of **A**.

4. Show that the determinant of a triangular matrix equals the product of its diagonal elements.

5. Show that (a) $\det(\mathbf{A}^*) = (\det \mathbf{A})^*$, (b) $\det \widetilde{\mathbf{A}} = \det \mathbf{A}$, (c) $\det \mathbf{A}^\dagger = (\det \mathbf{A})^*$.

*6. Show that the determinant of a hermitian matrix is real.

7. Show that the determinant of an antisymmetrix matrix of odd order is zero while that of an antihermitian matrix of odd order is purely imaginary or zero.

8. Find all the cofactors in the determinants

(a) $D=\begin{vmatrix} 1 & x & x^2 \\ x & x^2 & 1 \\ x^2 & 1 & x \end{vmatrix}$, (b) $D=\begin{vmatrix} \sin x & \cos x & e^{-x} \\ \cos x & -\sin x & -e^{-x} \\ -\sin x & -\cos x & e^{-x} \end{vmatrix}$,

and hence obtain dD/dx by using Eq. (6). Then develop the determinants and evaluate dD/dx explicitly. Verify that you get the same result.

5. Special matrices, II

In this section, we shall discuss some more special matrices which frequently occur in physics. Before we do that, it will be expedient to define the inverse of a matrix.

The inverse matrix

A matrix **A** is said to be *nonsingular* if there exists a matrix **B** such that

$$\mathbf{AB}=\mathbf{I}_n,\ \ \mathbf{BA}=\mathbf{I}_m, \tag{1}$$

where $\mathbf{I}_n$ and $\mathbf{I}_m$ are unit matrices, not necessarily of the same order. Then **A** is said to be the *inverse* of **B** and vice versa. For a given matrix **A**, if there exists no matrix **B** which satisfies Eqs. (1), then **A** has no inverse and is said to be *singular*.

The relationship between **A** and **B** is reciprocal, so that if **B** is the inverse of **A**, **A** is the inverse of **B**. The inverse matrix is also denoted by the symbol $\mathbf{A}^{-1}$, so that if Eqs. (1) are satisfied, we have

$$\mathbf{A}^{-1}\equiv\mathbf{B},\ \mathbf{B}^{-1}\equiv\mathbf{A}. \tag{2}$$

It is shown in Appendix II that a nonsingular matrix of finite order must be square, so that its inverse is also square and $\mathbf{I}_n$ and $\mathbf{I}_m$ in Eqs. (1) are of the same order. Further, in that case, the first of Eqs. (1) implies the second, and vice versa, i.e., $\mathbf{AB}=\mathbf{I}\Leftrightarrow\mathbf{BA}=\mathbf{I}$, so that any one of them is sufficient to define the inverse matrix. In this section, we shall deal with finite matrices only.[1] Then, taking the determinants of both sides of Eqs. (1), we have

$$(\det\mathbf{A})(\det\mathbf{B})=1. \tag{3}$$

If the determinant of either of the matrices is zero, Eq. (3) shows that the determinant of the other must be infinite, which means that the other matrix does not exist, so that the first matrix is singular. Thus, it follows that *a finite singular matrix has zero determinant.*

If **A** is a nonsingular matrix, it is possible to find a matrix **B** which satisfies Eqs. (1). The process of obtatning the inverse of a nonsingular matrix involves the following steps.

(a) Obtain the cofactor of every element of the matrix **A** and write the *matrix of cofactors* in the form

[1]For the inverse of an infinite matrix, see Section 14.

$$\mathbf{A}_c=\begin{bmatrix} A^{11} & A^{12} & \dots & A^{1n} \\ A^{21} & A^{22} & \dots & A^{2n} \\ \vdots & & & \\ A^{n1} & A^{n2} & \dots & A^{nn} \end{bmatrix}. \tag{4}$$

(b) Transpose the matrix of cofactors to obtain

$$\widetilde{\mathbf{A}}_c=\begin{bmatrix} A^{11} & A^{21} & \dots & A^{n1} \\ A^{12} & A^{22} & \dots & A^{n2} \\ \vdots & & & \\ A^{1n} & A^{2n} & & A^{nn} \end{bmatrix}. \tag{5}$$

This is called the *adjoint* of the matrix **A**.

(c) Divide this by det **A** to obtain the inverse of **A**:

$$\mathbf{B}\equiv\mathbf{A}^{-1}=\frac{\widetilde{\mathbf{A}}_c}{\det \mathbf{A}}=\frac{1}{\det \mathbf{A}}\begin{bmatrix} A^{11} & A^{21} & \dots & A^{n1} \\ A^{12} & A^{22} & \dots & A^{n2} \\ \vdots & & & \\ A^{1n} & A^{2n} & \dots & A^{nn} \end{bmatrix}. \tag{6a}$$

Alternatively,

$$\mathbf{A}\widetilde{\mathbf{A}}_c=(\det \mathbf{A})\mathbf{I}. \tag{6b}$$

We shall now prove that the matrix **B** obtained in Eq. (6a) satisfies the relation **AB**=**I**. For this, take the *ij*-th element of the product **AB** which, on using Eq. (6a), gives

$$(\mathbf{AB})_{ij}=\sum_{k=1}^{n}(\mathbf{A})_{ik}(\mathbf{B})_{kj}=\frac{1}{\det \mathbf{A}}\sum_{k=1}^{n} a_{ik}\mathbf{A}^{jk}$$
$$=\delta_{ij},\ 1\leqslant i, j\leqslant n, \tag{7}$$

where we have used the result of Eq. (4.5) in the last step. Eq. (7) shows that **AB**=**I**.

EXAMPLE 1: Find the inverse of the following matrix if it exists:

$$\mathbf{A}=\begin{bmatrix} -3 & 1 & -1 \\ 15 & -6 & 5 \\ -5 & 2 & -2 \end{bmatrix}. \tag{8}$$

Solution: The determinant of the given matrix is found to be -1. The matrix is therefore nonsingular and possesses an inverse. The matrix of cofactors is found to be

$$\mathbf{A}_c=\begin{bmatrix} 2 & 5 & 0 \\ 0 & 1 & 1 \\ -1 & 0 & 3 \end{bmatrix}. \tag{9}$$

The inverse of **A** is then obtained by transposing $\mathbf{A}_c$ and dividing it by -1 to give

$$A^{-1}=\begin{bmatrix} -2 & 0 & 1 \\ -5 & -1 & 0 \\ 0 & -1 & -3 \end{bmatrix}. \tag{10}$$

It can be verified explicitly by matrix multiplication that

$$AA^{-1}=A^{-1}A=\begin{bmatrix} 1 & 0 & 0 \\ 0 & 1 & 0 \\ 0 & 0 & 1 \end{bmatrix}. \tag{11}$$

Using Eq. (6) for **B**, it can also be shown that **BA**=**I**. Together with Eqs. (1), we then have

$$AA^{-1}=A^{-1}A=I. \tag{12}$$

Thus, *every matrix commutes with its inverse*.

The inverse of a matrix, if it exists, is unique. For, suppose that for a given matrix **A**, there are two matrices **B** and **C** which satisfy the relations

$$AB=I,\ AC=I. \tag{13}$$

If **AB**=**I**, we have seen that **BA**=**I**. Multiplying the second of Eqs. (13) from the left by **B**, we have

$$BAC=BI=B \Rightarrow C=B, \tag{14}$$

which proves our statement.

A result of considerable importance is proved in the following example.

EXAMPLE 2: Show that the inverse of the product of a number of matrices (all square and of the same order), none of which is singular, equals the product of the inverses taken in the reverse order.

Solution: Consider a particular case of three nonsingular matrices **A**, **B** and **C**. Then we have to show that

$$(ABC)^{-1}=C^{-1}B^{-1}A^{-1}. \tag{15}$$

If the product of two square matrices gives the unit matrix, the two matrices are said to be the inverses of each other. Taking the product of the matrix **ABC** with the matrix $C^{-1}B^{-1}A^{-1}$ and using the relations $CC^{-1}=I$, etc., we have

$$ABCC^{-1}B^{-1}A^{-1}=ABB^{-1}A^{-1}=AA^{-1}=I, \tag{16}$$

showing that $C^{-1}B^{-1}A^{-1}$ is the inverse of **ABC**. The method can be clearly generalized to the product of a finite number of nonsingular matrices. Thus, we have

$$(ABC\ldots X)^{-1}=X^{-1}\ldots C^{-1}B^{-1}A^{-1}. \tag{17}$$

EXAMPLE 3: Assuming that **A**, **I**−**A** and $I-A^{-1}$ are all nonsingular, show that

$$(\mathbf{I}-\mathbf{A})^{-1}+(\mathbf{I}-\mathbf{A}^{-1})^{-1}=\mathbf{I}. \tag{18}$$

Solution: We can write $(\mathbf{I}-\mathbf{A}^{-1})=\mathbf{A}^{-1}(\mathbf{A}-\mathbf{I})$, and hence

$$(\mathbf{I}-\mathbf{A}^{-1})^{-1}=[\mathbf{A}^{-1}(\mathbf{A}-\mathbf{I})]^{-1}=(\mathbf{A}-\mathbf{I})^{-1}(\mathbf{A}^{-1})^{-1}=-(\mathbf{I}-\mathbf{A})^{-1}\mathbf{A}. \tag{19}$$

Therefore

$$\begin{aligned}(\mathbf{I}-\mathbf{A})^{-1}+(\mathbf{I}-\mathbf{A}^{-1})^{-1}&=(\mathbf{I}-\mathbf{A})^{-1}-(\mathbf{I}-\mathbf{A})^{-1}\mathbf{A}\\ &=(\mathbf{I}-\mathbf{A})^{-1}(\mathbf{I}-\mathbf{A})=\mathbf{I}.\end{aligned} \tag{20}$$

Inverse transformation

Consider a linear transformation from a vector $\mathbf{x}$ to a vector $\mathbf{y}$, both of the same dimensions, of the form

$$y_i=\sum_{j=1}^{n} a_{ij}\,x_j,\ 1\leqslant i\leqslant n; \tag{21}$$

$$\mathbf{y}=\mathbf{A}\mathbf{x}. \tag{22}$$

If the matrix $\mathbf{A}$ of transformation is nonsingular, it is possible to invert the above transformation and express x_i in terms of y_i. If $\mathbf{B}=\mathbf{A}^{-1}$, multiply Eq. (22) from the left by $\mathbf{B}$ to obtain

$$\mathbf{B}\mathbf{y}=\mathbf{B}\mathbf{A}\mathbf{x}=\mathbf{x}. \tag{23}$$

Expressed in terms of components, this becomes

$$x_i=\sum_{j=1}^{n} b_{ij}\,y_j,\ 1\leqslant i\leqslant n, \tag{24}$$

which is also a linear transformation.

EXAMPLE 4: Find the inverse of the linear transformation

$$\begin{aligned}u&=-2x-2y+7z,\\ v&=4x+3y-12z,\\ w&=-x+2z,\end{aligned} \tag{25}$$

by obtaining the inverse of the matrix of transformation.

Solution: Eqs. (25) can be written in the matrix notation as

$$\begin{bmatrix} u\\ v\\ w\end{bmatrix}=\begin{bmatrix} -2 & -2 & 7\\ 4 & 3 & -12\\ -1 & 0 & 2\end{bmatrix}\begin{bmatrix} x\\ y\\ z\end{bmatrix}. \tag{26}$$

The determinant of the matrix of transformation is 1 and hence the inverse transformation exists. The inverse of the matrix of transformation is found to be

$$\begin{bmatrix} 6 & 4 & 3\\ 4 & 3 & 4\\ 3 & 2 & 2\end{bmatrix},$$

so that the inverse transformation becomes

$$\begin{bmatrix} 6 & 4 & 3 \\ 4 & 3 & 4 \\ 3 & 2 & 2 \end{bmatrix} \begin{bmatrix} u \\ v \\ w \end{bmatrix} = \begin{bmatrix} x \\ y \\ z \end{bmatrix}, \tag{27}$$

or

$$\begin{aligned} x &= 6u+4v+3w, \\ y &= 4u+3v+4w, \\ z &= 3u+2v+2w. \end{aligned} \tag{28}$$

Orthogonal matrix

A matrix **A** satisfying the relations

$$\mathbf{A}\,\widetilde{\mathbf{A}} = \mathbf{I}_n, \tag{29a}$$

$$\widetilde{\mathbf{A}}\,\mathbf{A} = \mathbf{I}_m, \tag{29b}$$

is called an *orthogonal matrix*. Again, it can be shown that if **A** is a finite matrix satisfying *both* Eqs. (29), then **A** must be square, and[2] $\mathbf{A}\,\widetilde{\mathbf{A}} = \mathbf{I} \Leftrightarrow \widetilde{\mathbf{A}}\,\mathbf{A} = \mathbf{I}$.

Let det $\mathbf{A} = d$. Taking the determinants of both sides of Eq. (29a), we have $d^2 = 1 \Rightarrow d = \pm 1$. This shows that *the determinant of an orthogonal matrix can only have values* $+1$ *or* -1.

At the same time, this shows that **A** is nonsingular, so that $\mathbf{A}^{-1}$ exists. Multiplying Eq. (29a) by $\mathbf{A}^{-1}$ from the right, we have

$$\mathbf{A}^{-1} = \widetilde{\mathbf{A}}. \tag{30}$$

This is an alternative way of defining an orthogonal matrix.

Equating the *ij*-th element of both sides of Eq. (29a), we find that

$$\sum_{k=1}^{n} a_{ik}\,a_{jk} = \delta_{ij}, \quad 1 \leqslant i, j \leqslant n, \tag{31a}$$

where n is the order of the matrix **A**, and a_{ik} is the *ik*-th element of **A**. Similarly, equating the *ij*-th element of both sides of Eq. (29b), we get

$$\sum_{k=1}^{n} a_{ki}\,a_{kj} = \delta_{ij}, \quad 1 \leqslant i, j \leqslant n. \tag{31b}$$

These are the conditions satisfied by the elements of an orthogonal matrix. However, the conditions of Eq. (31a) and Eq. (31b) are not independent for a finite matrix, because Eqs. (29a) and (29b) are

[2]If **A** is an infinite matrix, however, Eqs. (29a) and (29b) are independent of each other and **A** is orthogonal if and only if *both* Eqs. (29) are simultaneously satisfied.

themselves not independent. If any one of the Eqs. (29a) and (29b) holds good, the other also does.

EXAMPLE 5: Show that the matrix

$$\mathbf{A}=\begin{bmatrix} 1/\sqrt{3} & 1/\sqrt{6} & 1/\sqrt{2} \\ 1/\sqrt{3} & -\sqrt{2/3} & 0 \\ 1/\sqrt{3} & 1/\sqrt{6} & -1/\sqrt{2} \end{bmatrix} \tag{32}$$

is an orthogonal matrix and obtain its inverse.

Solution: The transpose of the given matrix is

$$\widetilde{\mathbf{A}}=\begin{bmatrix} 1/\sqrt{3} & 1/\sqrt{3} & 1/\sqrt{3} \\ 1/\sqrt{6} & -\sqrt{2/3} & 1/\sqrt{6} \\ 1/\sqrt{2} & 0 & -1/\sqrt{2} \end{bmatrix}. \tag{33}$$

Multiplying these two together, we find that

$$\mathbf{A}\widetilde{\mathbf{A}}=\begin{bmatrix} 1 & 0 & 0 \\ 0 & 1 & 0 \\ 0 & 0 & 1 \end{bmatrix}, \tag{34}$$

which is a unit matrix, showing that $\mathbf{A}$ is an orthogonal matrix. Now since for an orthogonal matrix, its inverse is equal to its transpose, $\mathbf{A}^{-1}$ is given by Eq. (33).

The elements of an orthogonal matrix may be real or complex. For example, according to the definition of Eqs. (29), $\begin{bmatrix} i & -\sqrt{2} \\ \sqrt{2} & i \end{bmatrix}$ is an orthogonal matrix However, in practice, we are always concerned with *real* orthogonal matrices and we shall employ the convention in this book that the phrase 'orthogonal matrix' will mean a real orthogonal matrix, unless otherwise stated.

Looking at Eq. (31a), we see that for fixed i and $1 \leqslant k \leqslant n$, a_{ik} are the elements of the i-th row of $\mathbf{A}$. Similarly, for fixed j and $1 \leqslant k \leqslant n$, a_{jk} are the elements of the j-th row of $\mathbf{A}$. Put in words, Eq. (31a) then means that *the sum of the products of the corresponding elements of two distinct rows of an orthogonal matrix is zero while the sum of the squares of the elements of any row is equal to unity.* If we regard the n rows of the matrix as n real vectors, this means that these n row vectors are orthogonal and normalized. Similarly, Eq. (31b) shows that all the columns of an orthogonal matrix are orthonormal.

Independent elements of an orthogonal matrix

For $1 \leqslant i,\ j \leqslant n$, Eq. (31a) stands for a set of n^2 equations. But these n^2 equations are not independent of each other, for, if we interchange i and j in Eq. (31a), we get the same equation. The independent equations among Eq. (31a) are therefore obtained for values of i and j given by $1 \leqslant i \leqslant j \leqslant n$. The number of these equations is $1+2+3+\ldots+n=n(n+1)/2$. Thus the n^2 elements of a real orthogonal matrix are constrained by $n(n+1)/2$ conditions. The number of independent parameters of an orthogonal matrix is therefore $n^2-n(n+1)/2=n(n-1)/2$. In other words, *every real orthogonal matrix of order n can be fully characterized by* $n(n-1)/2$ *real independent parameters*.

For $n=2$, we find that the number of independent parameters is one. The following example will help clarify the concepts.

EXAMPLE 6: Obtain the most general orthogonal matrix of order 2.

Solution: Let us start with an arbitrary matrix of order 2 which can be written as

$$\mathbf{A}=\begin{bmatrix} a & b \\ c & d \end{bmatrix}, \tag{35}$$

where a, b, c, d are any scalars, real or complex. If this is to be an orthogonal matrix, its elements must satisfy the conditions of Eq. (31a), which gives

$$a^2+b^2=1, \tag{36a}$$

$$c^2+d^2=1, \tag{36b}$$

$$ac+bd=0. \tag{36c}$$

The general solution of Eq. (36a) is $a=\cos\theta$, $b=\sin\theta$, where θ is real or complex. Similarly, Eq. (36b) can be satisfied if we choose $c=\cos\phi$, $d=\sin\phi$, where ϕ is a scalar. In order to satisfy Eq. (36c), we then find that θ and ϕ must be related by

$$\cos(\theta-\phi)=0 \Rightarrow \theta-\phi=\pm\pi/2. \tag{37}$$

The most general orthogonal matrix of order 2 then becomes

$$\mathbf{A}=\begin{bmatrix} \cos\theta & \sin\theta \\ \mp\sin\theta & \pm\cos\theta \end{bmatrix}. \tag{38}$$

Choosing the upper signs, we get the most general orthogonal matrix of order 2 with determinant $+1$, while the lower signs give the most general orthogonal matrix of order 2 with determinant -1. The matrix involves only one independent parameter as discussed above. Every orthogonal matrix of order 2 can be expressed in the form of Eq. (38) with some value of θ.

Unitary matrix

A matrix $\mathbf{U}$ satisfying the relations

$$\mathbf{UU}^{\dagger}=\mathbf{I}_n, \tag{39a}$$

$$\mathbf{U}^{\dagger}\mathbf{U}=\mathbf{I}_m, \tag{39b}$$

is called a *unitary matrix*. If $\mathbf{U}$ is a finite matrix satisfying *both* Eqs. (39), then $\mathbf{U}$ must be a square matrix, and[3] $\mathbf{UU}^{\dagger}=\mathbf{I} \Leftrightarrow \mathbf{U}^{\dagger}\mathbf{U}=\mathbf{I}$. The elements of a unitary matrix may be complex. In fact, it is evident from Eqs. (39) that a real unitary matrix is orthogonal.

Let det $\mathbf{U}=d$. Taking the determinants of both sides of Eq. (39a) and noting that det $\mathbf{U}^{\dagger}=d^*$, we have

$$dd^*=1 \Rightarrow |\, d\,|=1. \tag{40}$$

This shows that the determinant of a unitary matrix can be a complex number of unit magnitude, that is, a number of the form $e^{i\theta}$, where θ is a real number. It also shows that a unitary matrix is nonsingular and possesses an inverse. Multiplying Eq. (39a) from the left by $\mathbf{U}^{-1}$, we have

$$\mathbf{U}^{\dagger}=\mathbf{U}^{-1}. \tag{41}$$

Equating the ij-th element of both sides of Eq. (39a) and of Eq. (39b), we have

$$\sum_{k=1}^{n} u_{ik}u_{jk}^*=\delta_{ij}, \quad \sum_{k=1}^{n} u_{ki}u_{kj}^*=\delta_{ij}, \quad 1\leqslant i, j\leqslant n, \tag{42}$$

where u_{ik} is the ik-th element of $\mathbf{U}$ and n is its order. These are the conditions satisfied by the elements of a unitary matrix. As in the case of an orthogonal matrix, the sets of conditions of Eqs. (42) are not independent of each other for a finite matrix.

Independent elements of a unitary matrix

To obtain the number of real independent parameters of a unitary matrix, we first note that the total number of real parameters will be $2n^2$, where n is the order of the matrix. The factor 2 occurs because every element may be complex. However, these parameters will be related to each other through Eqs. (42). We must therefore determine the number of real independent conditions.

[3]However, if $\mathbf{U}$ is a matrix of infinite order, Eqs. (39a) and (39b) do not follow from each other. In that case, $\mathbf{U}$ will be unitary if and only if *both* Eqs. (39) are simultaneously satisfied.

As before, an interchange of i and j leaves Eqs. (42) unaffected. Therefore the independent conditions are obtained for the values of i and j given by $1 \leqslant i \leqslant j \leqslant n$. We consider the two cases $i=j$ and $i \neq j$ separately and write the first of Eqs. (42) in the form

$$\sum_{k=1}^{n} | u_{ik} |^2 = 1, \; 1 \leqslant i \leqslant n; \tag{43a}$$

$$\sum_{k=1}^{n} u_{ik} u_{jk}{}^* = 0, \; 1 \leqslant i < j \leqslant n. \tag{43b}$$

Eq. (43a) is an equation involving real numbers only and as i takes values from 1 to n, it stands for n real conditions. Eq. (43b), on the other hand, contains complex quantities on the left hand side and each equation is therefore equivalent to two real conditions. The number of equations in Eq. (43b) is $1+2+\ldots+(n-1)=n(n-1)/2$. Eq. (43b) is thus equivalent to $n(n-1)$ real conditions. The total number of real conditions is then $n+n(n-1)=n^2$.

A unitary matrix of order n thus involves $2n^2-n^2=n^2$ real independent parameters.

EXAMPLE 7: Obtain the most general unitary matrix of order 2.

Solution: An arbitrary matrix of order 2 can be written as

$$\mathbf{U} = \begin{bmatrix} a & b \\ c & d \end{bmatrix}, \tag{44}$$

where a, b, c and d may be complex scalars. If this matrix is to be unitary, it must satisfy the condition of Eq. (39a), which gives

$$\begin{bmatrix} a & b \\ c & d \end{bmatrix} \begin{bmatrix} a^* & c^* \\ b^* & d^* \end{bmatrix} = \begin{bmatrix} aa^*+bb^* & ac^*+bd^* \\ ca^*+db^* & cc^*+dd^* \end{bmatrix} = \begin{bmatrix} 1 & 0 \\ 0 & 1 \end{bmatrix}. \tag{45}$$

This gives the conditions

$$|a|^2+|b|^2=1, \; |c|^2+|d|^2=1, \; ac^*+bd^*=0. \tag{46}$$

These equations are equivalent to four real conditions, as they should be. From the first of Eqs. (46), we can choose a and b to be of the form

$$a = \cos\theta \, e^{i\alpha}, \; b = \sin\theta \, e^{i\gamma}, \tag{47}$$

where θ, α and γ are real numbers. Using this in the third of Eqs. (46), we find that

$$c^*/d^* = -\sin\theta \, e^{i\gamma}/(\cos\theta \, e^{i\alpha}). \tag{48}$$

In order to satisfy the second of Eqs. (46) and Eq. (48), we can then choose c and d to be

$$c=-\sin\theta\, e^{i(\beta-\gamma)},\ d=\cos\theta\, e^{i(\beta-\alpha)},\ \beta \text{ real.} \tag{49}$$

The most general unitary matrix of order 2 then becomes

$$\mathbf{U}=\begin{bmatrix} \cos\theta\, e^{i\alpha} & \sin\theta\, e^{i\gamma} \\ -\sin\theta\, e^{i(\beta-\gamma)} & \cos\theta\, e^{i(\beta-\alpha)} \end{bmatrix}, \tag{50}$$

which involves four real independent parameters α, β, γ and θ. The determinant of this matrix is $e^{i\beta}$ which is a complex number of unit magnitude.

The statement that this is the most general unitary matrix of order 2 means that (a) for any real values of α, β, γ and θ, the matrix of Eq. (50) is a unitary matrix, and conversely, (b) any unitary matrix of order 2 can be expressed in the form of Eq. (50) with some real values of α, β, γ and θ.

Orthogonal and unitary transformations

Consider a linear transformation in which the transformation matrix is square so that in the equation

$$\mathbf{y}=\mathbf{A}\mathbf{x}, \tag{51}$$

$\mathbf{x}$ and $\mathbf{y}$ are column vectors of order $n\times 1$ and $\mathbf{A}$ is of order $n\times n$. If the matrix $\mathbf{A}$ of transformation is unitary, it is said to be a *unitary transformation.* Using Eq. (39b) for a unitary matrix, we have, from Eq. (51), that

$$\mathbf{y}^\dagger\,\mathbf{y}=\mathbf{x}^\dagger\,\mathbf{A}^\dagger\,\mathbf{A}\mathbf{x}=\mathbf{x}^\dagger\,\mathbf{x}, \tag{52a}$$

or

$$\sum_{i=1}^{n} y_i{}^*y_i=\sum_{i=1}^{n} x_i{}^*x_i. \tag{52b}$$

This shows that *the norm of a vector remains invariant under a unitary transformation.* On the other hand, suppose $\mathbf{A}$ is a linear transformation which leaves the norm of every vector of a vector space unchanged, that is, $\|\mathbf{x}\|=\|\mathbf{A}\mathbf{x}\|$, or

$$\mathbf{x}^\dagger\,\mathbf{A}^\dagger\,\mathbf{A}\mathbf{x}=\mathbf{x}^\dagger\,\mathbf{x}. \tag{53}$$

Since this should be true for every vector $\mathbf{x}$ belonging to the vector space, it follows that $\mathbf{A}^\dagger\mathbf{A}=\mathbf{I}$, showing that $\mathbf{A}$ is unitary. Thus, *a linear transformation* (or *matrix*) *acting on a vector space is unitary if and only if it leaves the norm of every vector in the vector space unchanged.*

If the matrix $\mathbf{A}$ of transformation further happens to be real, it is called a *real orthogonal transformation.* In this case, Eqs. (52) reduce to

$$\tilde{\mathbf{y}}\,\mathbf{y}=\tilde{\mathbf{x}}\,\tilde{\mathbf{A}}\,\mathbf{A}\mathbf{x}=\tilde{\mathbf{x}}\,\mathbf{x}, \tag{54a}$$

$$\sum_{i=1}^{n} y_i^2=\sum_{i=1}^{n} x_i^2. \tag{54b}$$

Transformation of vectors and matrices

Consider a basis $\mathbf{e}=\{\mathbf{e}_1, \mathbf{e}_2,\ldots,\mathbf{e}_n\}$ in an n-dimensional vector space. The vectors of the set $\mathbf{e}$ are not necessarily orthonormal but are of course linearly independent in order to be a basis for the vector space. Let $\mathbf{e}'=\{\mathbf{e}_1', \mathbf{e}_2',\ldots,\mathbf{e}_n'\}$ be another basis (that is, another set of n linearly independent vectors) in the same vector space. Every vector of the set $\mathbf{e}'$ can be expressed as a linear combination of the set $\mathbf{e}$ in the form

$$\mathbf{e}_i'=\sum_{j=1}^{n} p_{ij}\mathbf{e}_j,\ 1\leqslant i\leqslant n, \tag{55a}$$

where p_{ij} are some coefficients. In the matrix form, we can write this as

$$\mathbf{e}'=\mathbf{P}\mathbf{e}, \tag{55b}$$

where $\mathbf{P}$ is a square matrix of order n whose ij-th element is p_{ij}. We say that $\mathbf{P}$ is the *matrix of transformation* from the basis $\mathbf{e}$ to the basis $\mathbf{e}'$.

Since $\mathbf{e}'$ is also a basis for the vector space, we can express the vectors of the set $\mathbf{e}$ as linear combinations of the vectors of the set $\mathbf{e}'$. We have

$$\mathbf{e}_i=\sum_{j=1}^{n} q_{ij}\mathbf{e}_j',\ 1\leqslant i\leqslant n; \tag{56a}$$

$$\mathbf{e}=\mathbf{Q}\mathbf{e}'. \tag{56b}$$

Substituting Eq. (55b) in Eq. (56b), we have

$$\mathbf{e}=\mathbf{Q}\mathbf{P}\mathbf{e}. \tag{57}$$

Since this must be true for any basis $\mathbf{e}$, we have

$$\mathbf{Q}\mathbf{P}=\mathbf{I},\ \mathbf{Q}=\mathbf{P}^{-1}, \tag{58}$$

showing that the inverse of the matrix $\mathbf{P}$ exists. This gives us the important result that *a matrix which transforms one set of linearly independent vectors into another set of linearly independent vectors* (in the same vector space) *must be a nonsingular matrix.* Further detail and a more rigorous proof are discussed in Example 8.6.

In Fig. 1, we have shown the effect of a transformation on vectors. In Fig. 1(a), the same vector is labeled **x** relative to the basis

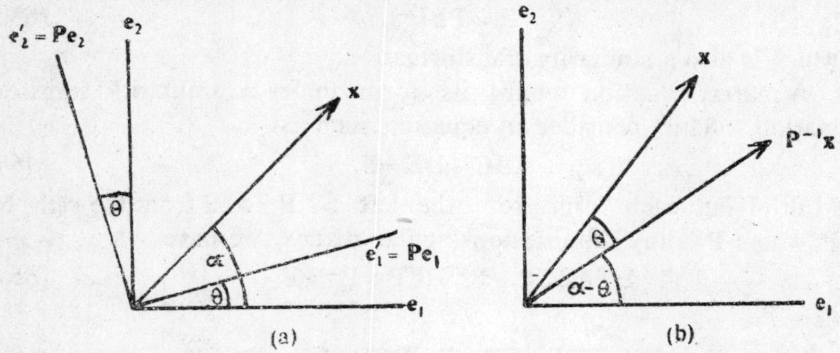

Figure 1. A transformation **P** on the basis vectors in (a) is equivalent to the inverse transformation $\mathbf{P}^{-1}$ on the vector **x** as in (b). **P** stands for an anticlockwise rotation through θ here, so that $\mathbf{P}^{-1}$ is a clockwise rotation through θ.

$\mathbf{e}=\{\mathbf{e}_1, \mathbf{e}_2\}$ and $\mathbf{x}'$ relative to the basis $\mathbf{e}'=\{\mathbf{e}_1', \mathbf{e}_2'\}$. Fig. 1 (b) shows the vector $\mathbf{P}^{-1}\mathbf{x}$ relative to the basis **e**, where **P** is the transformation taking **e** to $\mathbf{e}'$; Eq. (55b). It is clear from this figure that the components of the vector $\mathbf{x}'$ relative to $\mathbf{e}'$ are the same as those of $\mathbf{P}^{-1}\mathbf{x}$ relative to **e**. Hence we have

$$\mathbf{x}'=\mathbf{P}^{-1}\mathbf{x}, \ \mathbf{x}=\mathbf{P}\mathbf{x}'. \tag{59}$$

Consider now two vectors **x** and **y** relative to the basis **e**. Relative to the basis $\mathbf{e}'$, let the same vectors be denoted respectively by $\mathbf{x}'$ and $\mathbf{y}'$, so that

$$\mathbf{x}=\mathbf{P}\mathbf{x}', \ \mathbf{y}=\mathbf{P}\mathbf{y}'. \tag{60}$$

Suppose **x** and **y** are themselves related by the transformation **A** in the **e** coordinate system, so that

$$\mathbf{x}=\mathbf{A}\mathbf{y}. \tag{61}$$

We wish to find the matrix of transformation between the vectors $\mathbf{x}'$ and $\mathbf{y}'$ in the $\mathbf{e}'$ coordinate system. From Eqs. (60) and (61), we have

$$\mathbf{x}'=\mathbf{P}^{-1}\mathbf{x}=\mathbf{P}^{-1}\mathbf{A}\mathbf{y}=\mathbf{P}^{-1}\mathbf{A}\mathbf{P}\mathbf{y}' \equiv \mathbf{B}\mathbf{y}', \tag{62}$$

which defines the matrix **B** as

$$\mathbf{B}=\mathbf{P}^{-1}\mathbf{A}\mathbf{P}. \tag{63a}$$

This shows that if the matrix of transformation from **y** to **x** in the coordinate system **e** is **A**, the matrix of transformation for the *same* two vectors in the coordinate system $\mathbf{e}'=\mathbf{P}\mathbf{e}$ is $\mathbf{P}^{-1}\mathbf{A}\mathbf{P}$.

The matrices **A** and **B** of Eq. (63a) are said to be related by a *similarity transformation. They perform the same job in the same vector space on vectors referred to different coordinate systems.* The inverse transformation of Eq. (63a) is

$$\mathbf{A}=\mathbf{P}\mathbf{B}\mathbf{P}^{-1}, \tag{63b}$$

which is also a similarity transformation.

A matrix equation retains its form under a similarity transformation. Thus, consider an equation such as

$$\mathbf{ABC}+\mathbf{DE}=\mathbf{F}. \tag{64}$$

Multiplying each term from the left by $\mathbf{P}^{-1}$ and from the right by **P**, where **P** is any suitable nonsingular matrix, we have

$$\mathbf{P}^{-1}\mathbf{ABCP}+\mathbf{P}^{-1}\mathbf{DEP}=\mathbf{P}^{-1}\mathbf{FP}, \tag{65a}$$

or

$$\mathbf{P}^{-1}\mathbf{APP}^{-1}\mathbf{BPP}^{-1}\mathbf{CP}+\mathbf{P}^{-1}\mathbf{DPP}^{-1}\mathbf{EP}=\mathbf{P}^{-1}\mathbf{FP}. \tag{65b}$$

This is the same as

$$\mathbf{A}'\mathbf{B}'\mathbf{C}'+\mathbf{D}'\mathbf{E}'=\mathbf{F}', \tag{66}$$

where each primed matrix is related to the corresponding unprimed matrix by a similarity transformation through **P**. It is evident that Eq. (66) has the same form as Eq. (64). The advantage of this procedure is that a given matrix equation may be transformed by a similarity transformation until it appears in a simplified form.

If, in particular, the matrix **P** with which a similarity transformation is performed happens to be unitary, it is called a *unitary transformation.* If, further, **P** is a real orthogonal matrix, it is called an *orthogonal transformation.* We have

$$\mathbf{B}=\mathbf{P}^{-1}\mathbf{AP},\ \mathbf{P}^{-1}=\mathbf{P}^{\dagger}: \text{unitary transformation}; \tag{67a}$$

$$\mathbf{B}=\mathbf{P}^{-1}\mathbf{AP},\ \mathbf{P}^{-1}=\widetilde{\mathbf{P}}: \text{orthogonal transformation}. \tag{67b}$$

EXAMPLE 8: Find the matrix **A** of transformation on two-dimensional vectors which rotates them by 45° and doubles their lengths. Show that $\mathbf{A}^4=-16\mathbf{I}$.

Solution: Let us choose a two-dimensional cartesian coordinate system and let $\mathbf{x}=\{x_1, x_2\}$ be a vector. Let $\mathbf{y}=\{y_1, y_2\}$ be a vector making an angle of 45° with **x** and whose length is twice that of **x**. We shall perform the transformation in two steps—first rotate the vector by 45° and then double its length. Let **C** denote the first operation and let $\mathbf{z}=\{z_1, z_2\}=\mathbf{Cx}$. Let **x** make an angle θ with the basis vector $\mathbf{e}_1$ (see Fig. 2). Also, let $l=|\mathbf{x}|=|\mathbf{z}|$. From Fig. 2(a), we then have

$$x_1 = l\cos\theta,\ x_2 = l\sin\theta; \tag{68}$$

$$z_1 = l\cos(\theta+45°) = (l\cos\theta - l\sin\theta)/\sqrt{2} = (x_1 - x_2)/\sqrt{2},$$
$$z_2 = l\sin(\theta+45°) = (l\sin\theta + l\cos\theta)/\sqrt{2} = (x_1 + x_2)/\sqrt{2}. \tag{69}$$

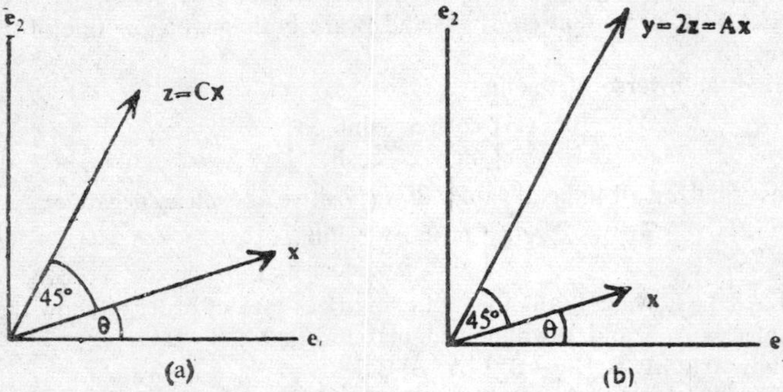

Figure 2. (a) The transformation **C** rotates the vector **x** by 45° to give the vector **z**=**Cx**. (b) The transformation **A** rotates **x** by 45° and doubles its length to give **y**=**Ax**.

Hence, in the matrix form, we have

$$\begin{bmatrix} z_1 \\ z_2 \end{bmatrix} = \begin{bmatrix} 1/\sqrt{2} & -1/\sqrt{2} \\ 1/\sqrt{2} & 1/\sqrt{2} \end{bmatrix} \begin{bmatrix} x_1 \\ x_2 \end{bmatrix}, \tag{70}$$

so that

$$\mathbf{C} = (1/\sqrt{2}) \begin{bmatrix} 1 & -1 \\ 1 & 1 \end{bmatrix}. \tag{71}$$

The matrix which doubles the length of a vector is simply 2**I**. Hence the desired matrix is

$$\mathbf{A} = 2\mathbf{IC} = \sqrt{2} \begin{bmatrix} 1 & -1 \\ 1 & 1 \end{bmatrix}. \tag{72}$$

It can be verified that

$$\mathbf{A}^2 = \begin{bmatrix} 0 & -4 \\ 4 & 0 \end{bmatrix},\ \mathbf{A}^4 = \begin{bmatrix} -16 & 0 \\ 0 & -16 \end{bmatrix} = -16\mathbf{I}. \tag{73}$$

Exercise 5

1. Work out explicitly the inverse of the matrix **A** of Eq. (32) and verify that you get the same result as that of Eq. (33).
2. Find the inverse of each of the following matrices :

(a) $\begin{bmatrix} 2 & 0 & -1 \\ 1 & 5 & -1 \\ 0 & 1 & 3 \end{bmatrix}$, *(b) $\begin{bmatrix} 0 & -1 & 3 \\ 4 & 2 & -2 \\ 1 & 1 & 1 \end{bmatrix}$,

(c) $\begin{bmatrix} 9 & 8 & 7 \\ 6 & 5 & 4 \\ 3 & 2 & 1 \end{bmatrix}$, *(d) $\begin{bmatrix} 1 & 0 & 1 & 1 \\ -1 & 2 & 3 & 4 \\ 6 & -2 & 1 & 0 \\ 3 & 0 & 1 & -1 \end{bmatrix}$.

3. If **A** is a nonsingular matrix, show that $\det(\mathbf{A}^{-1})=1/\det \mathbf{A}$.

*4. If $\mathbf{AB}=\mathbf{0}$, show that *either* **A** and **B** are both singular *or* one of them is a zero matrix.

5. Find the inverse of the matrix

$$\mathbf{A}=\begin{bmatrix} \cosh x & \sinh x \\ \sinh x & \cosh x \end{bmatrix},$$

and show that for all integral values of n (positive as well as negative),

$$\mathbf{A}^n=\begin{bmatrix} \cosh nx & \sinh nx \\ \sinh nx & \cosh nx \end{bmatrix}.$$

6. If $\mathbf{AXB}=\mathbf{I}$, where all the four matrices are of order n, show that the three matrices **A**, **X** and **B** are nonsingular and that $\mathbf{X}^{-1}=\mathbf{BA}$.

7. Show that $[\mathbf{A}, \mathbf{B}^{-1}]=-\mathbf{B}^{-1}[\mathbf{A}, \mathbf{B}]\mathbf{B}^{-1}$.

*8. Find the condition for which the matrix

$$\begin{bmatrix} q & p & p \\ p & q & p \\ p & p & q \end{bmatrix},$$

where p and q are numbers, is nonsingular. Show that the inverse, when it exists, is a matrix of the same form.

*9. If **P** is a nonsingular matrix and $\mathbf{P}^{-1}\mathbf{AP}$ and $\mathbf{P}^{-1}\mathbf{BP}$ are diagonal, prove that **A** and **B** commute.

10. Verify that

(a) $(\mathbf{A}^*)^{-1}=(\mathbf{A}^{-1})^*$, (b) $(\widetilde{\mathbf{A}})^{-1}=(\mathbf{A}^{-1})^T$, (c) $(\mathbf{A}^\dagger)^{-1}=(\mathbf{A}^{-1})^\dagger$.

*11. If **A** commutes with **B**, show that **A** also commutes with $\mathbf{B}^{-1}$.

12. Show that

$$\begin{bmatrix} \cos\alpha\cos\beta\cos\gamma-\sin\alpha\sin\gamma & -\cos\alpha\cos\beta\sin\gamma-\sin\alpha\cos\gamma & \cos\alpha\sin\beta \\ \sin\alpha\cos\beta\cos\gamma+\cos\alpha\sin\gamma & -\sin\alpha\cos\beta\sin\gamma+\cos\alpha\cos\gamma & \sin\alpha\sin\beta \\ -\sin\beta\cos\gamma & \sin\beta\sin\gamma & \cos\beta \end{bmatrix}$$

is an orthogonal matrix. [This is the most general orthogonal matrix of order 3 with determinant +1; see Eq. (14. 10).]

*13. Find the condition on the real scalars a and b for which the following matrix is orthogonal:

$$\begin{bmatrix} a+b & b-a \\ a-b & a+b \end{bmatrix}.$$

*14. If **x** is a nonzero real column vector, show that the matrix $\mathbf{I}-(2/\widetilde{\mathbf{x}}\,\mathbf{x})\,\widetilde{\mathbf{x}}\,\mathbf{x}$ is symmetric and orthogonal.

15. Show that the six matrices of Exercise 3.9 are orthogonal. Obtain the inverse of each and verify that $\mathbf{D}^{-1}=\mathbf{B}^{-1}\mathbf{A}^{-1}$.

16. Show that a unitary triangular matrix must be diagonal.

17. Show that hermitian and unitary matrices commute with their own hermitian conjugates.

*18. If **H** is a hermitian matrix, show that $\mathbf{U}=(\mathbf{H}-i\mathbf{I})(\mathbf{H}+i\mathbf{I})^{-1}$ is a unitary matrix.

6. Partitioning of matrices

It is often convenient to divide matrices into blocks, or to partition them, for the purposes of addition and multiplication. As we shall see below, this procedure also facilitates obtaining the inverse of a large matrix.

Consider a matrix **A** of order $m \times n$ with elements $(\mathbf{A})_{ij} = a_{ij}$. Let us partition this into four blocks as follows :

$$\mathbf{A} = \left[\begin{array}{c|c} \mathbf{A}_1 & \mathbf{A}_2 \\ \hline \mathbf{A}_3 & \mathbf{A}_4 \end{array}\right], \tag{1}$$

where $\mathbf{A}_1$, $\mathbf{A}_2$, $\mathbf{A}_3$ and $\mathbf{A}_4$ are rectangular matrices given by

$$\mathbf{A}_1 = \begin{bmatrix} a_{11} \dots a_{1l} \\ \vdots \\ a_{k1} \dots a_{kl} \end{bmatrix}, \; \mathbf{A}_2 = \begin{bmatrix} a_{1,l+1} \dots a_{1n} \\ \vdots \\ a_{k,l+1} \dots a_{kn} \end{bmatrix},$$

$$\mathbf{A}_3 = \begin{bmatrix} a_{k+1,1} & \dots a_{k+1,l} \\ \vdots & \\ a_{m1} & \dots a_{ml} \end{bmatrix}, \; \mathbf{A}_4 = \begin{bmatrix} a_{k+1,l+1} & \dots a_{k+1,n} \\ \vdots & \\ a_{m,l+1} & \dots a_{mn} \end{bmatrix}. \tag{2}$$

Here, k and l are arbitrary integers such that $1 \leqslant k \leqslant m$, $1 \leqslant l \leqslant n$. Now let **B** be some matrix of order $m \times n$ with elements $(\mathbf{B})_{ij} = b_{ij}$ and suppose that **B** is also divided into blocks in the same manner as **A**, i.e.,

$$\mathbf{B} = \left[\begin{array}{c|c} \mathbf{B}_1 & \mathbf{B}_2 \\ \hline \mathbf{B}_3 & \mathbf{B}_4 \end{array}\right], \tag{3}$$

where $\mathbf{B}_1$ is of order $k \times l$, $\mathbf{B}_2$ of order $k \times (n-l)$, $\mathbf{B}_3$ of order $(m-k) \times l$ and $\mathbf{B}_4$ of order $(m-k) \times (n-l)$. We say that **A** and **B** are *identically partitioned.*

Now it is very easy to see that

$$\mathbf{A}+\mathbf{B} = \left[\begin{array}{c|c} \mathbf{A}_1+\mathbf{B}_1 & \mathbf{A}_2+\mathbf{B}_2 \\ \hline \mathbf{A}_3+\mathbf{B}_3 & \mathbf{A}_4+\mathbf{B}_4 \end{array}\right]. \tag{4}$$

In the case of matrix addition, this is rather simple and almost trivial. Moreover, this method of partitioning matrices produces no simplification here. But we shall see that this method simplifies problems involving matrix multiplication and inversion.

Consider a matrix **A** of order $m \times n$ and a matrix **B** of order $n \times p$. Suppose both the matrices are divided into suitable blocks. Then their product can be written in the form

$$\mathbf{AB}=\begin{matrix} k\{ \\ m-k\{ \end{matrix}\left[\begin{array}{c:c} \overbrace{\mathbf{A}_1}^{l} & \overbrace{\mathbf{A}_2}^{n-l} \\ \hdashline \mathbf{A}_3 & \mathbf{A}_4 \end{array}\right]\left[\begin{array}{c:c} \overbrace{\mathbf{B}_1}^{t} & \overbrace{\mathbf{B}_2}^{p-t} \\ \hdashline \mathbf{B}_3 & \mathbf{B}_4 \end{array}\right]\begin{matrix} \}\, l \\ \}\, n-l \end{matrix}$$

$$= \begin{matrix} k\{ \\ m-k\{ \end{matrix}\left[\begin{array}{c:c} \overbrace{\mathbf{A}_1\mathbf{B}_1+\mathbf{A}_2\mathbf{B}_3}^{t} & \overbrace{\mathbf{A}_1\mathbf{B}_2+\mathbf{A}_2\mathbf{B}_4}^{p-t} \\ \hdashline \mathbf{A}_3\mathbf{B}_1+\mathbf{A}_4\mathbf{B}_3 & \mathbf{A}_3\mathbf{B}_2+\mathbf{A}_4\mathbf{B}_4 \end{array}\right]. \qquad (5)$$

It is not necessary to divide a matrix into four blocks only ; it may be divided into any number of blocks. A few examples of matrix multiplication by partitioning are given below.

(a) Let **A** be of order $m\times n$ and **B** of order $n\times p$. Then

$$\mathbf{AB}=\begin{matrix} r\{ \\ s\{ \\ t\{ \end{matrix}\left[\begin{array}{c} \overbrace{\mathbf{A}_1}^{n} \\ \hdashline \mathbf{A}_2 \\ \hdashline \mathbf{A}_3 \end{array}\right]\left[\begin{array}{c:c:c} \overbrace{\mathbf{B}_1}^{q} & \overbrace{\mathbf{B}_2}^{l} & \overbrace{\mathbf{B}_3}^{i} \end{array}\right]$$

$$=\begin{matrix} r\{ \\ s\{ \\ t\{ \end{matrix}\left[\begin{array}{c:c:c} \overbrace{\mathbf{A}_1\mathbf{B}_1}^{q} & \overbrace{\mathbf{A}_1\mathbf{B}_2}^{l} & \overbrace{\mathbf{A}_1\mathbf{B}_3}^{i} \\ \hdashline \mathbf{A}_2\mathbf{B}_1 & \mathbf{A}_2\mathbf{B}_2 & \mathbf{A}_2\mathbf{B}_3 \\ \hdashline \mathbf{A}_3\mathbf{B}_1 & \mathbf{A}_3\mathbf{B}_2 & \mathbf{A}_3\mathbf{B}_3 \end{array}\right], \qquad (6)$$

where $r+s+t=m$ and $q+l+i=p$.

(b) For the same matrix **AB**, we can also write

$$\mathbf{AB}=\begin{matrix} r\{ \\ s\{ \\ t\{ \end{matrix}\left[\begin{array}{c} \overbrace{\mathbf{A}_1}^{n} \\ \hdashline \mathbf{A}_2 \\ \hdashline \mathbf{A}_3 \end{array}\right]\left[\begin{array}{c:c} \overbrace{\mathbf{B}_1}^{l} & \overbrace{\mathbf{B}_2}^{i} \end{array}\right] = \begin{matrix} r\{ \\ s\{ \\ t\{ \end{matrix}\left[\begin{array}{c:c} \overbrace{\mathbf{A}_1\mathbf{B}_1}^{l} & \overbrace{\mathbf{A}_1\mathbf{B}_2}^{i} \\ \hdashline \mathbf{A}_2\mathbf{B}_1 & \mathbf{A}_2\mathbf{B}_2 \\ \hdashline \mathbf{A}_3\mathbf{B}_1 & \mathbf{A}_3\mathbf{B}_2 \end{array}\right], \qquad (7)$$

where $r+s+t=m$ and $l+i=p$.

(c) For the same matrix again, we could also write

$$\mathbf{AB}=m\{\left[\begin{array}{c:c:c} \overbrace{\mathbf{A}_1}^{r} & \overbrace{\mathbf{A}_2}^{s} & \overbrace{\mathbf{A}_3}^{t} \end{array}\right]\left[\begin{array}{c} \overbrace{\mathbf{B}_1}^{p} \\ \hdashline \mathbf{B}_2 \\ \hdashline \mathbf{B}_3 \end{array}\right]\begin{matrix} \}\, r \\ \}\, s \\ \}\, t \end{matrix} =[\mathbf{A}_1\mathbf{B}_1+\mathbf{A}_2\mathbf{B}_2+\mathbf{A}_3\mathbf{B}_3]_{m\times p}, \qquad (8)$$

where $r+s+t=n$.

Finally, we consider the use of the method of partitioning of matrices for finding the inverse of a matrix in the following example.

EXAMPLE 1: Let $\mathbf{A}$ be a matrix of order $m\times m$, $\mathbf{B}$ of order $m\times n$, $\mathbf{C}$ of order $n\times m$ and $\mathbf{D}$ of order $n\times n$. Assuming wherever necessary that the matrices are nonsingular, show that

$$\left[\begin{array}{c|c} \mathbf{A} & \mathbf{B} \\ \hline \mathbf{C} & \mathbf{D} \end{array}\right]^{-1} = \left[\begin{array}{c|c} (\mathbf{A}-\mathbf{B}\mathbf{D}^{-1}\mathbf{C})^{-1} & \mathbf{A}^{-1}\mathbf{B}(\mathbf{C}\mathbf{A}^{-1}\mathbf{B}-\mathbf{D})^{-1} \\ \hline \mathbf{D}^{-1}\mathbf{C}(\mathbf{B}\mathbf{D}^{-1}\mathbf{C}-\mathbf{A})^{-1} & (\mathbf{D}-\mathbf{C}\mathbf{A}^{-1}\mathbf{B})^{-1} \end{array}\right]. \quad (9)$$

Solution: To prove this, we note that the given matrix is of order $(m+n)\times(m+n)$. Let us assume that the required inverse matrix is

$$\mathbf{Z}=\left[\begin{array}{c|c} \mathbf{P} & \mathbf{Q} \\ \hline \mathbf{R} & \mathbf{S} \end{array}\right], \quad (10)$$

which is a square matrix of order $m+n$ and has been identically partitioned as the given matrix. Then we must have

$$\left[\begin{array}{c|c} \mathbf{A} & \mathbf{B} \\ \hline \mathbf{C} & \mathbf{D} \end{array}\right]\left[\begin{array}{c|c} \mathbf{P} & \mathbf{Q} \\ \hline \mathbf{R} & \mathbf{S} \end{array}\right] = \left[\begin{array}{c|c} \mathbf{I} & \mathbf{0} \\ \hline \mathbf{0} & \mathbf{I} \end{array}\right],$$

or

$$\left[\begin{array}{c|c} \mathbf{AP}+\mathbf{BR} & \mathbf{AQ}+\mathbf{BS} \\ \hline \mathbf{CP}+\mathbf{DR} & \mathbf{CQ}+\mathbf{DS} \end{array}\right] = \left[\begin{array}{c|c} \mathbf{I} & \mathbf{0} \\ \hline \mathbf{0} & \mathbf{I} \end{array}\right]. \quad (11)$$

This gives the four matrix equations

$$\mathbf{AP}+\mathbf{BR}=\mathbf{I}_m,\ \mathbf{CP}+\mathbf{DR}=\mathbf{0},\ \mathbf{AQ}+\mathbf{BS}=\mathbf{0},\ \mathbf{CQ}+\mathbf{DS}=\mathbf{I}_n. \quad (12)$$

In solving these equations for the four unknown matrices $\mathbf{P}$, $\mathbf{Q}$, $\mathbf{R}$ and $\mathbf{S}$, we must remember that $\mathbf{B}$ and $\mathbf{C}$ are rectangular matrices and hence their inverse is not defined. From the second of Eqs. (12), we have

$$\mathbf{R}=-\mathbf{D}^{-1}\mathbf{CP}. \quad (13)$$

Using this in the first of Eqs. (12), we get

$$\mathbf{AP}=\mathbf{I}_m-\mathbf{BR}=\mathbf{I}_m+\mathbf{BD}^{-1}\mathbf{CP}\Rightarrow(\mathbf{A}-\mathbf{BD}^{-1}\mathbf{C})\mathbf{P}=\mathbf{I}_m,$$

so that

$$\mathbf{P}=(\mathbf{A}-\mathbf{BD}^{-1}\mathbf{C})^{-1}. \quad (14a)$$

Putting this back in Eq. (13), we have

$$\mathbf{R}=-\mathbf{D}^{-1}\mathbf{C}(\mathbf{A}-\mathbf{BD}^{-1}\mathbf{C})^{-1}. \quad (14b)$$

Similarly, the third and the fourth of Eqs. (12) can be solved to obtain

$$\mathbf{S}=(\mathbf{D}-\mathbf{CA}^{-1}\mathbf{B})^{-1}, \quad \mathbf{Q}=-\mathbf{A}^{-1}\mathbf{B}(\mathbf{D}-\mathbf{CA}^{-1}\mathbf{B})^{-1}. \quad (14c)$$

Using these in Eq. (10), the desired result of Eq. (9) follows. Thus we see that the inverse of a large matrix can be obtained from the inverse of smaller matrices.

Submatrices and rank

Let **A** be a matrix of order $m \times n$. Any matrix obtained from **A** by omitting some of its rows or columns is called a *submatrix* of **A**. Thus, if

$$\mathbf{A}=\begin{bmatrix} 3 & 5 & 9 & 1 \\ 2 & 0 & -1 & 3 \\ -3 & -6 & 1 & 2 \end{bmatrix}, \tag{15}$$

some of its submatrices are

$$\begin{bmatrix} 3 & 5 & 9 \\ 2 & 0 & -1 \\ -3 & -6 & 1 \end{bmatrix}, \begin{bmatrix} 3 & 9 & 1 \\ 2 & -1 & 3 \\ -3 & 1 & 2 \end{bmatrix}, \begin{bmatrix} 5 & 9 & 1 \\ 0 & -1 & 3 \\ -6 & 1 & 2 \end{bmatrix},$$

$$\begin{bmatrix} 3 & 5 & 1 \\ 2 & 0 & 3 \end{bmatrix}, \begin{bmatrix} 2 & -1 & 3 \\ -3 & 1 & 2 \end{bmatrix}, \begin{bmatrix} 3 & 9 \\ 2 & -1 \end{bmatrix},$$

$$\begin{bmatrix} 5 & 1 \\ 0 & 3 \\ -6 & 2 \end{bmatrix}, \begin{bmatrix} 3 \\ 2 \\ -3 \end{bmatrix}, [\,3\;5\;1\,], [\,0\;3\,], [\,9\,], [\,-6\,], \text{etc.}$$

When a matrix is partitioned into a number of blocks, each block is a submatrix of the original matrix.

If a matrix **A** has at least one square nonsingular submatrix of order r but every square submatrix of **A** of order greater than r is singular, then r is said to be the *rank* of the matrix **A**. The rank of the matrix **A** of Eq. (15) is 3 because it has nonsingular square submatrices of order 3 but not higher than 3.

EXAMPLE 2: Find the rank of the matrix

$$\begin{bmatrix} 1 & 0 & 2 & -4 \\ 3 & -1 & 1 & 2 \\ 5 & -1 & 5 & -6 \\ 4 & -1 & 3 & -2 \end{bmatrix},$$

Solution: We note that the determinant of the given matrix is zero, so that its rank is less than 4. Next, we consider all possible submatrices of order 3 (there are 16 of them) and observe that all of them are also singular, so that the rank is less than 3. Finally, we note that the given matrix has nonsingular submatrices of order 2, so that its rank is 2.

This is evidently a lengthy and tedious procedure. In the above example, we had to find the determinants of a matrix of order four and sixteen matrices of order three before obtaining the rank. A simpler and more elegant method will be discussed in Example 8.7.

Since the determinant of a matrix remains unchanged on transposition, it follows that *the rank of the transpose of a matrix is the same as the rank of the original matrix.*

Exercise 6

*1. Obtain the inverse of the matrix

$$\left[\begin{array}{ccc|cc} 1 & 0 & 1 & 0 & 1 \\ 0 & 1 & 0 & 1 & 0 \\ 1 & 0 & 0 & 0 & 1 \\ \hline 0 & 1 & 0 & 0 & 1 \\ 1 & 0 & 1 & 1 & 1 \end{array}\right],$$

which has been partitioned into blocks as shown, by using the result of Eq. (9).

2. If **A**, **B**, **C** and **D** are nonsingular square matrices of order n, show that

$$\left[\begin{array}{c|c} \mathbf{A} & \mathbf{B} \\ \hline \mathbf{C} & \mathbf{D} \end{array}\right]^{-1} = \left[\begin{array}{c|c} (\mathbf{A}-\mathbf{B}\mathbf{D}^{-1}\mathbf{C})^{-1} & (\mathbf{C}-\mathbf{D}\mathbf{B}^{-1}\mathbf{A})^{-1} \\ \hline (\mathbf{B}-\mathbf{A}\mathbf{C}^{-1}\mathbf{D})^{-1} & (\mathbf{D}-\mathbf{C}\mathbf{A}^{-1}\mathbf{B})^{-1} \end{array}\right],$$

assuming that the large matrix is also nonsingular.

3. If **U** and **V** are unitary matrices of order n which commute with each other, show that

$$\frac{1}{\sqrt{2}}\begin{bmatrix} \mathbf{U} & -\mathbf{V} \\ \mathbf{V}^{\dagger} & \mathbf{U}^{\dagger} \end{bmatrix}$$

is a unitary matrix.

*4. A matrix is partitioned as

$$\left[\begin{array}{c|c|c} \mathbf{A} & \mathbf{B} & \mathbf{C} \\ \hline \mathbf{D} & \mathbf{E} & \mathbf{F} \end{array}\right].$$

Write down the complex conjugate, the transpose and the hermitian conjugate of the matrix.

5. Let

$$\mathbf{M} = \left[\begin{array}{c|c} \mathbf{A} & \mathbf{0} \\ \hline \mathbf{X} & \mathbf{B} \end{array}\right],$$

where **A** is of order $m \times m$, **B** of order $n \times n$, **X** of order $n \times m$ and **0** is the zero matrix of order $m \times n$. If **M** is unitary, show that **X** must be a zero matrix and **A** and **B** must be unitary.

6. Find the rank of each of the following matrices:

*(a) $\begin{bmatrix} 1 & 0 & 2 \\ 2 & 0 & 4 \\ 3 & 6 & -1 \end{bmatrix}$; (b) $\begin{bmatrix} 6 & 4 & 2 \\ 1 & -1 & 3 \\ 8 & -8 & 0 \\ 3 & -3 & 2 \end{bmatrix}$;

*(c) $\begin{bmatrix} 3 & 1 & -3 & 1 \\ 5 & 3 & 2 & -1 \\ 9 & 4 & 0 & 1 \\ -1 & 0 & 4 & -1 \end{bmatrix}$.

7. Systems of linear equations—particular cases

The methods of matrix algebra prove to be very useful in solving a system of linear equations in some unknowns. Let $x_1, x_2, \ldots, x_n$ be a set of n variables or parameters. An equation which does not contain any power of x_i higher than unity or their products with each other and involves constant coefficients (that is, independent of x_i) is called a *linear equation* in the variables x_i. Our object will be to study the most general system of m linear equations in n unknowns which can be written in the form

$$\begin{aligned} a_{11}x_1+a_{12}x_2+\ldots+a_{1n}x_n&=b_1,\\ a_{21}x_1+a_{22}x_2+\ldots+a_{2n}x_n&=b_2,\\ \ldots\ldots\ldots\ldots\ldots\ldots\ldots&\\ a_{m1}x_1+a_{m2}x_2+\ldots+a_{mn}x_n&=b_m. \end{aligned} \tag{1}$$

Here the coefficients a_{ij} and the right hand terms b_i are supposed to be known constants. The x_i's are unknown variables and we wish to determine how many solutions there are for them which satisfy Eqs. (1). This general problem is taken up in the next section. This section contains a preliminary discussion leading to the general problem, a few basic definitions and a number of examples.

Evidently, we can write Eqs. (1) in the matrix form as

$$\mathbf{Ax}=\mathbf{b}, \tag{2}$$

with $\mathbf{A}\equiv[a_{ij}]_{m\times o}$, $\mathbf{x}\equiv[x_i]_{n\times 1}$ and $\mathbf{b}\equiv[b_i]_{m\times 1}$. Note that $\mathbf{x}$ and $\mathbf{b}$ are column vectors of order n and m, respectively. Now, if $\mathbf{b}=\mathbf{0}$ (that is, each $b_i=0$ for $1\leqslant i\leqslant m$), the equations are called a set of *homogeneous equations.* If $\mathbf{b}\neq\mathbf{0}$ (which will be the case if even a single b_i is different from zero), they are called a set of *inhomogeneous equations.* It is clear that if $\mathbf{b}=\mathbf{0}$, the system possesses a solution $\mathbf{x}=\mathbf{0}$ (that is, $x_i=0$, $1\leqslant i\leqslant n$) which is known as the *trivial solution.* The equation

$$\mathbf{Ax}=\mathbf{0} \tag{3}$$

is called the *associated homogeneous equation.*

Linear equations in one unknown

As a first step, we consider the simplest situation, that is, one equation in one unknown of the form

$$ax=b. \tag{4}$$

We ask the question: What are the possible solutions of this equation? An elementary school-level answer would be to say that $x=b/a$. However, at the level at which one is reading this book, one cannot be so naive, because it is evident that if $a=0$, our simple answer fails. The complete answer is discussed below.

CASE (1): $a \neq 0$. Dividing Eq. (4) by a, we have $x=b/a$. We have two subcases; case (1a): $b=0$, $x=0$, the trivial solution; case (1b): $b \neq 0$, $x=b/a \neq 0$.

CASE (2): $a=0$. We have again two subcases; case (2a): $b \neq 0$, there is no solution for x; case (2b): $b=0$, x may have any finite value, or there is an infinite number of solutions, of which, the trivial solution $x=0$ is also one.

We can draw the following important conclusions from this simple example.

1. In case (1), the solution is unique.
2. In case (1a), the only solution is the trivial solution, $x=0$.
3. In case (2a), there is no solution, whereas in case (2b), there is an infinite number of solutions.
4. In case the equation is homogeneous [$b=0$; cases (1a) and (2b)], *a nontrivial solution exists if and only if* $a=0$.

Consider, now, a system of more than one equations in one unknown, such as

$$a_1 x=b_1,\ a_2 x=b_2, \ldots,\ a_m x=b_m. \tag{5}$$

If a single a_i is zero and the corresponding b_i is nonzero, it follows that the system has no solution. Let therefore $a_i \neq 0$, $1 \leqslant i \leqslant m$. The system of Eqs. (5) then has a solution if and only if $b_1/a_1=b_2/a_2 = \ldots = b_m/a_m$. In other words, we solve any one equation from the system. If this solution satisfies all the other equations of the system, the system has a solution, otherwise not.

Linear equations in two unknowns

Consider two linear equations in two unknowns:

$$\begin{aligned} a_{11} x_1 + a_{12} x_2 &= b_1, \\ a_{21} x_1 + a_{22} x_2 &= b_2. \end{aligned} \tag{6}$$

We have the following cases.

CASE (1): $\det \mathrm{A} \equiv \begin{vmatrix} a_{11} & a_{12} \\ a_{21} & a_{22} \end{vmatrix} = a_{11} a_{22} - a_{12} a_{21} \neq 0.$

We have

$$\mathbf{A}^{-1}=(1/\det \mathbf{A})\begin{bmatrix} a_{22} & -a_{12} \\ -a_{21} & a_{11} \end{bmatrix}. \tag{7}$$

The unique solution is therefore $\mathbf{x}=\mathbf{A}^{-1}\mathbf{b}$, or

$$x_1=(b_1 a_{22}-b_2 a_{12})/\det \mathbf{A},$$
$$x_2=(-b_1 a_{21}+b_2 a_{11})/\det \mathbf{A}. \tag{8}$$

We have the following two subcases.

CASE (1a): $\mathbf{b}=\mathbf{0}$ (that is, $b_1=b_2=0$); homogeneous equations. The only solution is the trivial solution $x_1=x_2=0$.

CASE (1b): $\mathbf{b}\neq\mathbf{0}$ (that is, at least one of the two numbers b_1 and b_2 is different from zero); inhomogeneous equations. The system has a unique and nontrivial solution given by Eqs. (8).

CASE (2): det $\mathbf{A}=0$, or

$$a_{11}a_{22}-a_{12}a_{21}=0. \tag{9}$$

We solve the first of Eqs. (6) for x_2 to give

$$x_2=(b_1-a_{11}x_1)/a_{12}, \tag{10}$$

assuming[1] $a_{12}\neq 0$. Using this, the left hand side of the second of Eqs. (6) becomes

$$a_{21}x_1+a_{22}(b_1-a_{11}x_1)/a_{12}=a_{22}b_1/a_{12}, \tag{11}$$

by virtue of Eq. (9). It is then evident that the question whether Eqs. (6) possess a solution or not depends on whether $a_{22}b_1/a_{12}$ equals b_2 or not. We now have the following three subcases.

CASE (2a): $\mathbf{b}\neq\mathbf{0}$ and $\mathbf{b}$ is such that $a_{22}b_1/a_{12}=b_2$. This together with Eq. (9) shows that

$$a_{21}/a_{11}=a_{22}/a_{12}=b_2/b_1. \tag{12}$$

This shows that the second of Eqs. (6) is just a multiple of the first and hence if one is satisfied, the other is also satisfied. It is therefore sufficient to solve one of them, say the first one, which gives the solution

$$x_2=(b_1-a_{11}x_1)/a_{12},\ x_1 \text{ arbitrary}. \tag{13}$$

There is thus an infinite number of solutions.

CASE (2b): $\mathbf{b}\neq\mathbf{0}$ and $\mathbf{b}$ is such that $a_{22}b_1/a_{12}\neq b_2$. In this case, Eqs. (6) have no solution.

[1] If $a_{12}=0$, Eq. (9) shows that either $a_{11}=0$ or $a_{22}=0$. In the first case, the first of Eqs. (6) becomes $0x_1+0x_2=b_1$ which is trivial if $b_1=0$ and has no solution if $b_1\neq 0$. In the second case, Eqs. (6) contain only one unknown and not two.

CASE (2c): $\mathbf{b}=\mathbf{0}$. In this case, the equation $a_{22}\, b_1/a_{12}=b_2$ is identically satisfied and hence Eqs. (6) once again have an infinite number of solutions given by

$$x_2=-a_{11}\, x_1/a_{12},\ x_1 \text{ arbitrary}. \tag{14}$$

All that we have done has a simple and elegant geometrical interpretation. If we regard x_1 and x_2 as the cartesian coordinates in a two-dimensional plane, a linear equation involving x_1 and x_2 denotes a straight line in the plane. The problem of solving Eqs. (6) is then equivalent to finding the point(s) of intersection of the two lines. We note that the slope of the line is determined by the ratio of the coefficients of x_1 and x_2 and the intersection with the x_2 axis is determined by the ratio of the inhomogeneous term b_i to the coefficient of x_2.

The two lines of Eqs. (6) will be parallel if

$$a_{11}/a_{12}=a_{21}/a_{22} \Rightarrow a_{11}a_{22}-a_{12}a_{21} \equiv \det \mathbf{A}=0. \tag{15}$$

Conversely, if $\det \mathbf{A}=0$, the lines will be parallel. The parallel lines will be *distinct* if $a_{22}/a_{12} \neq b_2/b_1$, in which case, there will be no point of intersection, that is, no solution. The parallel lines will be *the same* if $a_{22}/a_{12}=b_2/b_1$, in which case there is an infinite number of points common to both the lines, that is, an infinite of solutions. On the other hand, if $\det \mathbf{A} \neq 0$, the two lines have different slopes and it follows that they have one and only one point of intersection, that is, a unique solution. Notice that if the inhomogenous part of an equation is zero, the corresponding line passes through the origin. The five possible subcases discussed earlier are illustrated geometrically in Fig. 3.

Now, let us consider three linear equations in two unknowns, that is, the system

$$\begin{aligned} a_{11}x_1+a_{12}x_2&=b_1,\\ a_{21}x_1+a_{22}x_2&=b_2,\\ a_{31}x_1+a_{32}x_2&=b_3. \end{aligned} \tag{16}$$

It will be convenient to give both the algebraic and the geometrical interpretations simultaneously. The system under consideration will have a solution if and only if the three lines of Eqs. (16) meet at the same point. Now, it is well-known that the determinant

$$D=\begin{vmatrix} a_{11} & a_{12} & b_1 \\ a_{21} & a_{22} & b_2 \\ a_{31} & a_{32} & b_3 \end{vmatrix} \tag{17}$$

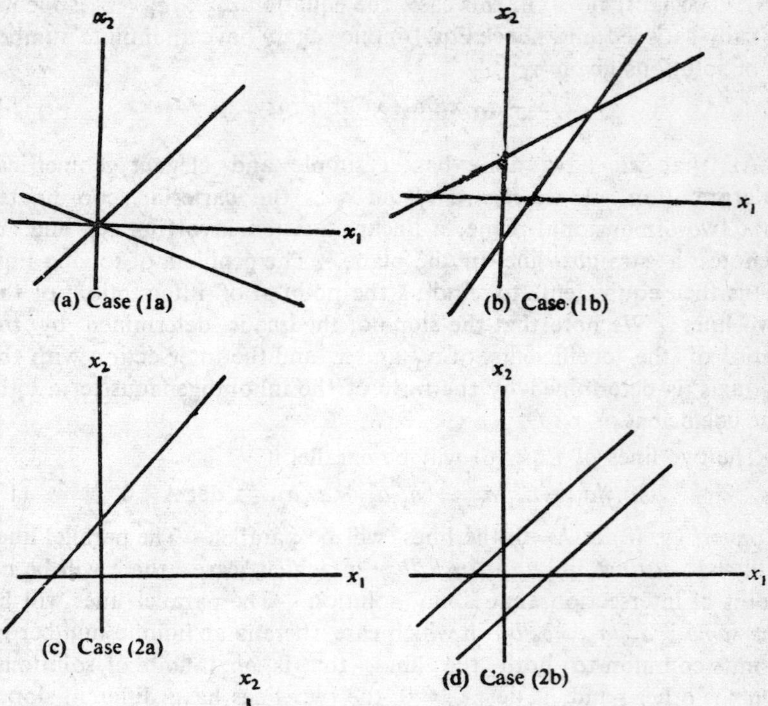

Figure 3. The solutions of Eqs. (6). In (a) and (b) there is a unique solution, in (c) and (e) an infinite number of solutions, and in (d) there is no solution.

is proportional to the area of the triangle enclosed by the three lines *provided no two of the three lines are parallel.* All the three lines would in that case meet at a common point if and only if $D=0$. From converse of Corollary 2 of Example 8.6, page 85, it follows that if $D=0$, at least one of its rows can be expressed as a linear combination of the remaining two rows. It means that, for example, we could express the coefficients of the third of Eqs. (16) as linear combinations of those of the first two in the form

$$a_{31}=ca_{11}+da_{21},\ a_{32}=ca_{12}+da_{22},\ b_3=cb_1+db_2, \tag{18}$$

where c and d are scalars. Borrowing the concept of linear dependence of vectors, we say that in this case the three equations are *linearly dependent*. If any two equations are satisfied, the third is identically satisfied.

Merely the vanishing of the determinant D does not guarantee a solution. Besides, it is necessary that no two lines be parallel to each other. Consider, for example, the situation when Eqs. (16) denote three parallel straight lines. In this case, the second column of D would be a multiple of the first and we would have $D=0$. But three parallel lines have no common point of intersection and hence there would be no solution. The assertion which can be made with certainty is the converse assertion that *if* $D \neq 0$, *there is no solution.* One final situation is that in which all three Eqs. (16) are multiples of each other. In this case, all the three equations denote the same straight line and there would be an infinite number of solutions. Some of the situations which arise with three equations in two unknowns are depicted geometrically in Fig. 4. We can summarize the conclusions as follows.

1. If $D\neq0$, the system has no solution.
2. If the three equations denote the same line, we would have $D=0$ and there is an infinite number of solutions.
3. If $D=0$ and not all the three equations denote the same straight line, there is one and only one solution provided no two lines are parallel, and there is no solution if any two lines are parallel.

It is evident that the method can be generalized to any number of equations in two unknowns.

Linear equations in three unknowns

Now that we have sufficient experience with equations involving one and two unknowns, we will discuss only briefly the geometrical interpretation of linear equations in three unknowns. The detailed algebraic treatment of the general case is then taken up in the next section.

Consider linear equations in three variables. If we regard x_1, x_2, x_3 as the cartesian coordinates in a three-dimensional vector space, any linear equation of the form

$$a_{11}x_1+a_{12}x_2+a_{13}x_3=b_1 \tag{19}$$

represents a plane (or a two-dimensional *subspace*). If the inhomogeneous part b_1 vanishes, the plane passes through the origin of

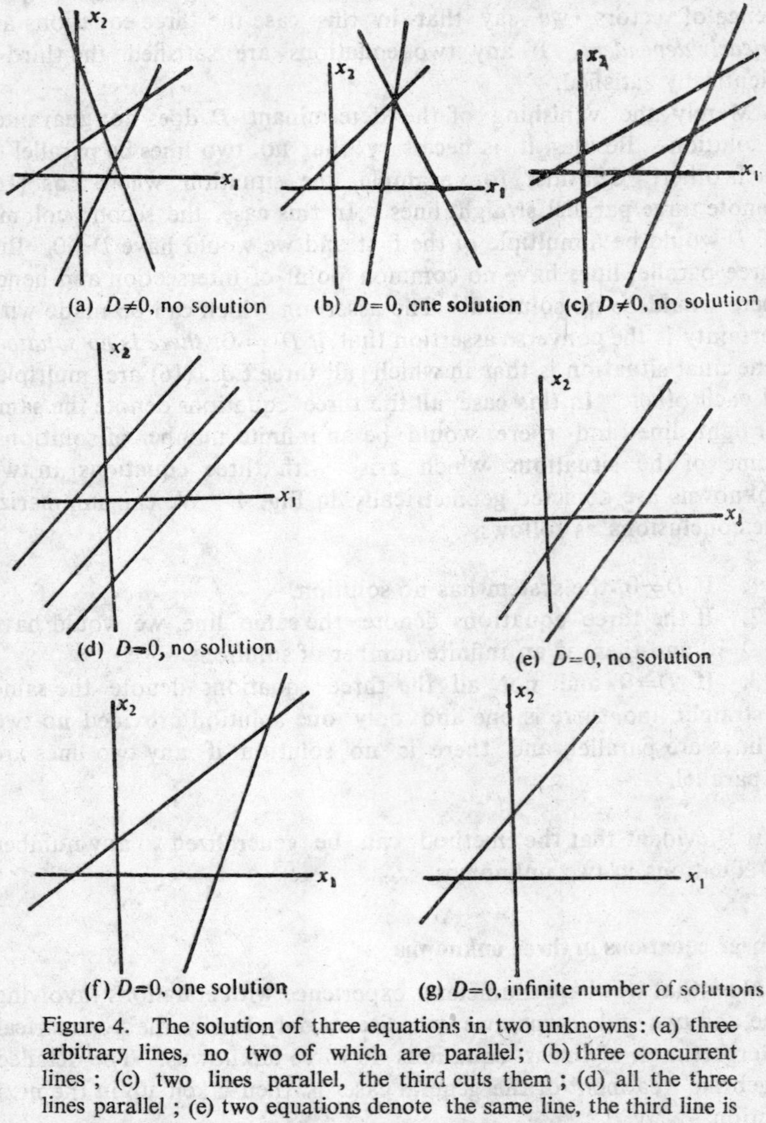

(a) $D \neq 0$, no solution (b) $D=0$, one solution (c) $D \neq 0$, no solution

(d) $D=0$, no solution (e) $D=0$, no solution

(f) $D=0$, one solution (g) $D=0$, infinite number of solutions

Figure 4. The solution of three equations in two unknowns: (a) three arbitrary lines, no two of which are parallel; (b) three concurrent lines ; (c) two lines parallel, the third cuts them ; (d) all the three lines parallel ; (e) two equations denote the same line, the third line is parallel ; (f) two equations denote the same line, the third line cuts it ; (g) all the three equations denote the same straight line.

coordinates. If we have a single equation as above to solve, there will be an infinite number of solutions, for the coordinates of every point in the plane satisfy the equation. Algebraically, the solution can be written as

$$x_3=(b_1-a_{11}x_1-a_{12}x_2)/a_{13},\ x_1,\ x_2 \text{ arbitrary},\ a_{13}\neq 0. \tag{20}$$

There is thus a two-fold arbitrariness.

Consider now two equations in three variables:

$$\begin{aligned} a_{11}x_1+a_{12}x_2+a_{13}x_3&=b_1,\\ a_{21}x_1+a_{22}x_2+a_{23}x_3&=b_2. \end{aligned} \tag{21}$$

Each of the two equations denotes a plane and the two planes will meet in a straight line except when they are parallel. We have therefore the following cases.

CASE (1): The planes are not parallel and hence meet in a straight line. There will be an infinite number of solutions, for the coordinates of every point on the line of intersection will satisfy both the equations. There is one-fold arbitrariness.

CASE (2): The planes will be parallel if and only if the coefficients of the homogeneous parts of the two equations are respectively proportional to each other, that is,

$$a_{21}/a_{11}=a_{22}/a_{12}=a_{23}/a_{13}\neq b_2/b_1. \tag{22}$$

In this case, there is no solution.

CASE (3): In the case (2), it may further happen that the inhomogeneous parts also bear the same ratio to each other as the coefficients of the homogeneous parts, that is,

$$a_{21}/a_{11}=a_{22}/a_{12}=a_{23}/a_{13}=b_2/b_1. \tag{23}$$

In this case, both the equations denote the same plane and there will be an infinite number of solutions, with a two-fold arbitrariness.

Consider the following two matrices :

$$\mathbf{A}=\begin{bmatrix} a_{11} & a_{12} & a_{13}\\ a_{21} & a_{22} & a_{23}\end{bmatrix},\quad \mathbf{B}=\begin{bmatrix} a_{11} & a_{12} & a_{13} & b_1\\ a_{21} & a_{22} & a_{23} & b_2\end{bmatrix}, \tag{24}$$

the first of which involves the coefficients of the homogeneous parts alone while the second has an additional column which is the inhomogeneous part of the equations. **A** is known as the *matrix of coefficients* and **B** as the *augmented matrix*. Let us try to study the ranks of these two matrices in the three cases discussed above. It is easy to see that we shall have the following results:

CASE (1): rank of $\mathbf{A}=2$, rank of $\mathbf{B}=2$;
CASE (2): rank of $\mathbf{A}=1$, rank of $\mathbf{B}=2$;

CASE (3): rank of $\mathbf{A}=1$, rank of $\mathbf{B}=1$.

It is then evident that if we determine the ranks of **A** and **B**, we know how many solutions the system has. Note that the rank of **B** gives us the number of independent equations. In the case (3), there is only one independent equation; the other is just a multiple of the first. The rank of **A** gives the number of independent associated homogeneous equations obtained by equating the inhomogeneous part to zero.

If we have three equations in three variables, we have a case of three planes. All the three planes may meet at a point whose coordinates will then be the unique solution of the system. The planes may pass through the same straight line in which case there is an infinite number of solutions with one arbitrary parameter. Finally, all the three planes may be identical when again there is an infinite number of solutions with two arbitrary parameters. In all other cases, there is no solution[2].

If we have more than three equations in three unknowns, it is convenient to solve any three of these equations first and to verify whether this solution satisfies the remaining equations also.

Cramer's rule

Consider a system of three equations in three unknowns:

$$\begin{aligned} a_{11}x_1+a_{12}x_2+a_{13}x_3&=b_1,\\ a_{21}x_1+a_{22}x_2+a_{23}x_3&=b_2,\\ a_{31}x_1+a_{32}x_2+a_{33}x_3&=b_3. \end{aligned} \tag{25}$$

Let $\mathbf{A}\equiv[a_{ij}]$ be the matrix of coefficients and let A^{ij} denote the cofactor of a_{ij} in det **A**. To solve the above system, let us multiply the first of Eqs. (25) by A^{1k}, the second by A^{2k}, the third by A^{3k} $(1\leqslant k\leqslant 3)$, add, and rearrange terms on the left hand side to obtain

$$\begin{aligned} x_1\,(a_{11}A^{1k}+a_{21}A^{2k}+a_{31}A^{3k})+x_2\,(a_{12}A^{1k}+a_{22}A^{2k}+a_{32}A^{3k})\\ +x_3\,(a_{13}A^{1k}+a_{23}A^{2k}+a_{33}A^{3k})=b_1A^{1k}+b_2A^{2k}+b_3A^{3k}. \end{aligned} \tag{26}$$

It follows from Eq. (4.5) that on the left hand side of Eq. (26), only the coefficient of x_k will survive and will be equal to det **A**, while the other two coefficients will vanish, leading to

[2]Note that the system may not have a solution even if no two of the three planes are parallel. This case corresponds to that of three straight lines in a plane; Fig. 4 (a).

$$x_k \det \mathbf{A}=b_1 A^{1k}+b_2 A^{2k}+b_3 A^{3k}, \tag{27}$$

or, if $\det \mathbf{A} \neq 0$,

$$x_k=(b_1 A^{1k}+b_2 A^{2k}+b_3 A^{3k})/\det \mathbf{A},\ k=1, 2, 3. \tag{28}$$

Thus, if $\det \mathbf{A} \neq 0$, the system of Eqs. (25) has a unique solution given by Eq. (28).

It is seen that the expression on the right hand side of Eq. (27) is the determinant of the matrix obtained by replacing the k-th column of $\mathbf{A}$ by the elements b_1, b_2, b_3. Let us denote these determinants by

$$D_1=\begin{vmatrix} b_1 & a_{12} & a_{13} \\ b_2 & a_{22} & a_{23} \\ b_3 & a_{32} & a_{33} \end{vmatrix},\ D_2=\begin{vmatrix} a_{11} & b_1 & a_{13} \\ a_{21} & b_2 & a_{23} \\ a_{31} & b_3 & a_{33} \end{vmatrix},\ D_3=\begin{vmatrix} a_{11} & a_{12} & b_1 \\ a_{21} & a_{22} & b_2 \\ a_{31} & a_{32} & b_3 \end{vmatrix}. \tag{29}$$

Then the solution of Eq. (28) can be written as

$$x_1=D_1/\det \mathbf{A},\ x_2=D_2/\det \mathbf{A},\ x_3=D_3/\det \mathbf{A}. \tag{30}$$

This procedure is known as the *Cramer's rule* which is based on the *Cramer's theorem* which states, in general, that *if the determinant of the matrix* **A** *of coefficients of a system of n linear equations in the same number of unknowns is not zero, the system has precisely one solution given by*

$$x_k=D_k/\det \mathbf{A},\ 1 \leqslant k \leqslant n, \tag{31}$$

where D_k is the determinant obtained by replacing the k-th column of **A** *by the inhomogeneous terms.*

The above treatment also gives two results as corollaries:

(a) If $\det \mathbf{A}=0$, but $D_k \neq 0$ for at least one value of k, $1 \leqslant k \leqslant n$, the system has no solution.

(b) If $\det \mathbf{A} \neq 0$ and the system is homogeneous $[b_i=0,\ 1 \leqslant i \leqslant n]$, the system has only the trivial solution $x_1=x_2=\ldots=x_n=0$.

EXAMPLE 1: Solve the system of equations

$$\begin{aligned} 4a+2b+\ c&=-8, \\ 3a-\ b+2c&=\ \ 4, \\ -a-\ b-\ c&=\ \ 2. \end{aligned} \tag{32}$$

Solution: The determinant of the system is

$$\det \mathbf{A}=\begin{vmatrix} 4 & 2 & 1 \\ 3 & -1 & 2 \\ -1 & -1 & -1 \end{vmatrix}=10. \tag{33}$$

The determinants D_k are found to be

$$D_1=\begin{vmatrix} -8 & 2 & 1 \\ 4 & -1 & 2 \\ 2 & -1 & -1 \end{vmatrix}=-10,\ D_2=\begin{vmatrix} 4 & -8 & 1 \\ 3 & 4 & 2 \\ -1 & 2 & -1 \end{vmatrix}=-30, \tag{34}$$

$$D_3=\begin{vmatrix} 4 & 2 & -8 \\ 3 & -1 & 4 \\ -1 & -1 & 2 \end{vmatrix}=20.$$

By Cramer's rule, the solution is therefore

$$a=D_1/\det \mathbf{A}=-1,\ b=D_2/\det \mathbf{A}=-3,\ c=D_3/\det \mathbf{A}=2. \qquad (35)$$

Exercise 7

*1. By finding the inverse of the matrix of coefficients, invert the transformation

$$\begin{aligned} u&=19x+2y-9z, \\ v&=-4x-y+2z, \\ w&=-4x+z. \end{aligned}$$

2. Solve the equations *(a) $\mathbf{Ax}=2\mathbf{x}$, (b) $\mathbf{Ax}=-\mathbf{x}$, *(c) $\mathbf{Ax}=3\mathbf{x}$, where

$$\mathbf{A}=\begin{bmatrix} 0 & 1 & 1 \\ 1 & 0 & 1 \\ 1 & 1 & 0 \end{bmatrix},\ \mathbf{x}=\begin{bmatrix} x_1 \\ x_2 \\ x_3 \end{bmatrix}.$$

*3. Solve the following systems of linear equations:

(a) $$\begin{aligned} 2x+y&=3, \\ -x+4y&=0, \\ x+3y&=-1; \end{aligned}$$

(b) $$\begin{aligned} x+y+3z&=6, \\ 2x+3y-4z&=6, \\ 3x+2y+7z&=0. \end{aligned}$$

4. Using Kirchhoff's laws, show that the currents in the electrical network of Fig. 5 are given by

$$\begin{aligned} I_1-I_2-I_3&=0, \\ R_2 I_2-R_3 I_3&=0, \\ R_1 I_1+R_2 I_2&=E_0. \end{aligned}$$

Hence solve for the unknown currents I_1, I_2, I_3 in terms of R_1, R_2, R_3 and E_0.

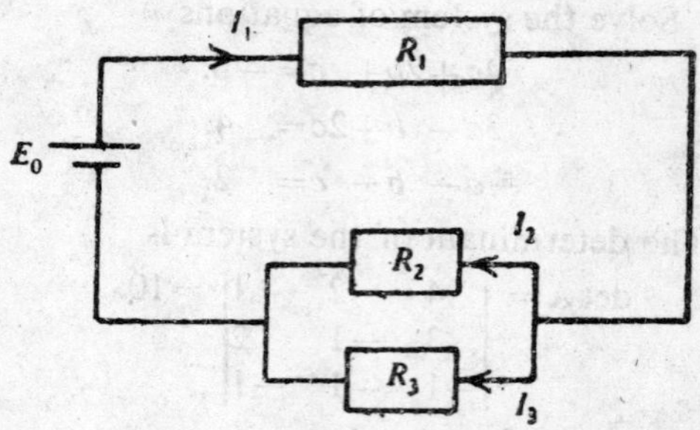

Figure 5. Electrical network for Exercise 4.

*5. By using Kirchhoff's laws, obtain the currents in the network shown in Fig. 6.

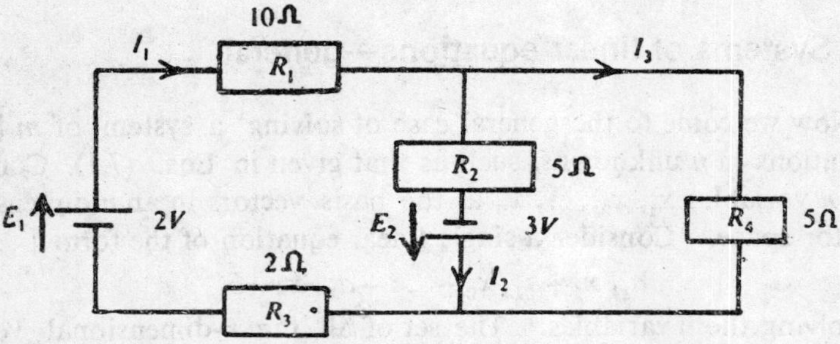

Figure 6. Electrical network for Exercise 5.

*6. Given

$$A = \begin{bmatrix} 1 & 2 & 3 \\ 4 & 5 & 6 \end{bmatrix},$$

find the general matrix **B** which satisfies the equation **AB**=**I**, where **I** is the unit matrix. Show that **B** contains two arbitrary parameters. Show that there is no matrix **B** which satisfies the equation **BA**=**I**. (The unit matrices appearing in the two equations may be of different orders.) [See discussions on the inverse matrix in Section 5 and Appendix II.]

8. Systems of linear equations—general

Now we come to the general case of solving a system of m linear equations in n unknowns, such as that given in Eqs. (7.1). Consider the n variables $x_1, x_2, \ldots, x_n$ as the basis vectors in an n-dimensional vector space. Consider a single linear equation of the form

$$a_{11} x_1 + a_{12} x_2 + \ldots + a_{1n} x_n = b_1 \tag{1}$$

involving the n variables. The set of all the n-dimensional vectors subject to Eq. (1) is an $(n-1)$-dimensional subspace of the full space. If there are two linear equations connecting the n variables, the possible vectors subject to these two equations lie in an $(n-2)$-dimensional vector space. In general, with certain exceptions as we shall discuss below, if the n variables are subject to m constraints $(m<n)$, the possible solutions lie in an $(n-m)$-dimensional subspace. If the number of constraints is equal to the number of variables, there will generally be a unique solution, while if the number of conditions exceeds the number of variables, there will, in general, be no solution. We shall discuss these three cases separately.

m equations in n unknowns, $m < n$

In general, a system of m linear equations in n unknowns $(m < n)$ would have an infinite number of solutions with an $(n-m)$-fold arbitrariness. However, there is an exceptional case when the system has no solution, in spite of the fact that the number of constraints is less than the number of variables.

Consider the system of Eqs. (7.1) and define the matrix $\mathbf{A}$ of order $m \times n$ as in Eq. (7.2). Further, the augmented matrix for the system is given by

$$\mathbf{B} = \begin{bmatrix} a_{11} & a_{12} & \cdots & a_{1n} & b_1 \\ a_{21} & a_{22} & \cdots & a_{2n} & b_2 \\ \vdots & & & & \\ a_{m1} & a_{m2} & \cdots & a_{mn} & b_m \end{bmatrix}, \tag{2}$$

which has for its last column the inhomogeneous part of Eqs. (7.1) and which is therefore of order $m \times (n+1)$. Let

$$r = \text{rank of } \mathbf{A}, \quad s = \text{rank of } \mathbf{B}. \tag{3}$$

For the case $m < n$, it is evident that $r \leqslant m$ and $s \leqslant m$; s gives the number of linearly independent equations from among the system of m equations, while r gives the number of linearly independent equations from among the system of m associated homogeneous equations. Moreover, it is also clear that $s=r$ or $r+1$. Now we state without proof the following results:

(a) If $s=r$, the system has an infinite number of solutions with $(n-r)$-fold arbitrariness.

(b) If $s>r$, the system has no solution. In this case it is said that the equations are *inconsistent*.

EXAMPLE 1: (a) Solve the following system of linear equations:

$$\begin{aligned} x+2y+2z-3w&=5,\\ 2x-y+z+2w&=1,\\ 3x+y+3z-w&=6. \end{aligned} \tag{4}$$

Solution: The matrix of coefficients and the augmented matrix are in this case given by

$$\mathbf{A}=\begin{bmatrix} 1 & 2 & 2 & -3 \\ 2 & -1 & 1 & 2 \\ 3 & 1 & 3 & -1 \end{bmatrix}, \mathbf{B}=\begin{bmatrix} 1 & 2 & 2 & -3 & 5 \\ 2 & -1 & 1 & 2 & 1 \\ 3 & 1 & 3 & -1 & 6 \end{bmatrix}. \tag{5}$$

Calculating the ranks of these matrices, we find that

$$r\equiv\text{rank of } \mathbf{A}=2, \quad s\equiv\text{rank of } \mathbf{B}=2. \tag{6}$$

Since $s=r$, the system would have an infinite number of solutions with $(4-2)=2$-fold arbitrariness. It is evident that the three equations are linearly dependent; the third equation is in fact the sum of the first two equations. The solution is found to be

$$x=(7-4z-w)/5, \; y=(9-3z+8w)/5, \; z, w \text{ arbitrary}, \tag{7}$$

(b) Solve the system of Eqs. (4) with the inhomogeneous part of the third equation replaced by 7, that is, the system

$$\begin{aligned} x+2y+2z-3w&=5,\\ 2x-y+z+2w&=1,\\ 3x+y+3z-w&=7. \end{aligned} \tag{8}$$

Solution: It is evident that with this change the three equations are no longer dependent although their homogeneous parts are dependent. We now have

$$\mathbf{B}=\begin{bmatrix} 1 & 2 & 2 & -3 & 5 \\ 2 & -1 & 1 & 2 & 1 \\ 3 & 1 & 3 & -1 & 7 \end{bmatrix}, \tag{9}$$

so that

$$r=2, s=3. \tag{10}$$

Since $s>r$, the system has no solution. This can be explicitly verified. The first two equations have the solution given in Eqs. (7).

Substituting this in the left hand side of the third equation, we get

$$3x+y+3z-w=6\neq 7. \tag{11}$$

Thus, Eqs. (8) are inconsistent and there are no values of x, y, z and w which satisfy the three equations simultaneously.

n equations in n unknowns

Our second case arises when $m=n$. The matrix $\mathbf{A}$ in this case reduces to a square matrix of order n. Since a square matrix possesses an inverse except when its determinant vanishes, the discussion of this case follows along the same lines as in the case of one equation in one unknown; Eq. (7.3). We have the following situation.

CASE (1): det $\mathbf{A}\neq 0$. If r=rank of $\mathbf{A}$, then $r=n$, $\mathbf{A}^{-1}$ exists, and multiplying the equation

$$\mathbf{Ax}=\mathbf{b} \tag{12}$$

from the left by $\mathbf{A}^{-1}$, we find that

$$\mathbf{x}=\mathbf{A}^{-1}\mathbf{b} \tag{13}$$

is the unique solution of the given system. If $\mathbf{b}=\mathbf{0}$ [case (1a)], the solution is trivial, whereas if $\mathbf{b}\neq\mathbf{0}$ [case (1b)], the system has a nontrivial solution.

CASE (2): det $\mathbf{A}=0$. The rank of $\mathbf{A}$ is now $r<n$. Let s be the rank of the augmented matrix $\mathbf{B}$. We have the following subcases.

CASE (2a): $\mathbf{b}\neq\mathbf{0}$. In this case, we have $s=r$ or $s>r$. If $s=r$, the system has an infinite number of solutions with an $(n-r)$-fold arbitrariness. If $s>r$, the system has no solution.

CASE (2b): $\mathbf{b}=\mathbf{0}$. In this case, we have $s=r$, and hence the system has an infinite number of solutions with an $(n-r)$-fold arbitrariness.

As a corollary, we note that if $\mathbf{b}=\mathbf{0}$, a case of n linear homogeneous equations in n variables [cases (1a) and (2b)], the system has a nontrivial solution if and only if det $\mathbf{A}=0$. Expressed differently, *the necessary and sufficient condition for a system of n linear homogeneous equations in n variables to possess a nontrivial solution is that the determinant of the coefficient matrix vanish.*

EXAMPLE 2: Solve the system of equations

$$\begin{aligned} x-y+2z+w&=a,\\ x+z-w&=b,\\ -2x+y+z&=c,\\ x+w&=d. \end{aligned} \tag{14}$$

Solution: The matrix of coefficients is

$$A=\begin{bmatrix} 1 & -1 & 2 & 1 \\ 1 & 0 & 1 & -1 \\ -2 & 1 & 1 & 0 \\ 1 & 0 & 0 & 1 \end{bmatrix}. \tag{15}$$

It can be found that det A=8, so that A is nonsingular. The inverse matrix is found to be

$$A^{-1}=\frac{1}{8}\begin{bmatrix} -1 & 3 & -1 & 4 \\ -4 & 4 & 4 & 8 \\ 2 & 2 & 2 & 0 \\ 1 & -3 & 1 & 4 \end{bmatrix} \tag{16}$$

Writing the system of Eqs. (14) in the matrix form

$$\begin{bmatrix} 1 & -1 & 2 & 1 \\ 1 & 0 & 1 & -1 \\ -2 & 1 & 1 & 0 \\ 1 & 0 & 0 & 1 \end{bmatrix}\begin{bmatrix} x \\ y \\ z \\ w \end{bmatrix}=\begin{bmatrix} a \\ b \\ c \\ d \end{bmatrix}, \tag{17}$$

and multiplying from the left by A^{-1}, we find the solution to be

$$\begin{bmatrix} x \\ y \\ z \\ w \end{bmatrix}=\frac{1}{8}\begin{bmatrix} -1 & 3 & -1 & 4 \\ -4 & 4 & 4 & 8 \\ 2 & 2 & 2 & 0 \\ 1 & -3 & 1 & 4 \end{bmatrix}\begin{bmatrix} a \\ b \\ c \\ d \end{bmatrix}, \tag{18a}$$

or

$$x=(-a+3b-c+4d)/8, \quad y=(-a+b+c+2d)/2,$$
$$z=(a+b+c)/4, \quad w=(a-3b+c+4d)/8. \tag{18b}$$

Note that if the system of equations is homogeneous, that is, $a=b=c=d=0$, the only solution is the trivial solution $x=y=z=w=0$.

EXAMPLE 3: For which value of k does the system of equations

$$\begin{aligned} x-\ y+2z&=0, \\ 3x+2y+\ z&=0, \\ x+2y+kz&=0, \end{aligned} \tag{19}$$

possess a nontrivial solution? Find this solution.

Solution: We note that the given system of equations is homogeneous. If we put $x=y=z=0$, Eqs. (19) are identically satisfied irrespective of what value k takes. However, to obtain a nontrivial solution, we must impose the condition that the determinant of the coefficient matrix be zero. We have

$$\det \mathbf{A} = \begin{vmatrix} 1 & -1 & 2 \\ 3 & 2 & 1 \\ 1 & 2 & k \end{vmatrix} = 5k+5, \tag{20}$$

so that $\det \mathbf{A}=0$ when $k=-1$. Thus the system has a nontrivial solution if and only if $k=-1$.

Once det **A** has been made to vanish, the three equations are no longer independent. In fact, it can be checked that with $k=-1$, the third row in det **A** is equal to 3/5 times the second row minus 4/5 times the first row. Thus, only two equations are independent, rank of $\mathbf{A}=2$, and the desired solution is

$$x=-z, \; y=z, \; z \text{ arbitrary}. \tag{21}$$

m equations in *n* unknowns, $m>n$

If the number of equations exceeds the number of variables, we consider the cases of homogeneous and inhomogeneous systems separately.

In the case of homogeneous equations, consider the matrix of coefficients with rank r. It is evident that $r \leqslant n$. If $r=n$, the system has one and only one solution which is the trivial solution, $x_i=0$ for $1 \leqslant i \leqslant n$, irrespective of the value of m. If $r<n$, the system has an infinite number of solutions with an $(n-r)$-fold arbitrariness (one of which is the trivial solution), irrespective of the value of m.

In the case of inhomogeneous equations, we consider the augmented matrix **B** whose order is $m\times(n+1)$ and whose rank is s. Once again, it is evident that $s \leqslant n+1$. If $s=n+1$, it implies that there are $n+1$ independent constraints on n variables and, as such, the system cannot have any solution. If $s \leqslant n$, it means that the number of independent equations is less than or equal to the number of variables. We can then pick out s independent equations from the given system and solve them by methods discussed earlier in this section.

Another alternative in the case of inhomogeneous equations is to pick out *any* n equations and solve them and to substitute this solution in the remaining $m-n$ equations. If these are identically satisfied, the equations are consistent and the solution is final; if they are not, the system has no solution.

EXAMPLE 4: Obtain a solution of the system of equations

$$\begin{aligned} 7x+4y&=3, \\ -2x+3y&=4, \end{aligned} \tag{22}$$

$$2x+\ y=-2.$$

Solution: Here we have three inhomogeneous equations in two unknowns. Hence we consider the augmented matrix

$$\mathbf{B}=\begin{bmatrix} 7 & 4 & 3 \\ -2 & 3 & 4 \\ 2 & 1 & -2 \end{bmatrix}. \tag{23}$$

Its rank is found to be three. Thus, there are three independent equations in two variables, so that the system has no solution.

EXAMPLE 5: Solve the simultaneous equations

$$\begin{aligned} 2x+\ y+3z&=7, \\ 3x-2y+8z&=7, \\ x-2y+4z&=1, \\ -2x+\ y-4z&=-3. \end{aligned} \tag{24}$$

Solution: Here we have four inhomogeneous equations in three unknowns. Let us pick out any three equations and solve them. The first three equations are found to have the solution

$$x=3-2z,\ y=z+1,\ z \text{ arbitrary}. \tag{25}$$

Substituting this in the fourth equation, it reduces to $z=2$. Hence the final solution is

$$x=-1,\ y=3,\ z=2. \tag{26}$$

To summarize, we can say that a system of m linear equations in n unknowns has zero, one or an infinite number of solutions. In the case of inhomogeneous equations, there is no solution if the rank of the augmented matrix is greater than that of the coefficient matrix, and one or more solutions if the two matrices have the same rank. In the case of homogeneous equations, the augmented matrix and the coefficient matrix do have the same rank, so that there is at least one solution. In either case, when a solution exists, the multiplicity (arbitrariness) of the solution is $n-r$, where n is the number of unknowns and r the rank of the coefficient matrix.

Number of independent vectors

We have earlier discussed how to determine whether a given set of vectors is linearly dependent or independent. In case the set is linearly dependent, how to find *how many* vectors of the set are linearly independent?

Suppose we have n vectors $\mathbf{x}_1, \mathbf{x}_2, \ldots, \mathbf{x}_n$, each of dimension m

(that is, having m components). To determine the number of linearly independent vectors from the given set, we solve the equation

$$a_1 \mathbf{x}_1 + a_2 \mathbf{x}_2 + \ldots + a_n \mathbf{x}_n = \mathbf{0} \tag{27}$$

for the unknowns a_i. If the solution involves k arbitrary constants, the number of independent vectors is $n-k$.

The following examples contain a few important results regarding the linear dependence and independence of vectors.

EXAMPLE 6: Let $\mathbf{x}_1, \mathbf{x}_2, \ldots, \mathbf{x}_n$ be n column vectors of dimension m, the i-th vector being denoted by $\mathbf{x}_i \equiv \{x_{1i}\ x_{2i} \ldots x_{mi}\}$. Let $\mathbf{P}$ be a matrix whose columns are the column vectors $\mathbf{x}_i$, that is,

$$\mathbf{P} \equiv [\mathbf{x}_1\ \mathbf{x}_2 \ldots \mathbf{x}_n] = \begin{bmatrix} x_{11} & \ldots & x_{1i} & \ldots & x_{1n} \\ x_{21} & \ldots & x_{2i} & \ldots & x_{2n} \\ \vdots & & & & \\ x_{m1} & \ldots & x_{mi} & \ldots & x_{mn} \end{bmatrix}. \tag{28}$$

Then prove that *the n vectors are linearly independent if and only if the matrix $\mathbf{P}$ has rank n.*

Solution: Consider Eq. (27) which becomes

$$a_1 \begin{bmatrix} x_{11} \\ x_{21} \\ \vdots \\ x_{m1} \end{bmatrix} + a_2 \begin{bmatrix} x_{12} \\ x_{22} \\ \vdots \\ x_{m2} \end{bmatrix} + \ldots + a_n \begin{bmatrix} x_{1n} \\ x_{2n} \\ \vdots \\ x_{mn} \end{bmatrix} = \begin{bmatrix} 0 \\ 0 \\ \vdots \\ 0 \end{bmatrix}. \tag{29}$$

Written out separately, these equations take the form

$$\begin{aligned} x_{11} a_1 + x_{12} a_2 + \ldots + x_{1n} a_n &= 0, \\ x_{21} a_1 + x_{22} a_2 + \ldots + x_{2n} a_n &= 0, \\ \ldots\ldots\ldots\ldots\ldots\ldots\ldots & \\ x_{m1} a_1 + x_{m2} a_2 + \ldots + x_{mn} a_n &= 0. \end{aligned} \tag{30}$$

This is a set of m linear homogeneous equations in the n unknowns $a_1, a_2, \ldots, a_n$, with coefficients x_{ij}. Whatever be the value of m in relation to n, if the rank of the coefficient matrix equals n, the only solution of Eqs. (30) is the trivial solution $a_i = 0$, $1 \leqslant i \leqslant n$, showing that the vectors $\mathbf{x}_i$ are linearly independent. The converse is also true: if the vectors are linearly independent, the only solution of Eqs. (30) must be $a_i = 0$, $1 \leqslant i \leqslant n$, which is possible only if the rank of the coefficient matrix in Eqs. (30) is n. Hence the result follows.

The above result has the following two corollaries which can be treated as special cases of the result.

COROLLARY 1: If $m < n$, the rank of the coefficient matrix in Eqs. (30) cannot be equal to n, and the equations would have nontrivial solutions showing that the vectors $\mathbf{x}_i$ must be linearly dependent. This leads to the conclusion that *n vectors of dimension less than n must be linearly dependent.*

COROLLARY 2: If $m=n$, the coefficient matrix is square, and if the vectors are linearly independent, the rank of the coefficient matrix is n, that is, equal to the order of the matrix, so that the determinant of the matrix cannot be zero. This leads to the conclusion that *the matrix constructed with n linearly independent vectors of dimension n as its columns is nonsingular.*

EXAMPLE 7: Show that *the rank of a matrix equals the maximum number of linearly independent row vectors or column vectors of the matrix.*

Solution: This is a consequence of the theorem proved in Example 6 above. Consider n column vectors of dimension m (that is, of order $m \times 1$). Let k of these vectors be linearly independent; then $k \leqslant \min(n, m)$, where $\min(n, m)$ means the smaller of the two integers n and m. Let us pick up k linearly independent vectors from the given n vectors and arrange them in columns to obtain a matrix of order $m \times k$. Since the columns are independent, the rank of this matrix, by the theorem of Example 6, is k. Now, the remaining $n-k$ vectors are dependent on these k vectors and if we put these $n-k$ vectors as additional columns to obtain a bigger matrix of order $m \times n$, the rank of the resulting matrix will still be k, which is the number of linearly independent column vectors. (The rank of a matrix remains unchanged on interchanging any of its columns or rows.)

Finally, since the rank of a matrix remains the same on transposition, the number of linearly independent rows of the matrix must be the same as the number of linearly independent columns. Hence the result follows.

Exercise 8

1. Find the possible values of λ for which the following systems of homogeneous equations possess nontrivial solutions. Obtain the corresponding solutions.

*(a) $(2-\lambda)x+4y=0,$
$x+(3+\lambda)y=0.$

(b) $(1-\lambda)x+2y=0,$
$2x+(1-\lambda)y+z=0,$
$5y+(1-\lambda)z=0.$

2. Find the ranks of the following matrices by the method of Example 7:

(a) $\begin{bmatrix} 2 & 2 & 2 & 3 \\ -2 & 8 & 2 & -5 \\ 0 & 10 & 4 & 9 \end{bmatrix}$; *(b) $\begin{bmatrix} 3 & 5 & 9 & 1 \\ 1 & 3 & 4 & 0 \\ -3 & 2 & 0 & 4 \\ 1 & -1 & 1 & -1 \\ 0 & 3 & 3 & 1 \end{bmatrix}$.

*3. Solve the simultaneous equations

$$u+v+w+x=1,$$
$$u+v+w-x=2,$$
$$u+v-w-x=3,$$
$$u-v-w-x=4.$$

4. Show that if $\mathbf{Ax}=\mathbf{0}$ for all n-dimensional vectors $\mathbf{x}$, then $\mathbf{A}$ must be a zero matrix.

5. (a) Show that the equation $\mathbf{Ax}=\mathbf{0}$, where

$$\mathbf{A}=\begin{bmatrix} 0 & c & -b \\ -c & 0 & a \\ b & -a & 0 \end{bmatrix},$$

has the solution $\mathbf{x}=\{at \quad bt \quad ct\}$, where t is arbitrary. (b) If

$$\mathbf{B}=\begin{bmatrix} a^2 & ab & ac \\ ab & b^2 & bc \\ ac & bc & c^2 \end{bmatrix}, \quad \mathbf{y}=\begin{bmatrix} y_1 \\ y_2 \\ y_3 \end{bmatrix},$$

show that $\mathbf{By}$ is a vector of the kind of $\mathbf{x}$ of part (a). (c) Hence show, without computing the product, that $\mathbf{AB}=\mathbf{0}$.

*6. Find nontrivial solutions of the following systems in parametric form:

(a) $4x+2y+z+5t=0,$
$x-7y+4z+5t=0,$
$6x+12y-3z+3t=0,$
$3x+7y-2z+t=0.$

(b) $x_1+x_2+x_3+x_4+x_5=0,$
$x_1+2x_2+3x_3+4x_4+5x_5=0,$
$5x_1+4x_2+3x_3+2x_4+x_5=0,$
$x_1-x_2+x_3-x_4+x_5=0.$

9. The eigenvalue problem, I

Let $\mathbf{A} \equiv [a_{ij}]$ be a square matrix of order n. Suppose there is an n-dimensional nonzero column vector $\mathbf{x}$ such that the action of $\mathbf{A}$ on $\mathbf{x}$ (that is, the matrix product $\mathbf{Ax}$) gives a vector which is just a multiple of $\mathbf{x}$, that is

$$\mathbf{Ax} = \lambda \mathbf{x}, \tag{1}$$

where λ is a scalar. In other words, the transformation represented by the matrix $\mathbf{A}$ just multiplies the vector $\mathbf{x}$ by a scalar λ. The vector $\mathbf{x}$ is then called an *eigenvector*[1] of the matrix $\mathbf{A}$ and λ is called an *eigenvalue* of $\mathbf{A}$ corresponding to the eigenvector $\mathbf{x}$. The problem of finding the eigenvectors and the eigenvalues of a matrix is called the *eigenvalue problem.*

The problem can be expressed in an alternative form as follows. Whatever value λ may take, Eq. (1) is satisfied if $\mathbf{x}=\mathbf{0}$, the zero vector. But the problem is to find those values of λ for which Eq. (1) may be satisfied by a nonzero vector also. These values of λ are the eigenvalues of $\mathbf{A}$.

Determination of eigenvalues

Let us write the system of Eq. (1) explicitly. If we denote the column vector $\mathbf{x}$ by $\{x_1, x_2, \ldots, x_n\}$, Eq. (1) becomes

$$\begin{bmatrix} a_{11} & a_{12} & \cdots & a_{1n} \\ a_{21} & a_{22} & \cdots & a_{2n} \\ \vdots & & & \\ a_{n1} & a_{n2} & \cdots & a_{nn} \end{bmatrix} \begin{bmatrix} x_1 \\ x_2 \\ \vdots \\ x_n \end{bmatrix} = \lambda \begin{bmatrix} x_1 \\ x_2 \\ \vdots \\ x_n \end{bmatrix}, \tag{2}$$

or, writing the equations separately, we have

$$\begin{aligned} (a_{11}-\lambda)x_1 + a_{12}x_2 + \ldots + a_{1n}x_n &= 0, \\ a_{21}x_1 + (a_{22}-\lambda)x_2 + \ldots + a_{2n}x_n &= 0, \\ \ldots\ldots\ldots\ldots\ldots\ldots\ldots\ldots & \\ a_{n1}x_1 + a_{n2}x_2 + \ldots + (a_{nn}-\lambda)x_n &= 0. \end{aligned} \tag{3a}$$

In the matrix form, this could be written as[2]

$$(\mathbf{A} - \lambda \mathbf{I})\mathbf{x} = \mathbf{0}. \tag{3b}$$

[1] '*Eigen*' is a German word meaning *proper* or *characteristic*. Eigenvalues and eigenvectors are also known as *characteristic values* and *characteristic vectors*, respectively.

[2] It is necessary to introduce the unit matrix $\mathbf{I}$ in writing $\mathbf{A}-\lambda\mathbf{I}$. It would be wrong to write $\mathbf{A}-\lambda$ because $\mathbf{A}$ is of order $n \times n$ while λ is of order 1×1.

We note that this is a system of n simultaneous linear homogeneous equations in the n parameters x_i, and always possesses the trivial solution $x_i=0$, $1\leqslant i\leqslant n$. However, the system will also possess nontrivial solutions *if, and only if*, the determinant of the coefficient matrix vanishes, that is, if, and only if,

$$D(\lambda)\equiv\det(\mathbf{A}-\lambda\mathbf{I})=\begin{vmatrix} a_{11}-\lambda & a_{12} & \ldots & a_{1n} \\ a_{21} & a_{22}-\lambda & \ldots & a_{2n} \\ \vdots & & & \\ a_{n1} & a_{n2} & \ldots & a_{nn}-\lambda \end{vmatrix}=0. \quad (4)$$

It is evident that the determinant $D(\lambda)$ is a polynomial of degree n in λ. It is called the *characteristic polynomial* of the matrix **A**. Let us write it as

$$D(\lambda)=c_0\lambda^n+c_1\lambda^{n-1}+c_2\lambda^{n-2}+\ldots+c_{n-1}\lambda+c_n, \quad (5)$$

where the coefficients c_i are functions of the elements of **A**. Some of them can be easily determined; in fact, it can be verified by an inspection of Eq. (4) that

$$c_0=(-1)^n,\ c_1=(-1)^{n-1}(a_{11}+a_{22}+\ldots+a_{nn}),$$
$$c_n=D(0)=\det\mathbf{A}. \quad (6)$$

The determinantal equation (4) then reduces to the polynomial equation

$$c_0\lambda^n+c_1\lambda^{n-1}+c_2\lambda^{n-2}+\ldots+c_{n-1}\lambda+c_n=0. \quad (7)$$

Eq. (4) or Eq. (7) is known as the *characteristic equation* for **A**. From the theory of polynomials, it is known that every polynomial of degree n has exactly n roots which make the polynomial vanish. Let us denote the roots of Eq. (7) by $\lambda_1, \lambda_2, \ldots, \lambda_n$. These are the values of λ for which $\det(\mathbf{A}-\lambda\mathbf{I})$ vanishes and hence Eq. (3b) will possess nontrivial solutions. In other words, we have the result that *every square matrix has exactly as many eigenvalues as its order and these eigenvalues are the roots of its characteristic polynomial.*

Evidently, we can write the characteristic polynomial as

$$c_0\lambda^n+c_1\lambda^{n-1}+\ldots+c_n=(\lambda_1-\lambda)(\lambda_2-\lambda)\ldots(\lambda_n-\lambda). \quad (8)$$

It follows that

$$c_1=(-1)^{n-1}(\lambda_1+\lambda_2+\ldots+\lambda_n),\ c_n=\lambda_1\lambda_2\ldots\lambda_n. \quad (9)$$

Comparing with Eqs. (6), we have the following two important results regarding the eigenvalues of a matrix.

1. *The sum of the eigenvalues equals the trace of the matrix*, i.e.,

$$\lambda_1+\lambda_2+\ldots+\lambda_n=a_{11}+a_{22}+\ldots+a_{nn}\equiv\mathrm{Tr}\,\mathbf{A}. \quad (10a)$$

2. *The product of the eigenvalues equals the determinant of the matrix*, i.e.,

$$\lambda_1\lambda_2\ldots\lambda_n=\det \mathbf{A}. \tag{10b}$$

The set of the eigenvalues of a matrix is called its *eigenvalue spectrum*.

EXAMPLE 1: Find the eigenvalues of the matrix

$$\mathbf{A}=\begin{bmatrix} 2 & -2 & 3 \\ 2 & -1 & 0 \\ -1 & 1 & 1 \end{bmatrix}. \tag{11}$$

Solution: The characteristic polynomial is

$$D(\lambda)=\det(\mathbf{A}-\lambda\mathbf{I})=\begin{vmatrix} 2-\lambda & -2 & 3 \\ 2 & -1-\lambda & 0 \\ -1 & 1 & 1-\lambda \end{vmatrix}$$

$$=-\lambda^3+2\lambda^2-6\lambda+5=(1-\lambda)(\lambda^2-\lambda+5).$$

The eigenvalues of $\mathbf{A}$, which are the solutions of the equation $D(\lambda)=0$, are therefore 1, $(1+i\sqrt{19})/2$ and $(1-i\sqrt{19})/2$. It is left to the reader to verify Eqs. (10) in this case.

EXAMPLE 2: If $\mathbf{A}$ is nonsingular, show that the eigenvalues of $\mathbf{A}^{-1}$ are the reciprocals of those of $\mathbf{A}$ and every eigenvector of $\mathbf{A}$ is also an eigenvector of $\mathbf{A}^{-1}$.

Solution: Let λ be an eigenvalue of $\mathbf{A}$ corresponding to the eigenvector $\mathbf{x}$, so that

$$\mathbf{A}\mathbf{x}=\lambda\mathbf{x}. \tag{12}$$

Since $\mathbf{A}^{-1}$ exists, multiply Eq. (12) from the left by $\mathbf{A}^{-1}$ to obtain

$$\mathbf{A}^{-1}\mathbf{A}\mathbf{x}=\mathbf{A}^{-1}\lambda\mathbf{x}\Rightarrow\mathbf{x}=\lambda\mathbf{A}^{-1}\mathbf{x}. \tag{13}$$

Since $\mathbf{A}$ is nonsingular, Exercise 2 shows that λ must be nonzero. Hence dividing Eq. (13) by λ, we finally have

$$\mathbf{A}^{-1}\mathbf{x}=(1/\lambda)\mathbf{x}. \tag{14}$$

Since this must be true for every eigenvalue of $\mathbf{A}$, the result follows.

EXAMPLE 3: Show that all the eigenvalues of a unitary matrix have unit magnitude.

Solution: Let $\mathbf{U}$ be a unitary matrix and let $\mathbf{x}$ be an eigenvector of $\mathbf{U}$ with the eigenvalue λ, so that

$$\mathbf{U}\mathbf{x}=\lambda\mathbf{x}. \tag{15}$$

Taking the hermitian conjugate of both sides, we have

$$x^\dagger U^\dagger = \lambda^* x^\dagger . \tag{16}$$

Multiplying Eq. (15) from the left by Eq. (16), we get

$$x^\dagger U^\dagger U x = \lambda\lambda^* x^\dagger x. \tag{17}$$

Since **U** is unitary, $U^\dagger U = I$, so that Eq. (17) reduces to $x^\dagger x(|\lambda|^2 - 1) = 0$. Now, $x^\dagger x$ is the square of the norm of **x** and hence cannot vanish unless **x** is a null vector. Since we are interested in nonzero vectors **x** which satisfy Eq. (15), we must have $|\lambda|^2 = 1$, or $|\lambda| = 1$, proving the desired result.

Eigenvectors and their properties

Although every matrix of order n has exactly n eigenvalues, it does not necessarily have n linearly independent eigenvectors. In this section, we shall deal only with those matrices which have n linearly independent eigenvectors.[3] For reasons which will become clear in the following pages, we shall refer to these matrices as *diagonalizable matrices*.

If **x** is an eigenvector of **A**, it is easy to see that any multiple of **x** is also an eigenvector of **A**. This can be shown by multiplying Eq. (1) by an arbitrary constant c, so that we have $c\mathbf{A}\mathbf{x} = c\lambda\mathbf{x}$. Now since every matrix commutes with a scalar, we have $c\mathbf{A} = \mathbf{A}c$, giving $\mathbf{A}(c\mathbf{x}) = \lambda(c\mathbf{x})$, showing that $c\mathbf{x}$ is also an eigenvector of **A** with the same eigenvalue λ.

However, $c\mathbf{x}$ is linearly dependent on **x** and if we were to count all such eigenvectors separately, we would clearly have an infinite number of them, for, we could multiply **x** by *any* arbitrary constant and get a new eigenvector. Such eigenvectors are therefore not counted separately.

The eigenvalues of a matrix may not all be distinct; some of them may be repeated more than once. This happens when the characteristic polynomial has two or more identical roots. If an eigenvalue occurs k times, the matrix has at most k *linearly independent* eigenvectors all corresponding to the same eigenvalue; k is then called the *multiplicity* of the eigenvalue.

When λ is chosen to be equal to one of the eigenvalues of **A**, say λ_i, then $\det(\mathbf{A} - \lambda_i\mathbf{I}) = 0$ by design. Therefore it is clear that $s \equiv$ rank of $(\mathbf{A} - \lambda_i\mathbf{I}) < n$. From the system of Eqs. (3a), only s equations

[3]See Section 10 for a discussion of matrices which do not possess n linearly independent eigenvectors.

will be linearly independent and the system will have a solution with $(n-s)$-fold arbitrariness, or, in other words, the system will have $n-s$ linearly independent solutions.

Now if an eigenvalue λ_i has multiplicity k, the matrix $\mathbf{A}-\lambda_i\mathbf{I}$ will have rank $s=n-k$, so that the system of Eqs. (3a) has exactly $n-s=k$ linearly independent solutions. That is, *an eigenvalue with multiplicity k of a diagonalizable matrix* **A** *has associated with it k linearly independent eigenvectors of* **A**. Such linearly independent eigenvectors having the same eigenvalue are said to be *degenerate eigenvectors.*

Once the eigenvalues have been determined, the eigenvectors can be easily determined. We replace λ by one of the eigenvalues of **A** in Eqs. (3a) and solve them for $x_1, x_2, \ldots, x_n$; these are then just the components of the required eigenvector. The procedure can be repeated for every distinct eigenvalue of **A**.

EXAMPLE 4: Find the eigenvalues and the eigenvectors of the matrix

$$\mathbf{A}=\begin{bmatrix} q & p & p \\ p & q & p \\ p & p & q \end{bmatrix}, \tag{18}$$

where p and q are scalars, $p\neq 0$.

Solution: The characteristic polynomial of the given matrix is

$$\begin{aligned}\det(\mathbf{A}-\lambda\mathbf{I})&=\begin{vmatrix} q-\lambda & p & p \\ p & q-\lambda & p \\ p & p & q-\lambda \end{vmatrix}\\ &=(q-p-\lambda)\,[\lambda^2-(2q+p)\lambda+(q^2+qp-2p^2)]\\ &=(q-p-\lambda)^2\,(q+2p-\lambda).\end{aligned} \tag{19}$$

The eigenvalues of **A** are therefore

$$\lambda_1=q-p,\ \lambda_2=q-p,\ \lambda_3=q+2p. \tag{20}$$

We note that the eigenvalue $q-p$ has multiplicity 2, while the eigenvalue $q+2p$ has multiplicity 1.

To obtain the eigenvectors, we first take $\lambda=\lambda_3=q+2p$ and solve the equation $(\mathbf{A}-\lambda_3\mathbf{I})\,\mathbf{x}=\mathbf{0}$. We have

$$\begin{bmatrix} -2p & p & p \\ p & -2p & p \\ p & p & -2p \end{bmatrix}\begin{bmatrix} x_1 \\ x_2 \\ x_3 \end{bmatrix}=\begin{bmatrix} 0 \\ 0 \\ 0 \end{bmatrix}, \tag{21a}$$

or

$$-2px_1+px_2+px_3=0,$$

$$px_1 - 2px_2 + px_3 = 0, \tag{21b}$$
$$px_1 + px_2 - 2px_3 = 0.$$

The rank of the matrix in Eq. (21a)i s found to be 2, so that only two of the Eqs. (21b) will be independent; the third equation does not give any additional constraint. Solving any two of Eqs. (21b), we obtain

$$x_3 = x_2 = x_1,\ x_1 \text{ arbitrary.} \tag{22}$$

Taking $x_1 = a$, the eigenvector corresponding to the eigenvalue $\lambda_3 = q+2p$ is found to be

$$\mathbf{x}_3' = \{a \quad a \quad a\}, \tag{23a}$$

or, choosing $a=1$,

$$\mathbf{x}_3 = \{1 \quad 1 \quad 1\}. \tag{23b}$$

Next we take $\lambda = q-p$ and solve the equation $[\mathbf{A}-(q-p)\mathbf{I}]\mathbf{x}=\mathbf{0}$. We have

$$\begin{bmatrix} p & p & p \\ p & p & p \\ p & p & p \end{bmatrix} \begin{bmatrix} x_1 \\ x_2 \\ x_3 \end{bmatrix} = \begin{bmatrix} 0 \\ 0 \\ 0 \end{bmatrix} \tag{24}$$

It is easily seen that the matrix of Eq. (24) has rank 1, which is consistent with the fact that the eigenvalue $q-p$ has multiplicity 2. There will be only one independent equation and it is of no avail to write all the three equations. The equation to be solved is

$$px_1 + px_2 + px_3 = 0, \tag{25a}$$

whose solution, since $p \neq 0$, is

$$x_3 = -x_1 - x_2, \quad x_1 \text{ and } x_2 \text{ arbitrary,} \tag{25b}$$

which involves a two-fold arbitrariness. x_1 and x_2 may be given any values whatsoever according to our choice. Let us first choose $x_1=1$, $x_2=0$ to obtain the first eigenvector

$$\mathbf{x}_1 = \{1 \quad 0 \quad -1\}. \tag{26}$$

To obtain the second eigenvector, we might choose $x_1=0$, $x_2=1$, so that $x_3=-1$, which yields the eigenvector

$$\mathbf{x}_2 = \{0 \quad 1 \quad -1\}. \tag{27}$$

Note that the most general eigenvector having the eigenvalue $q-p$ is $\{a \quad b \quad -a-b\}$ with any arbitrary values of a and b. However, this will depend linearly on $\mathbf{x}_1$ and $\mathbf{x}_2$; in fact, it can be seen that

$$\{a \quad b \quad -a-b\} = a\mathbf{x}_1 + b\mathbf{x}_2. \tag{28}$$

The answer, finally, is therefore :

(i) eigenvalue $q+2p$, eigenvector $\{1 \quad 1 \quad 1\}$;

(ii) eigenvalue $q-p$ with multiplicity 2; a set of two linearly independent eigenvectors is $\{1 \ 0 \ -1\}$, $\{0 \ 1 \ -1\}$.

If normalized eigenvectors are desired, the above eigenvectors can be normalized to give

$$\mathbf{x}_1=\frac{1}{\sqrt{2}}\{1 \ 0 \ -1\},\ \mathbf{x}_2=\frac{1}{\sqrt{2}}\{0 \ 1 \ -1\},\ \mathbf{x}_3=\frac{1}{\sqrt{3}}\{1 \ 1 \ 1\}. \tag{29}$$

It can be explicitly verified that the operation of the matrix **A** of Eq. (18) on any of its eigenvectors gives a constant multiple of the same eigenvector. Thus,

$$\begin{bmatrix} q & p & p \\ p & q & p \\ p & p & q \end{bmatrix}\begin{bmatrix} a \\ a \\ a \end{bmatrix}=\begin{bmatrix} (q+2p)a \\ (q+2p)a \\ (q+2p)a \end{bmatrix}=(q+2p)\begin{bmatrix} a \\ a \\ a \end{bmatrix}, \tag{30a}$$

$$\begin{bmatrix} q & p & p \\ p & q & p \\ p & p & q \end{bmatrix}\begin{bmatrix} a \\ b \\ -a-b \end{bmatrix}=\begin{bmatrix} (q-p)a \\ (q-p)b \\ (p-q)(a+b) \end{bmatrix}=(q-p)\begin{bmatrix} a \\ b \\ -a-b \end{bmatrix}. \tag{30b}$$

EXAMPLE 5: Show that (a) all the eigenvalues of a hermitian matrix are real, and (b) the eigenvectors corresponding to distinct eigenvalues are orthogonal.

Solution: (a) Let **H** be a hermitian matrix and let **x** be an eigenvector of **H** with the eigenvalue λ, so that

$$\mathbf{Hx}=\lambda\mathbf{x}. \tag{31a}$$

Taking the hermitian conjugate of this equation and noting that $\mathbf{H}^\dagger=\mathbf{H}$, we have

$$\mathbf{x}^\dagger\mathbf{H}=\lambda^*\mathbf{x}^\dagger. \tag{31b}$$

Multiplying Eq. (31a) from the left by $\mathbf{x}^\dagger$, Eq. (31b) from the right by **x** and subtracting one from the other, we have

$$(\lambda-\lambda^*)\mathbf{x}^\dagger\mathbf{x}=0. \tag{32}$$

Now, since $\mathbf{x}^\dagger\mathbf{x}$ cannot be zero, we have

$$\lambda^*=\lambda, \tag{33}$$

showing that the eigenvalue λ is real.

(b) Let $\mathbf{x}_1$ and $\mathbf{x}_2$ be two eigenvectors of **H** corresponding to the distinct eigenvalues λ_1 and λ_2, respectively, so that

$$\mathbf{Hx}_1=\lambda_1\mathbf{x}_1, \tag{34a}$$

$$\mathbf{Hx}_2=\lambda_2\mathbf{x}_2. \tag{34b}$$

Taking the hermitian conjugate of Eq. (34b), we have

$$\mathbf{x}_2^\dagger\mathbf{H}=\lambda_2\mathbf{x}_2^\dagger, \tag{35}$$

where we have used the fact that λ_2 is real. Multiplying Eq. (34a) from the left by $\mathbf{x}_2^\dagger$, Eq. (35) from the right by $\mathbf{x}_1$ and subtracting, we get

$$(\lambda_1-\lambda_2)\mathbf{x}_2^\dagger\,\mathbf{x}_1=0. \tag{36}$$

Since $\lambda_1\neq\lambda_2$, we have $\mathbf{x}_2^\dagger\,\mathbf{x}_1=0$, showing that the vectors $\mathbf{x}_1$ and $\mathbf{x}_2$ are orthogonal.

Diagonalization of a matrix

Let $\mathbf{A}$ be a square matrix of order n with elements a_{ij}, which has n linearly independent eigenvectors which may be denoted by $\mathbf{x}_i$. Let λ_i be the corresponding eigenvalues so that $\mathbf{A}\mathbf{x}_i=\lambda_i\mathbf{x}_i$. If we denote the eigenvectors $\mathbf{x}_i$ by column vectors in the form $\mathbf{x}_i=\{\mathbf{x}_{1i}\ \mathbf{x}_{2i}\ \ldots\ \mathbf{x}_{ni}\}$, then the eigenvalue equation can be written in the matrix form

$$\begin{bmatrix} a_{11} & a_{12} & \cdots & a_{1n} \\ a_{21} & a_{22} & \cdots & a_{2n} \\ \vdots & & & \\ a_{n1} & a_{n2} & \cdots & a_{nn} \end{bmatrix}\begin{bmatrix} x_{1i} \\ x_{2i} \\ \vdots \\ x_{ni} \end{bmatrix}=\lambda_i\begin{bmatrix} x_{1i} \\ x_{2i} \\ \vdots \\ x_{ni} \end{bmatrix}. \tag{37a}$$

Writing explicitly the j-th equation from the above, we have

$$\sum_{k=1}^{n} a_{jk}x_{ki}=\lambda_i x_{ji}. \tag{37b}$$

Now construct a matrix $\mathbf{P}$ of order $n\times n$ whose columns are the vectors $\mathbf{x}_i$, i.e.,

$$\mathbf{P}=[\mathbf{x}_1\ \ \mathbf{x}_2\ldots\mathbf{x}_n]=\begin{bmatrix} x_{11}\ldots x_{1i}\ldots x_{1n} \\ x_{21}\ldots x_{2i}\ldots x_{2n} \\ \vdots \\ x_{n1}\ldots x_{ni}\ldots x_{nn} \end{bmatrix};\ (\mathbf{P})_{ji}=x_{ji}. \tag{38}$$

Since the vectors $\mathbf{x}_i$ are linearly independent, the matrix $\mathbf{P}$, by Corollary 2 of Example 8.6, will be nonsingular; therefore $\mathbf{P}^{-1}$ exists. Now we state that the matrix $\mathbf{P}^{-1}\mathbf{A}\mathbf{P}$ *is diagonal whose diagonal elements are the eigenvalues of* $\mathbf{A}$.

To prove this, let us define a diagonal matrix Λ whose diagonal elements are λ_i, i.e.,

$$\Lambda=\begin{bmatrix} \lambda_1 & & & & \\ & \lambda_2 & & & 0 \\ & & \cdot & & \\ & & & \cdot & \\ & 0 & & & \cdot \\ & & & & \lambda_n \end{bmatrix}. \tag{39}$$

We wish to prove that

$$\mathbf{P}^{-1}\mathbf{A}\mathbf{P}=\mathbf{\Lambda}, \tag{40a}$$

or, multiplying from the left by **P**, that

$$\mathbf{A}\mathbf{P}=\mathbf{P}\mathbf{\Lambda}. \tag{40b}$$

Taking the ji-th element of the left hand side of the above equation, we get

$$(\mathbf{AP})_{ji}=\sum_{k=1}^{n}(\mathbf{A})_{jk}\,(\mathbf{P})_{ki}=\sum_{k=1}^{n}a_{jk}\,x_{ki}=\lambda_i\,x_{ji}, \tag{41a}$$

where we have used Eq. (37b). Now the ji-th element of the right hand side of Eq. (40b) is

$$(\mathbf{P\Lambda})_{ji}=\sum_{k=1}^{n}(\mathbf{P})_{jk}\,(\mathbf{\Lambda})_{ki}=\sum_{k=1}^{n}x_{jk}\,\lambda_i\,\delta_{ki}=\lambda_i\,x_{ji}. \tag{41b}$$

Eqs. (41a) and (41b) clearly prove the validity of Eqs. (40). This proves our statement.

This process is called the *diagonalization of a matrix.*

EXAMPLE 6: Diagonalize the matrix of Eq. (18).

Solution: Let us construct a matrix whose columns are the eigenvectors of the given matrix **A** of Eq. (18). From Eqs. (23b), (26) and (27), we have

$$\mathbf{P}=\begin{bmatrix} 1 & 0 & 1 \\ 0 & 1 & 1 \\ -1 & -1 & 1 \end{bmatrix}. \tag{42}$$

The inverse is found to be

$$\mathbf{P}^{-1}=\tfrac{1}{3}\begin{bmatrix} 2 & -1 & -1 \\ -1 & 2 & -1 \\ 1 & 1 & 1 \end{bmatrix}. \tag{43}$$

Then we have

$$\mathbf{P}^{-1}\mathbf{A}\mathbf{P}=\tfrac{1}{3}\begin{bmatrix} 2 & -1 & -1 \\ -1 & 2 & -1 \\ 1 & 1 & 1 \end{bmatrix}\begin{bmatrix} q & p & p \\ p & q & p \\ p & p & q \end{bmatrix}$$

$$\times\begin{bmatrix} 1 & 0 & 1 \\ 0 & 1 & 1 \\ -1 & -1 & 1 \end{bmatrix}=\begin{bmatrix} q-p & 0 & 0 \\ 0 & q-p & 0 \\ 0 & 0 & q+2p \end{bmatrix}. \tag{44}$$

Notice that $\mathbf{P}^{-1}\mathbf{A}\mathbf{P}$ is a diagonal matrix whose diagonal elements are just the eigenvalues of **A**.

Thus, we can state the following theorem:

If a matrix of order n has n linearly independent eigenvectors, then it is related through a similarity transformation to a diagonal matrix whose diagonal elements are the eigenvalues of the matrix.

It is important to note that the matrix **P** which is able to diagonalize a given matrix is not unique. This is because we could arrange the eigenvectors $\mathbf{x}_1, \mathbf{x}_2, \ldots, \mathbf{x}_n$ *in any order* in the construction of **P**. In Example 6, for instance, instead of choosing $\mathbf{P}=[\mathbf{x}_1\ \mathbf{x}_2\ \mathbf{x}_3]$, we could have chosen

$$\mathbf{Q}=[\mathbf{x}_3\ \mathbf{x}_1\ \mathbf{x}_2]=\begin{bmatrix} 1 & 1 & 0 \\ 1 & 0 & 1 \\ 1 & -1 & -1 \end{bmatrix}. \tag{45}$$

We would then find that

$$\mathbf{Q}^{-1}\mathbf{A}\mathbf{Q}=\begin{bmatrix} q+2p & 0 & 0 \\ 0 & q-p & 0 \\ 0 & 0 & q-p \end{bmatrix}, \tag{46}$$

which is again a diagonal matrix whose diagonal elements are the eigenvalues of **A**. We note that the order of the eigenvalues in Eq. (46) is different from that in Eq. (44) and that, in each case, the order of the eigenvalues corresponds to the order of the eigenvectors in the construction of the diagonalizing matrix. We can therefore state the general rule that *in the process of diagonalization* $\mathbf{P}^{-1}\mathbf{A}\mathbf{P}=\mathbf{\Lambda}$, *the order of the eigenvalues in* $\mathbf{\Lambda}$ *corresponds to the order of the eigenvectors of* **A** *in constructing* **P**.

Eigenvectors of commuting matrices

We shall now prove a theorem of great importance in matrix algebra as well as in quantum mechanics. The theorem reads as follows.[4]

It is possible to find a common set of eigenvectors for two commuting matrices.

Let **A** and **B** be two square matrices, each of order *n*, which commute with each other, so that

$$\mathbf{AB}-\mathbf{BA}=[\mathbf{A}, \mathbf{B}]=0. \tag{47}$$

[4] The theorem also holds good in quantum mechanics on replacing the word 'matrix' by 'operator'. The proof also holds good with suitable modifications.

First, let λ be an eigenvalue of **A** with multiplicity 1, corresponding to the eigenvector **x**. Consider the eigenvalue equation

$$\mathbf{A}\mathbf{x}=\lambda\mathbf{x}, \tag{48}$$

and multiply both sides from the left by **B**. Using Eq. (47), this becomes

$$\mathbf{B}\mathbf{A}\mathbf{x}=\mathbf{A}(\mathbf{B}\mathbf{x})=\lambda(\mathbf{B}\mathbf{x}). \tag{49}$$

Now **B** is an $n\times n$ matrix and **x** is an $n\times 1$ vector; therefore **Bx** is also an $n\times 1$ vector. Eq. (49) shows that **Bx** is also an eigenvector of **A** with the same eigenvalue λ. But since **x** is a nondegenerate eigenvector of **A**, any other vector which is an eigenvector of **A** with the same eigenvalue as that of **x** must be a multiple of **x**, that is,

$$\mathbf{B}\mathbf{x}=\mu\,\mathbf{x}, \tag{50}$$

where μ is a scalar. This shows that **x** is also an eigenvector of **B** with the eigenvalue μ. In this first part, we have thus proved that *if two matrices commute, every nondegenerate eigenvector of one is also an eigenvector of the other, and vice versa.*

Next, let λ be an eigenvalue of **A** with multiplicity k. This means that **A** has k linearly independent eigenvectors, say $\mathbf{x}_1, \mathbf{x}_2, \ldots, \mathbf{x}_k$, each corresponding to the eigenvalue λ; or

$$\mathbf{A}\mathbf{x}_i=\lambda\,\mathbf{x}_i,\ 1\leqslant i\leqslant k. \tag{51}$$

Again, multiply Eq. (51) from the left by **B** to obtain

$$\mathbf{A}(\mathbf{B}\mathbf{x}_i)=\lambda(\mathbf{B}\mathbf{x}_i), \tag{52}$$

showing that **Bx** is also an eigenvector of **A** with the same eigenvalue λ. Now the most general eigenvector of **A** having the eigenvalue λ must be a linear combination of the degenerate eigenvectors $\mathbf{x}_1, \mathbf{x}_2, \ldots, \mathbf{x}_k$, so that

$$\mathbf{B}\mathbf{x}_i=\sum_{j=1}^{k} c_{ij}\mathbf{x}_i,\ 1\leqslant i\leqslant k, \tag{53}$$

where c_{ij} are certain scalar coefficients. Define a matrix $\mathbf{C}\equiv[c_{ij}]$ of order k and let the matrix **C** be diagonalized in the form

$$\mathbf{D}\mathbf{C}\mathbf{D}^{-1}=\begin{bmatrix} \mu_1 & & & & \\ & \mu_2 & & \mathbf{0} & \\ & & \cdot & & \\ & & & \cdot & \\ & \mathbf{0} & & & \cdot \\ & & & & \mu_k \end{bmatrix}\equiv\mathbf{Q},\ \mathbf{D}\mathbf{C}=\mathbf{Q}\mathbf{D}, \tag{54}$$

which defines **Q** as the diagonal matrix with diagonal elements μ_i.

Let the ij-th element of $\mathbf{D}$ be d_{ij}; then equating the ij-th element of both sides of Eq. (54), we have $(\mathbf{DC})_{ij}=(\mathbf{QD})_{ij}$, or

$$\sum_{l=1}^{k} d_{il}c_{lj}=\sum_{l=1}^{k} \mu_i\delta_{il}d_{lj}=\mu_i d_{ij}. \tag{55}$$

Define new vectors $\mathbf{y}_i$'s by

$$\mathbf{y}_i=\sum_{l=1}^{k} d_{il}\mathbf{x}_l,\ 1\leqslant i\leqslant k, \tag{56}$$

which are linear combinations of the $\mathbf{x}_i$'s. Since the matrix $\mathbf{D}$ is nonsingular, the k vectors $\mathbf{y}_i$ will be linearly independent. Moreover, since they are linear combinations of the degenerate eigenvectors, each $\mathbf{y}_i$ will also be an eigenvector of $\mathbf{A}$ with the same eigenvalue λ. Now consider the operation of $\mathbf{B}$ on $\mathbf{y}_i$:

$$\begin{aligned}\mathbf{By}_i&=\sum_{l=1}^{k} d_{il}\mathbf{Bx}_l\\&=\sum_{l,j=1}^{k} d_{il}c_{lj}\mathbf{x}_j \quad \text{[by Eq. (53)]}\\&=\sum_{j=1}^{k} \mu_i d_{ij}\mathbf{x}_j \quad \text{[by Eq. (55)]}\\&=\mu_i\mathbf{y}_i,\ 1\leqslant i\leqslant k \quad \text{[by Eq. (56)]}.\end{aligned} \tag{57}$$

This shows that $\mathbf{y}_i$ $(1\leqslant i\leqslant k)$ are also eigenvectors of $\mathbf{B}$ associated respectively with the eigenvalues μ_i. The k vectors $\mathbf{y}_i$ are therefore common eigenvectors of $\mathbf{A}$ and $\mathbf{B}$.

We have now proved that *if* $\mathbf{A}$ *has* k *degenerate eigenvectors* $\mathbf{x}_i$, *then it is possible to find a set of* k *independent linear combinations of* $\mathbf{x}_i$'s *which are also eigenvectors of* $\mathbf{B}$.

Continuing this process for every distinct eigenvalue of $\mathbf{A}$, we see that finally it would be possible to find a common set of eigenvectors for the two commuting matrices, thus proving the theorem. Note that Exercise 5.9 is the converse of this theorem.

The theorem can obviously be extended to any number of mutually commuting matrices, that is, a set of matrices every one of which commutes with every other member of the set. We therefore have the generalized theorem that *it is possible to find a common set of eigenvectors for any number of mutually commuting matrices.* Alternatively, we can say that *all mutually commuting matrices can be simultaneously diagonalized.*

EXAMPLE 7: Find a common set of eigenvectors for the two matrices

$$\mathbf{A}=\begin{bmatrix} -1 & \sqrt{6} & \sqrt{2} \\ \sqrt{6} & 0 & \sqrt{3} \\ \sqrt{2} & \sqrt{3} & -2 \end{bmatrix}, \quad \mathbf{B}=\begin{bmatrix} 10 & \sqrt{6} & -\sqrt{2} \\ \sqrt{6} & 9 & \sqrt{3} \\ -\sqrt{2} & \sqrt{3} & 11 \end{bmatrix}. \tag{58}$$

Solution: It is easily verified that **A** and **B** commute; in fact,

$$\mathbf{AB}=\mathbf{BA}=\begin{bmatrix} -6 & 9\sqrt{6} & 15\sqrt{2} \\ 9\sqrt{6} & 9 & 9\sqrt{3} \\ 15\sqrt{2} & 9\sqrt{3} & -11 \end{bmatrix}. \tag{59}$$

The eigenvalues of **A** are found to be 3, —3, —3. The eigenvalue 3 has multiplicity 1 while the eigenvalue —3 has multiplicity 2.

The eigenvector corresponding to $\lambda=3$ is found to be

$$\mathbf{x}_1=\{\sqrt{2} \quad \sqrt{3} \quad 1\}. \tag{60}$$

Since $\mathbf{x}_1$ is a nondegenerate eigenvector of **A**, it must also be an eigenvector of **B**, as can be readily verified:

$$\begin{bmatrix} 10 & \sqrt{6} & -\sqrt{2} \\ \sqrt{6} & 9 & \sqrt{3} \\ -\sqrt{2} & \sqrt{3} & 11 \end{bmatrix}\begin{bmatrix} \sqrt{2} \\ \sqrt{3} \\ 1 \end{bmatrix}=\begin{bmatrix} 12\sqrt{2} \\ 12\sqrt{3} \\ 12 \end{bmatrix}=12\begin{bmatrix} \sqrt{2} \\ \sqrt{3} \\ 1 \end{bmatrix}. \tag{61}$$

Thus 12 is one of the eigenvalues of **B** associated with the eigenvector $\mathbf{x}_1$.

For the doubly degenerate eigenvalue $\lambda=-3$, we shall proceed along a slightly different path than that followed in the proof of the theorem. The most general eigenvector of **A** having the eigenvalue $\lambda=-3$ is found to be

$$\mathbf{x}=\{a \quad b \quad -\sqrt{2}a-\sqrt{3}b\}. \tag{62}$$

which involves two arbitrary constants. We wish to obtain two linearly independent eigenvectors of the form of Eq. (62) which will also be eigenvectors of **B**. We shall therefore find the conditions on a and b for which **x** is an eigenvector of **B**. We have

$$\mathbf{Bx}=\begin{bmatrix} 10 & \sqrt{6} & -\sqrt{2} \\ \sqrt{6} & 9 & \sqrt{3} \\ -\sqrt{2} & \sqrt{3} & 11 \end{bmatrix}\begin{bmatrix} a \\ b \\ -\sqrt{2}a-\sqrt{3}b \end{bmatrix}$$

$$=\begin{bmatrix} 12a+2\sqrt{6}b \\ 6b \\ -12\sqrt{2}a-10\sqrt{3}b \end{bmatrix}. \tag{63}$$

For **x** to be an eigenvector of **B**, we must therefore have

$$
\begin{aligned}
12a+2\sqrt{6}b&=\mu a,\\
6b&=\mu b,\\
12\sqrt{2}a+10\sqrt{3}b&=\mu(\sqrt{2}a+\sqrt{3}b).
\end{aligned}
\tag{64}
$$

The second of the three equations gives *either* $\mu=6$ *or* $b=0$. Taking $b=0$, we find that the first and the third equations are satisfied if $\mu=12$. The corresponding eigenvector, with $a=1$, is

$$\mathbf{x}_2=\{1 \quad 0 \quad -\sqrt{2}\}. \tag{65}$$

Taking $\mu=6$, the first and the third of Eqs. (64) are satisfied if $b=-\sqrt{3}a/\sqrt{2}$. Choosing $a=\sqrt{2}$, the corresponding eigenvector is found to be

$$\mathbf{x}_3=\{\sqrt{2} \quad -\sqrt{3} \quad 1\}. \tag{66}$$

The three eigenvectors $\mathbf{x}_1$, $\mathbf{x}_2$ and $\mathbf{x}_3$, given respectively by Eqs. (60), (65) and (66), are a common set of eigenvectors for **A** and **B**. Their eigenvalues with respect to **A** are respectively 3, −3, −3, while those with respect to **B** are respectively 12, 12, 6. Notice that any linear combination of $\mathbf{x}_2$ and $\mathbf{x}_3$ will be an eigenvector of **A** but not of **B**. Similarly, any linear combination of $\mathbf{x}_1$ and $\mathbf{x}_2$ will be an eigenvector of **B** but not of **A**.

Reduction of coupled differential equations to eigenvalue problem

It is often possible, by a simple substitution, to reduce a system of coupled differential equations to an eigenvalue problem and then solve it. If the system has boundary conditions, these can be incorporated in the final solution. Instead of developing any general theory, we shall illustrate this by means of a few examples.

EXAMPLE 8: Solve the following coupled differential equations:

$$
\begin{aligned}
\dot{y}_1&=-11y_1-10y_2+5y_3,\\
\dot{y}_2&=\ \ \ 5y_1+\ 4y_2-5y_3,\\
\dot{y}_3&=-20y_1-20y_2+4y_3,
\end{aligned}
\tag{67}
$$

where the y_i are functions of a parameter t and the dots denote differentiation with respect to t.

Solution: Define a column vector **y** by

$$\mathbf{y}(t)=\{y_1(t) \quad y_2(t) \quad y_3(t)\}. \tag{68}$$

Then Eq. (67) can be written in the matrix form

$$\frac{d\mathbf{y}}{dt}=\mathbf{A}\mathbf{y}, \text{ where } \mathbf{A}=\begin{bmatrix} -11 & -10 & 5\\ 5 & 4 & -5\\ -20 & -20 & 4 \end{bmatrix}. \tag{69}$$

Let us try a solution $\mathbf{y}=\mathbf{x}e^{wt}$, where $\mathbf{x}=\{x_1 \quad x_2 \quad x_3\}$, $\mathbf{x}$ and w being independent of t. Then $d\mathbf{y}/dt=w\mathbf{x}e^{wt}$ and Eq. (69) becomes

$$w\mathbf{x}e^{wt}=\mathbf{A}\mathbf{y}=\mathbf{A}\mathbf{x}e^{wt}\Rightarrow\mathbf{A}\mathbf{x}=w\mathbf{x}, \tag{70}$$

which is an eigenvalue equation. Thus the possible values of w are the eigenvalues of $\mathbf{A}$ and the corresponding eigenvectors will give the solutions $\mathbf{y}=\mathbf{x}e^{wt}$.

We find that the eigenvalues of $\mathbf{A}$ are $w=4$, -6, -1 and the corresponding eigenvectors are, respectively,

$$\mathbf{x}_1=\{1 \quad -1 \quad 1\},\ \mathbf{x}_2=\{0 \quad 1 \quad 2\},\ \mathbf{x}_3=\{1 \quad -1 \quad 0\}. \tag{71}$$

The most general solution for $\mathbf{y}$ is therefore

$$\mathbf{y}=a\mathbf{x}_1e^{4t}+b\mathbf{x}_2e^{-6t}+c\mathbf{x}_3e^{-t}, \tag{72a}$$

giving

$$\begin{aligned} y_1(t)&=ae^{4t}+ce^{-t},\\ y_2(t)&=-ae^{4t}+be^{-6t}-ce^{-t},\\ y_3(t)&=ae^{4t}+2be^{-6t}. \end{aligned} \tag{72b}$$

The solution involves three arbitrary constants as it should, for we have three differential equations, each of the first order. It can be verified explicitly by substituting Eqs. (72b) that Eqs. (67) are satisfied.

Suppose we desire a solution of the given system satisfying the initial conditions $y_1(0)=5$, $y_2(0)=-4$, $y_3(0)=4$. Putting $t=0$ in Eqs. (72b) and using the initial values, we get

$$a+c=5,\ -a+b-c=-4,\ a+2b=4. \tag{73}$$

yielding $a=2$, $b=1$, $c=3$. The desired solution is thus

$$y_1(t)=2e^{4t}+3e^{-t},\ y_2(t)=-2e^{4t}+e^{-6t}-3e^{-t},\ y_3(t)=2e^{4t}+2e^{-6t}. \tag{74}$$

EXAMPLE 9: Discuss the vibrations of the two coupled springs shown in Fig. 7. The springs are massless and have spring constants $k_1=18$, $k_2=6$, respectively. The masses attached at their endpoints are $m_1=3$, $m_2=2$.

Solution: As shown in the figure, let y_1 and y_2 be the respective displacements of the two masses, measured positive downward. The elongation of the first spring is y_1 while that of the second spring is y_2-y_1. The mass m_1 is acted upon by two forces: the force due to the first spring is opposite to y_1 while that due to the second spring is parallel to y_2-y_1, force also being measured positive downward. The mass m_2 is acted upon by one force, that due to the second spring, which is opposite to y_2-y_1. The restoring force equals the

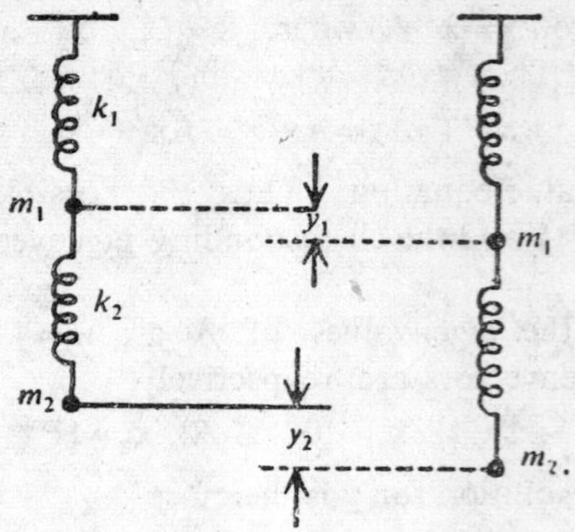

In equilibrium In motion

Figure 7. Vibrations of a system of coupled springs.

spring constant times the elongation. The equations of motion for the two masses are therefore

$$m_1\ddot{y}_1 = -k_1 y_1 + k_2(y_2 - y_1),$$
$$m_2\ddot{y}_2 = -k_2(y_2 - y_1), \tag{75}$$

where the dots denote time derivatives.

Here we have a system of two coupled differential equations, each of the second order. If we define

$$\mathbf{y} = \begin{bmatrix} y_1 \\ y_2 \end{bmatrix}, \quad \mathbf{A} = \begin{bmatrix} -(k_1+k_2)/m_1 & k_2/m_1 \\ k_2/m_2 & -k_2/m_2 \end{bmatrix}, \tag{76}$$

Eqs. (75) can be written as

$$\ddot{\mathbf{y}} = \mathbf{A}\mathbf{y}. \tag{77}$$

Assume a solution of the form

$$\mathbf{y} = \mathbf{x}e^{wt}, \tag{78}$$

so that Eq. (77) reduces to

$$\mathbf{A}\mathbf{x} = w^2\mathbf{x}. \tag{79}$$

Thus the possible values of w are the square roots of the eigenvalues of **A** with the corresponding solutions being the eigenvectors of **A**.

With the particular values of m_1, m_2, k_1 and k_2 given in the problem, the matrix **A** becomes

$$\mathbf{A} = \begin{bmatrix} -8 & 2 \\ 3 & -3 \end{bmatrix}. \tag{80}$$

The eigenvalues and the eigenvectors of this matrix are found to be

$$w^2 = -9, \ \mathbf{x}_1 = \{2 \quad -1\} \ ; \ w^2 = -2, \ \mathbf{x}_2 = \{1 \quad 3\}. \tag{81}$$

The most general solution of the system is therefore

$$\mathbf{y}(t)=b_1\mathbf{x}_1e^{3it}+b_2\mathbf{x}_1e^{-3it}+b_3\mathbf{x}_2e^{\sqrt{2}it}+b_4\mathbf{x}_2e^{-\sqrt{2}it}, \tag{82a}$$

or

$$\mathbf{y}(t)=a_1\mathbf{x}_1\cos 3t+a_2\mathbf{x}_1\sin 3t+a_3\mathbf{x}_2\cos\sqrt{2}t+a_4\mathbf{x}_2\sin\sqrt{2}t, \tag{82b}$$

where b_i and a_i $(1\leqslant i\leqslant 4)$ are arbitrary constants. The general solution contains four arbitrary constants which is consistent with the fact that we have two differential equations, each of order two. Writing out the components of $\mathbf{y}(t)$ separately from Eq. (82b), we have

$$\begin{aligned} y_1(t)&=\ \ 2a_1\cos 3t+2a_2\sin 3t+a_3\cos\sqrt{2}t+a_3\sin\sqrt{2}t,\\ y_2(t)&=-a_1\cos 3t-a_2\sin 3t+3a_3\cos\sqrt{2}t+3a_4\sin\sqrt{2}t. \end{aligned} \tag{83}$$

It can be verified by direct substitution of these in Eqs. (75) that they are satisfied.

Initial conditions can be easily incorporated in the general solution to determine the arbitrary constants. Suppose at $t=0$, the lower mass is pulled down a certain distance d and then released. The initial velocities of the masses are zero so that $\dot{y}_1(0)=0$, $\dot{y}_2(0)=0$. Moreover, $y_2(0)=d$. To determine $y_1(0)$, we consider the equilibrium of mass m_1 at $t=0$, which gives

$$k_1y_1(0)=k_2[y_2(0)-y_1(0)]\Rightarrow y_1(0)=[k_2/(k_1+k_2)]y_2(0). \tag{84}$$

Choosing $d=4$ and using the specified values of k_1 and k_2, we have $y_1(0)=1$, $y_2(0)=4$. Putting these four initial conditions in Eqs. (83), we obtain

$$2a_1+a_3=1,\ -a_1+3a_3=4,\ 6a_2+\sqrt{2}a_4=0,\ -3a_2+3\sqrt{2}a_4=0, \tag{85}$$

which have the solution

$$a_1=-1/7,\ a_3=9/7,\ a_2=a_4=0. \tag{86}$$

The displacements of the two masses under the given conditions are therefore given by

$$\begin{aligned} y_1(t)&=(9\cos\sqrt{2}t-2\cos 3t)/7,\\ y_2(t)&=(27\cos\sqrt{2}t+\cos 3t)/7. \end{aligned} \tag{87}$$

Exercise 9

*1. Show that the eigenvalues of a triangular matrix are its diagonal elements.

2. Show that a matrix is singular *if and only if* at least one of its eigenvalues is zero.

3. Prove the result of Example 3 by using the result of Example 2 and noting that $\mathbf{U}^{-1}=\mathbf{U}^{\dagger}$ for a unitary matrix.

*4. What are the possible eigenvalues of an orthogonal matrix?

5. Let **H** and **U** be the matrices defined in Exercise 5.18. Assuming that **U** does not have the eigenvalue unity, solve for **H** and show that $\mathbf{H}=i(\mathbf{I}+\mathbf{U})(\mathbf{I}-\mathbf{U})^{-1}$. [This is known as the *Cayley transform* connecting a unitary and a hermitian matrix].

6. Prove that a real symmetric matrix is orthogonal if and only if its eigenvalues are all ± 1.

*7. Show that the eigenvalues of a real matrix are either real or appear in pairs of complex conjugates. Deduce that if the order of a real matrix is odd, it has at least one real eigenvalue.

*8. If two matrices **A** and **B** are related by a similarity transformation $\mathbf{B}=\mathbf{P}^{-1}\mathbf{A}\mathbf{P}$, show that **A** and **B** have the same characteristic polynomial, and hence the same eigenvalue spectrum.

9. Show that the eigenvalues of $\mathbf{A}^k$, where k is a positive integer, are the k-th powers of those of **A**. If **A** is nonsingular, show that this result holds good for negative integral k also.

10. If λ is an eigenvalue of **A**, show that the matrix polynomial

$$f(\mathbf{A})=a_0\mathbf{I}+a_1\mathbf{A}+a_2\mathbf{A}^2+\ldots+a_k\mathbf{A}^k$$

has $f(\lambda)=a_0+a_1\lambda+a_2\lambda^2+\ldots+a_k\lambda^k$ as one of its eigenvalues.

11. Show that any column vector of dimension n is an eigenvector of a constant matrix of order n.

12. Given a matrix **A**, show that it is possible to find a nonzero vector **x** such that **Ax** is the null vector if and only if **A** is singular.

*13. Find the eigenvalues of the matrix

$$\mathbf{A}=\begin{bmatrix} 3 & 2 & 2 \\ 4 & 3 & -3 \\ 0 & 0 & 1 \end{bmatrix}.$$

Then determine $\mathbf{A}^{-1}$, find its eigenvalues and verify that they are the reciprocals of those of **A**. Also show that **A** and $\mathbf{A}^{-1}$ have the same eigenvalues.

14. Find the eigenvectors of the matrix of Eq. (11).

15. Find the eigenvalues and the eigenvectors of the following matrices :

*(a) $\begin{bmatrix} \cos\theta & \sin\theta \\ -\sin\theta & \cos\theta \end{bmatrix}$; (b) $\begin{bmatrix} 0 & 1 & 0 \\ 0 & 0 & 1 \\ 1 & 0 & 0 \end{bmatrix}$;

(c) $\begin{bmatrix} q & p & p & p \\ p & q & p & p \\ p & p & q & p \\ p & p & p & q \end{bmatrix}$, $p\neq 0$.

*16 Let **x** be an eigenvector of a hermitian matrix **H**. If **y** is any vector orthogonal to **x**, show that **Hy** is also orthogonal to **x**.

17. Prove that the eigenvectors of a unitary matrix corresponding to distinct eigenvalues are orthogonal.

18. If a matrix **P** diagonalizes a matrix **A** (that is, $\mathbf{P}^{-1}\mathbf{A}\mathbf{P}$ is diagonal), prove that the columns of **P** are the eigenvectors of **A**. [This is the converse of what is proved in Eqs. (40).]

19. Find a common set of eigenvectors for the following commuting matrices:

(a) $\begin{bmatrix} 1 & -3 & 6 \\ 1 & 5 & -2 \\ 2 & 2 & 0 \end{bmatrix}$ and $\begin{bmatrix} 8 & -4 & -4 \\ -2 & -2 & 4 \\ -3 & -9 & 12 \end{bmatrix}$;

(b) $\begin{bmatrix} 26 & -36 & 0 \\ 18 & -25 & 0 \\ -30 & 44 & 3 \end{bmatrix}$ and $\begin{bmatrix} 1 & 0 & 0 \\ 0 & 1 & 0 \\ 24 & -34 & -1 \end{bmatrix}$

*20. Find the solution of the equation

$$\frac{d\mathbf{x}}{dt} = \begin{bmatrix} 3 & 2 \\ 2 & 3 \end{bmatrix} \mathbf{x}$$

which has the value $\mathbf{x}(0) = \{1 \quad 1\}$ at $t=0$.

21. Show that the matrices of Eqs. (58) satisfy the relation $(\mathbf{A}+3\mathbf{I})(\mathbf{B}-12\mathbf{I}) = \mathbf{0}$. Hence show that (a) if $\mathbf{x}$ is any vector which is not an eigenvector of $\mathbf{B}$ with the eigenvalue 12, then $(\mathbf{B}-12\mathbf{I})\mathbf{x}$ is an eigenvector of $\mathbf{A}$; (b) If $\mathbf{y}$ is any vector which is not an eigenvector of $\mathbf{A}$ with the eigenvalue -3, then $(\mathbf{A}+3\mathbf{I})\mathbf{y}$ is an eigenvector of $\mathbf{B}$.

22. Discuss the vibrations of the coupled springs of Fig. 7 for the following sets of values:

(a) $m_1=2,\ m_2=3,\ k_1=8,\ k_2=6,\ y_2(0)=7,\ \dot{y}_1(0)=\dot{y}_2(0)=0$;

(b) $m_1=4,\ m_2=2,\ k_1=36,\ k_2=8,\ y_2(0)=11,\ \dot{y}_1(0)=\dot{y}_2(0)=0$.

*23. Find the eigenvalues and the eigenvectors of the $n\times n$ matrix

$$\begin{bmatrix} q & p & p \dots p \\ p & q & p \dots p \\ p & p & q \dots p \\ \dots & \dots & \dots \\ p & p & p \dots q \end{bmatrix},$$

where p and q are real scalars, $p\neq 0$. [This is a generalization of Example 4.]

24. If λ_i, $1 \leqslant i \leqslant n$, are the eigenvalues of a matrix $\mathbf{A}$ of order n, show that

$$\mathrm{Tr}(\mathbf{A}^k) = \sum_{i=1}^{n} \lambda_i^k.$$

10. The eigenvalue problem, II

In this section,[1] we shall discuss some of the advanced topics concerning the eigenvalue problem such as the conditions for diagonalizability, nondiagonalizable matrices, Cayley-Hamilton theorem, etc.

Cayley-Hamilton theorem

We have seen in Section 3 that there are at most n^2 linearly independent matrices of order $n\times n$. Therefore if $\mathbf{A}$ is a square matrix of order n, the set of n^2+1 matrices $\mathbf{I}$, $\mathbf{A}$, $\mathbf{A}^2$, $\mathbf{A}^3$, ..., $\mathbf{A}^{n^2}$ must be linearly dependent. In other words, for any square matrix $\mathbf{A}$, it is possible to satisfy the equation

$$c_0\mathbf{I}+c_1\mathbf{A}+c_2\mathbf{A}^2+\ldots+c_N\mathbf{A}^N=\mathbf{0}, \tag{1}$$

with at least some nonzero coefficients c_i, where we have used the index $N=n^2$ for convenience. We thus see that there is a matrix polynomial of at most order n^2 which identically equals the zero matrix.

However, the importance of the Cayley-Hamilton theorem is that, in fact, for any square matrix of order n, there is a matrix polynomial of degree not more than n which equals the null matrix. If $\mathbf{A}$ is a square matrix whose characteristic polynomial is

$$D(\lambda)=c_0\lambda^n+c_1\lambda^{n-1}+c_2\lambda^{n-2}+\ldots+c_{n-1}\lambda+c_n, \tag{9.5}$$

the Cayley-Hamilton theorem states that $D(\mathbf{A})=\mathbf{0}$, or

$$c_0\mathbf{A}^n+c_1\mathbf{A}^{n-1}+c_2\mathbf{A}^{n-2}+\ldots+c_{n-1}\mathbf{A}+c_n\mathbf{I}=\mathbf{0}. \tag{2}$$

It is often expressed by saying that *every square matrix satisfies its own characteristic equatian.* In short, if $D(\lambda)=\det(\mathbf{A}-\lambda\mathbf{I})$, then $D(\mathbf{A})=\mathbf{0}$.

Consider the matrix $\mathbf{A}-\lambda\mathbf{I}$ and let us define $\mathbf{B}(\lambda)$ as the matrix of cofactors of $\mathbf{A}-\lambda\mathbf{I}$. Then it follows from Eq. (5.6b) that

$$(\mathbf{A}-\lambda\mathbf{I})\widetilde{\mathbf{B}}(\lambda)=D(\lambda)\mathbf{I}. \tag{3}$$

Since $\mathbf{A}-\lambda\mathbf{I}$ is a matrix of order n, each element of $\mathbf{B}(\lambda)$, being a cofactor of some element of $\mathbf{A}-\lambda\mathbf{I}$, will be a polynomial in λ of degree not more than n$-$1. $\widetilde{\mathbf{B}}(\lambda)$ can therefore be expressed as a polynomial in λ of degree $n-1$ [(see Eq. (3.49)] in the form

$$\widetilde{\mathbf{B}}(\lambda)=\lambda^{n-1}\mathbf{B}_0+\lambda^{n-2}\mathbf{B}_1+\ldots+\lambda\mathbf{B}_{n-2}+\mathbf{B}_{n-1}, \tag{4}$$

[1]This section may be skipped without loss of continuity.

where each $\mathbf{B}_i$ is a matrix of order $n\times n$. Then the left hand side of Eq. (3) becomes

$$(\mathbf{A}-\lambda\mathbf{I})\widetilde{\mathbf{B}}(\lambda)=(\mathbf{A}-\lambda\mathbf{I})\ (\lambda^{n-1}\mathbf{B}_0+\lambda^{n-1}\mathbf{B}_1+\ldots+\mathbf{B}_{n-1})$$
$$=-[\lambda^n\mathbf{B}_0+\lambda^{n-1}(\mathbf{B}_1-\mathbf{A}\mathbf{B}_0)+\lambda^{n-2}(\mathbf{B}_2-\mathbf{A}\mathbf{B}_1)+$$
$$\ldots+\lambda(\mathbf{B}_{n-1}-\mathbf{A}\mathbf{B}_{n-2})-\mathbf{A}\mathbf{B}_{n-1}], \qquad (5)$$

while the right hand side of Eq. (3) is

$$D(\lambda)\mathbf{I}=c_0\lambda^n\mathbf{I}+c_1\lambda^{n-1}\mathbf{I}+\ldots+c_{n-1}\lambda\mathbf{I}+c_n\mathbf{I}. \qquad (6)$$

Since the two polynomial expansions of Eqs. (5) and (6) must be identical, the coefficients of the same powers of λ must be the same, so that we have

$$\begin{aligned} -\mathbf{B}_0&=c_0\mathbf{I},\\ \mathbf{A}\mathbf{B}_0-\mathbf{B}_1&=c_1\mathbf{I},\\ \mathbf{A}\mathbf{B}_1-\mathbf{B}_2&=c_2\mathbf{I}, \\ &\ldots\ldots\ldots\ldots\\ \mathbf{A}\mathbf{B}_{n-2}-\mathbf{B}_{n-1}&=c_{n-1}\mathbf{I},\\ \mathbf{A}\mathbf{B}_{n-1}&=c_n\mathbf{I}. \end{aligned} \qquad (7)$$

Multiplying the first of Eqs. (7) from the left by $\mathbf{A}^n$, the second by $\mathbf{A}^{n-1}$, etc., until the last but one by $\mathbf{A}$ and adding all up, we have

$$0=c_0\mathbf{A}+c_1\mathbf{A}^{n-1}+c_2\mathbf{A}^{n-2}+\ldots+c_{n-1}\mathbf{A}+c_n\mathbf{I}, \qquad (2)$$

which is the desired result. If λ_i are the eigenvalues of $\mathbf{A}$, Eq. (2) can also be written as

$$(\mathbf{A}-\lambda_1\mathbf{I})\ (\mathbf{A}-\lambda_2\mathbf{I})\ldots(\mathbf{A}-\lambda_n\mathbf{I})=0. \qquad (8)$$

EXAMPLE 1: Consider the matrix $\mathbf{A}$ of Eq. (9.69) :

$$\mathbf{A}=\begin{bmatrix} -11 & -10 & 5 \\ 5 & 4 & -5 \\ -20 & -20 & 4 \end{bmatrix}. \qquad (9)$$

Show that this matrix satisfies the Cayley-Hamilton theorem.

Solution : The eigenvalues of $\mathbf{A}$ are 4, -6, -1. We have

$$(\mathbf{A}-4\mathbf{I})\ (\mathbf{A}+6\mathbf{I})\ (\mathbf{A}+\mathbf{I})$$
$$=\begin{bmatrix} -15 & -10 & 5 \\ 5 & 0 & -5 \\ -20 & -20 & 0 \end{bmatrix}\begin{bmatrix} -5 & -10 & 5 \\ 5 & 10 & -5 \\ -20 & -20 & 10 \end{bmatrix}$$
$$\times\begin{bmatrix} -10 & -10 & 5 \\ 5 & 5 & -5 \\ -20 & -20 & 5 \end{bmatrix}=0. \qquad (10)$$

Minimal polynomial

It turns out that in the case of some matrices having eigenvalues with multiplicity greater than unity, there may exist polynomials of degree less than n which equal the zero matrix. This will be clear from the following example.

EXAMPLE 2: If $\mathbf{A}$ is the matrix of Eq. (9.18), show that

$$[\mathbf{A}-(q+2p)\mathbf{I}]\,[\mathbf{A}-(q-p)\,\mathbf{I}]=\mathbf{0}.$$

Solution: The eigenvalues of the given matrix were found to be $q+2p$ with multiplicity one and $q-p$ with multiplicity two. The Cayley-Hamilton theorem asserts that

$$[\mathbf{A}-(q+2p)\,\mathbf{I}]\,[\mathbf{A}-(q-p)\,\mathbf{I}]^2=\mathbf{0}. \tag{11}$$

However, consider the given matrix polynomial:

$$[\mathbf{A}-(q+2p)\,\mathbf{I}]\,[\mathbf{A}-(q-p)\,\mathbf{I}]$$
$$=\begin{bmatrix} -2p & p & p \\ p & -2p & p \\ p & p & -2p \end{bmatrix}\begin{bmatrix} p & p & p \\ p & p & p \\ p & p & p \end{bmatrix}$$
$$=\begin{bmatrix} 0 & 0 & 0 \\ 0 & 0 & 0 \\ 0 & 0 & 0 \end{bmatrix}. \tag{12}$$

This shows that for the matrix under consideration, there exists a polynomial of degree 2 which vanishes, and this is

$$\mathbf{A}^2-(2q+p)\mathbf{A}+(q^2+pq-2p^2)\,\mathbf{I}=\mathbf{0}. \tag{13}$$

Let λ_μ, $1\leqslant\mu\leqslant l$, denote the l distinct eigenvalues of a matrix $\mathbf{A}$ of order n. Let d_μ be the multiplicity of the eigenvalue λ_μ, so that

$$\sum_{\mu=1}^{l} d_\mu=n. \tag{14}$$

The charactersitic polynomial of $\mathbf{A}$ can then be denoted by

$$D(\lambda)=(\lambda_1-\lambda)^{d_1}(\lambda_2-\lambda)^{d_2}\ldots(\lambda_l-\lambda)^{d_l}. \tag{15}$$

The Cayley-Hamilton theorem states that

$$D(\mathbf{A})=(\lambda_1\mathbf{I}-\mathbf{A})^{d_1}\,(\lambda_2\mathbf{I}-\mathbf{A})^{d_2}\ldots(\lambda_l\mathbf{I}-\mathbf{A})^{d_l}=\mathbf{0}. \tag{16}$$

But as we have seen in Example 2 above, there may exist a matrix polynomial of degree less than n which equals the zero matrix. If $r_1, r_2, \ldots, r_l$ are the smallest positive integers for which

$$p(\mathbf{A})\equiv(\lambda_1\mathbf{I}-\mathbf{A})^{r_1}(\lambda_2\mathbf{I}-\mathbf{A})^{r_2}\ldots(\lambda_l\mathbf{I}-\mathbf{A})^{r_l}=\mathbf{0}, \tag{17}$$

where $r_\mu\leqslant d_\mu$ $(1\leqslant\mu\leqslant l)$, then

$$p(\lambda)=(\lambda_1-\lambda)^{r_1}\,(\lambda_2-\lambda)^{r_2}\ldots(\lambda_l-\lambda)^{r_l} \tag{18}$$

is called the *minimal polynomial* of the matrix. The highest degree of the minimal polynomial is evidently n, and this happens when the minimal polynomial equals the characteristic polynomial, as in Example 1. It is also clear that the minimal polynomial divides the characteristic polynomial exactly.

Now we shall state without proof that *each factor is contained in the minimal polynomial at least once*, i.e., $r_\mu > 0$, $1 \leqslant \mu \leqslant l$. It follows at once that if all the eigenvalues of a matrix are distinct, its minimal polynomial equals its characteristic polynomial.

Condition for diagonalizability

The condition for the diagonalizability of a matrix can be stated in the following different ways. Once again, we shall merely state it without proof. *A matrix of order n is diagonalizable* (or *has a set of n linearly independent eigenvectors*) *if and only if its minimal polynomial is of the form*

$$p(\lambda) = (\lambda_1 - \lambda)(\lambda_2 - \lambda) \ldots (\lambda_l - \lambda),$$

where λ_μ are the *distinct* eigenvalues of the matrix. In other words, *a matrix is diagonalizable if and only if its minimal polynomial has distinct roots.* In the following example, we have discussed the case of a nondiagonalizable matrix.

EXAMPLE 3: Discuss the diagonalizability of the matrix

$$\mathbf{A} = \begin{bmatrix} 3 & 1 & -1 \\ 2 & 2 & -1 \\ 2 & 2 & 0 \end{bmatrix}. \tag{19}$$

Solution: The characteristic polynomial of the given matrix is found to be

$$D(\lambda) = \begin{bmatrix} 3-\lambda & 1 & -1 \\ 2 & 2-\lambda & -1 \\ 2 & 2 & -\lambda \end{bmatrix} = (1-\lambda)(2-\lambda)^2. \tag{20}$$

The eigenvalues are therefore 1, 2, 2. To find the minimal polynomial, consider

$$(\mathbf{A}-\mathbf{I})(\mathbf{A}-2\mathbf{I}) = \begin{bmatrix} 2 & 1 & -1 \\ 2 & 1 & -1 \\ 2 & 2 & -1 \end{bmatrix} \begin{bmatrix} 1 & 1 & -1 \\ 2 & 0 & -1 \\ 2 & 2 & -2 \end{bmatrix}$$

$$= \begin{bmatrix} 2 & 0 & -1 \\ 2 & 0 & -1 \\ 4 & 0 & -2 \end{bmatrix}, \tag{21}$$

which is not a zero matrix. On the other hand, $(\mathbf{A}-\mathbf{I})(\mathbf{A}-2\mathbf{I})^2$ must be a zero matrix by the Cayley-Hamilton theorem. The minimal polynomial is therefore $(1-\lambda)(2-\lambda)^2$, which has a multiple root and hence the matrix is not diagonalizable.

Let us see what happens when we try to obtain the eigenvectors of **A**. For the simple root $\lambda=1$, the eigenvector is easily found to be

$$\mathbf{x}_1=\{1 \quad 0 \quad 2\}. \tag{22}$$

For the double root $\lambda=2$, we have

$$(\mathbf{A}-2\mathbf{I})\mathbf{x}=\begin{bmatrix} 1 & 1 & -1 \\ 2 & 0 & -1 \\ 2 & 2 & -2 \end{bmatrix}\begin{bmatrix} a \\ b \\ c \end{bmatrix}=\begin{bmatrix} 0 \\ 0 \\ 0 \end{bmatrix}. \tag{23}$$

Here we find a peculiarity. Since the eigenvalue 2 has multiplicity 2, we expect, according to our earlier experience, that the matrix $\mathbf{A}-2\mathbf{I}$ would have rank $3-2=1$. However, it has rank 2 here, so that Eq. (23) can have only one linearly independent solution. The solution is $b=a$, $c=2a$, which gives the second eigenvector

$$\mathbf{x}_2=\{1 \quad 1 \quad 2\}. \tag{24}$$

Thus there is only one eigenvector associated with the eigenvalue $\lambda=2$ which has multiplicity 2.

Let us take any vector **u** linearly independent of $\mathbf{x}_1$ and $\mathbf{x}_2$. We could take a simple vector for **u**, say $\mathbf{u}=\{0 \quad 0 \quad 1\}$. Define a matrix $\mathbf{P}=[\mathbf{x}_1 \quad \mathbf{x}_2 \quad \mathbf{u}]$ and consider the similarity transformation

$$\mathbf{P}^{-1}\mathbf{A}\mathbf{P}=\begin{bmatrix} 1 & -1 & 0 \\ 0 & 1 & 0 \\ -2 & 0 & 1 \end{bmatrix}\begin{bmatrix} 3 & 1 & -1 \\ 2 & 2 & -1 \\ 2 & 2 & 0 \end{bmatrix} \times\begin{bmatrix} 1 & 1 & 0 \\ 0 & 1 & 0 \\ 2 & 2 & 1 \end{bmatrix}=\left[\begin{array}{c|cc} 1 & 0 & 0 \\ \hline 0 & 2 & -1 \\ 0 & 0 & 2 \end{array}\right]. \tag{25}$$

The similarity transformation **P** thus reduces **A** only to a triangular form, not to a diagonal form. Whatever components we choose for the vector **u** (linearly independent of $\mathbf{x}_1$ and $\mathbf{x}_2$), the matrix **P** will only reduce **A** to a triangular form. There exists no matrix **P** for which $\mathbf{P}^{-1}\mathbf{A}\mathbf{P}$ is diagonal, or, **A** is not similar to any diagonal matrix.

We can conclude from the discussion of this section that *a matrix which is similar to a triangular matrix at least two of whose diagonal elements, say i-th and j-th, are equal and the ij-th or ji-th element is nonvanishing, is nondiagonalizable.*

Diagonalization of hermitian matrices

A theorem of great importance in physics is that *every hermitian matrix can be diagonalized by a unitary matrix.* Although we shall not give a rigorous proof of this theorem, we shall point out the arguments one must follow.

We have seen in this section that every matrix can be brought to a triangular form by a similarity transformation. Let **H** be a hermitian matrix of order n and let us assume that it has only k linearly independent eigenvectors, $k \leqslant n$. If the eigenvectors correspond to distinct eigenvalues of **H**, they will be orthogonal to each other by virtue of Example 9.5. In case a number of eigenvectors correspond to the same eigenvalue, they can be chosen to be orthogonal by the Schmidt's process. Further, these eigenvectors can be normalized to obtain a set of k orthonormal eigenvectors of **H.** Now it is possible to choose a suitable set of $n-k$ orthonormal vectors each of which is orthogonal to the k eigenvectors, such that a matrix **U** constructed with these n vectors as column vectors brings **H** to a triangular form, say **T**. Since the columns of **U** are orthonormal vectors, **U** is clearly a unitary matrix. We have

$$\mathbf{U}^{-1}\mathbf{H}\mathbf{U}=\mathbf{T}. \tag{26}$$

Taking the hermitian conjugate of Eq. (26), it is easy to see that $\mathbf{T}^{\dagger}=\mathbf{T}$, so that **T** is also hermitian. But **T** is also triangular, and a matrix can be simultaneously hermitian and triangular only if it is diagonal with real elements. It therefore follows that it is possible to diagonalize a hermitian matrix by a unitary matrix. All the column vectors of **U** will be eigenvectors of **H**.

Note that by virtue of Exercise 9.17, this theorem applies to unitary matrices also. In fact, it applies to a more general class of matrices known as *normal matrices. A matrix is said to be normal if it commutes with its own hermitian conjugate* : $\mathbf{N}\mathbf{N}^{\dagger}=\mathbf{N}^{\dagger}\mathbf{N}$. The general theorem, whose proof is beyond the scope of this book, states that *a matrix of order n possesses a complete set of n orthogonal eigenvectors if and only if it commutes with its own hermitian conjugate.* It is easy to see that hermitian and unitary matrices are normal matrices.

Since a real symmetric matrix is also hermitian and a real unitary matrix is orthogonal, it follows, as a special case of the above theorem, that *every real symmetric matrix can be diagonalized by an orthogonal matrix.*

Matrix polynomials

An expression of the form

$$f(\mathbf{A})=a_k\mathbf{A}^k+a_{k-1}\mathbf{A}^{k-1}+\ldots+a_1\mathbf{A}+a_0\mathbf{I}, \tag{27}$$

where a_i are scalar coefficients, is called a *matrix polynomial* in **A** of degree k. The existence of a minimal polynomial for every matrix provides a great simplification in evaluating matrix polynomials. In fact, if m is the degree of the minimal polynomial of a matrix, it can be shown that every polynomial of the matrix of degree greater than $m-1$ is identically equal to a polynomial of degree $m-1$. This will be best understood by the following example.

EXAMPLE 4: If **A** is the matrix of Eq. (19), show that

$$2\mathbf{A}^5+4\mathbf{A}^4-6\mathbf{A}^3+\mathbf{A}^2-3\mathbf{A}-3\mathbf{I}=137\mathbf{A}^2-331\mathbf{A}+189\mathbf{I}. \tag{28}$$

Solution: We have seen in Example 3 that

$$(\mathbf{A}-\mathbf{I})\,(\mathbf{A}-2\mathbf{I})^2=\mathbf{A}^3-5\mathbf{A}^2+8\mathbf{A}-4\mathbf{I}=\mathbf{0}. \tag{29}$$

Multiplying Eq. (29) by $2\mathbf{A}^2$, we have

$$2\mathbf{A}^5-10\mathbf{A}^4+16\mathbf{A}^3-8\mathbf{A}^2=\mathbf{0}. \tag{30}$$

Let us denote the given polynomial by

$$f(\mathbf{A})=2\mathbf{A}^5+4\mathbf{A}^4-6\mathbf{A}^3+\mathbf{A}^2-3\mathbf{A}-3\mathbf{I}, \tag{31}$$

which, on using Eq. (30), reduces to

$$f(\mathbf{A})=14\mathbf{A}^4-22\mathbf{A}^3+9\mathbf{A}^2-3\mathbf{A}-3\mathbf{I}. \tag{32}$$

Again, multiplying Eq. (29) by $14\mathbf{A}$ and subtracting from Eq. (32), we have

$$f(\mathbf{A})=48\mathbf{A}^3-103\mathbf{A}^2+53\mathbf{A}-3\mathbf{I}. \tag{33}$$

Finally, multiplying Eq. (29) by 48 and subtracting from Eq. (33), we have

$$f(\mathbf{A})=137\mathbf{A}^2-331\mathbf{A}+189\mathbf{I}, \tag{34}$$

which proves the desired result.

Exercise 10

1. Determine the characteristic polynomial and minimal polynomial of each of the following matrices. Find which of them are diagonalizable and which are not. Find a set of linearly independent eigenvectors in each case. Reduce each matrix to a diagonal or a triangular form.

(a) $\begin{bmatrix} 2 & 3 & 5 \\ 1 & -1 & 0 \\ 2 & 1 & -1 \end{bmatrix}$; *(b) $\begin{bmatrix} 0 & -1 & -1 \\ 2 & 4 & 1 \\ 0 & -2 & 2 \end{bmatrix}$;

(c) $\begin{bmatrix} 3 & 4 & 0 \\ 2 & 2 & 1 \\ -10 & -12 & -2 \end{bmatrix}$; (d) $\begin{bmatrix} 2 & 2 & 1 \\ 5 & -1 & 1 \\ 1 & -2 & 2 \end{bmatrix}$.

2. Show that a matrix $\mathbf{A}=\mathbf{HU}$, where $\mathbf{H}$ is hermitian and $\mathbf{U}$ is unitary, is normal if and only if $\mathbf{H}^2$ commutes with $\mathbf{U}$.

3. Let $\mathbf{A}$ be the matrix of Eq. (9). Reduce the polynomial $2\mathbf{A}^4+5\mathbf{A}^3-47\mathbf{A}^2-28\mathbf{A}+21\mathbf{I}$ to a polynomial at most of degree 2.

*4. Consider a general matrix $\begin{bmatrix} a & b \\ c & d \end{bmatrix}$ of order two with real elements which is not diagonal (that is, at least one of the two elements b and c is nonzero). Show that the matrix is nondiagonalizable *if and only if* $(a-d)^2+4bc=0$.

*5. If $\mathbf{A}$ is the matrix of Eq. (9.18), express $\mathbf{A}^3$ and $\mathbf{A}^{-1}$ as linear combinations of $\mathbf{A}$ and $\mathbf{I}$. Verify the results by explicit calculation.

6. Let $\mathbf{A}$ be the matrix of Eq. (9). Construct the matrices

$$\mathbf{P}_1=(\mathbf{A}+6\mathbf{I})(\mathbf{A}+\mathbf{I})/50,$$
$$\mathbf{P}_2=(\mathbf{A}-4\mathbf{I})(\mathbf{A}+\mathbf{I})/50,$$
$$\mathbf{P}_3=-(\mathbf{A}-4\mathbf{I})(\mathbf{A}+6\mathbf{I})/25.$$

Show that these matrices satisfy the equations

(a) $\mathbf{P}_1+\mathbf{P}_2+\mathbf{P}_3=\mathbf{I}$; (b) $4\mathbf{P}_1-6\mathbf{P}_2-\mathbf{P}_3=\mathbf{A}$;

(c) $16\mathbf{P}_1+36\mathbf{P}_2+\mathbf{P}_3=\mathbf{A}^2$; (d) $\mathbf{P}_i^2=\mathbf{P}_i$, $i=1, 2, 3$.

*7. This is a generalization of Exercise 6. Let $\mathbf{A}$ be any matrix of order 3 with distinct eigenvalues $\lambda_1, \lambda_2, \lambda_3$. Let

$$\mathbf{P}_1=(\mathbf{A}-\lambda_2\mathbf{I})(\mathbf{A}-\lambda_3\mathbf{I})/[(\lambda_1-\lambda_2)(\lambda_1-\lambda_3)],$$
$$\mathbf{P}_2=(\mathbf{A}-\lambda_1\mathbf{I})(\mathbf{A}-\lambda_3\mathbf{I})/[(\lambda_2-\lambda_1)(\lambda_2-\lambda_3)],$$
$$\mathbf{P}_3=(\mathbf{A}-\lambda_1\mathbf{I})(\mathbf{A}-\lambda_2\mathbf{I})/[(\lambda_3-\lambda_1)(\lambda_3-\lambda_2)].$$

(a) Show that they satisfy the equations

(i) $\mathbf{P}_1+\mathbf{P}_2+\mathbf{P}_3=\mathbf{I}$; (ii) $\lambda_1\mathbf{P}_1+\lambda_2\mathbf{P}_2+\lambda_3\mathbf{P}_3=\mathbf{A}$;

(iii) $\lambda_1^2\mathbf{P}_1+\lambda_2^2\mathbf{P}_2+\lambda_3^2\mathbf{P}_3=\mathbf{A}^2$; (iv) $\mathbf{P}_i^2=\mathbf{P}_i$, $1\leq i\leq 3$.

(b) Express $\mathbf{A}^3$ as a linear combination of $\mathbf{P}_1$, $\mathbf{P}_2$ and $\mathbf{P}_3$.

11. Bilinear and quadratic forms

If $x_1, x_2, \ldots, x_n$ and $y_1, y_2, \ldots, y_n$ are a set of $2n$ variables, real or complex, the expression

$$B = \sum_{i,j=1}^{n} a_{ij} x_i^* y_j, \tag{1}$$

where a_{ij} are scalar coefficients, is called a *bilinear form* in the $2n$ variables. If we define the column vectors **x** and **y** and the matrix **A** by

$$\mathbf{x} \equiv \{x_i\},\ \mathbf{y} \equiv \{y_i\},\ \mathbf{A} \equiv [a_{ij}], \tag{2}$$

then it is easy to see that the bilinear form B can be written as

$$B = \mathbf{x}^\dagger \mathbf{A} \mathbf{y}. \tag{3}$$

If we replace **y** by **x** in the above expression, the resulting expression

$$Q = \mathbf{x}^\dagger \mathbf{A} \mathbf{x} = \sum_{i,j=1}^{n} a_{ij} x_i^* x_j \tag{4}$$

is called a *quadratic form* in the n variables x_i. The value of a bilinear form or a quadratic form is a number.

EXAMPLE 1: Obtain the bilinear form $\mathbf{x}^\dagger \mathbf{A}\mathbf{y}$ and the quadratic form $\mathbf{x}^\dagger \mathbf{A}\mathbf{x}$ if

$$\mathbf{x} = \begin{bmatrix} 1+i \\ 3 \\ 2i \end{bmatrix},\quad \mathbf{y} = \begin{bmatrix} 2-i \\ 0 \\ 2+2i \end{bmatrix},\quad \mathbf{A} = \begin{bmatrix} 3+i & 0 & 9 \\ 6i & 2+3i & 0 \\ -i & 2i & 1-i \end{bmatrix}.$$

Solution: We have

$$B = \mathbf{x}^\dagger \mathbf{A} \mathbf{y} = [1-i \quad 3 \quad -2i]$$

$$\times \begin{bmatrix} 3+i & 0 & 9 \\ 6i & 2+3i & 0 \\ -i & 2i & 1-i \end{bmatrix} \begin{bmatrix} 2-i \\ 0 \\ 2+2i \end{bmatrix} = 56+22i, \tag{5a}$$

and

$$Q = \mathbf{x}^\dagger \mathbf{A} \mathbf{x} = [1-i \quad 3 \quad -2i]$$

$$\times \begin{bmatrix} 3+i & 0 & 9 \\ 6i & 2+3i & 0 \\ -i & 2i & 1-i \end{bmatrix} \begin{bmatrix} 1+i \\ 3 \\ 2i \end{bmatrix} = 38+59i. \tag{5b}$$

If **x** is a real vector and **A** is any real matrix, the quadratic form $\mathbf{x}^\dagger \mathbf{A}\mathbf{x} = \tilde{\mathbf{x}} \mathbf{A}\mathbf{x}$ can *always* be written as $\tilde{\mathbf{x}} \mathbf{S} \mathbf{x}$ where **S** is a real symmetric matrix. For, the quadratic form, being only a number, equals its own transpose, so that we have

$$(\tilde{\mathbf{x}}\mathbf{A}\mathbf{x})^T=\tilde{\mathbf{x}}\mathbf{A}\mathbf{x}\Rightarrow\tilde{\mathbf{x}}\,\tilde{\mathbf{A}}\,\mathbf{x}=\tilde{\mathbf{x}}\,\mathbf{A}\mathbf{x}. \tag{6}$$

Therefore we have

$$\tilde{\mathbf{x}}\mathbf{A}\mathbf{x}=\tfrac{1}{2}(\tilde{\mathbf{x}}\mathbf{A}\mathbf{x}+\tilde{\mathbf{x}}\tilde{\mathbf{A}}\mathbf{x})=\tfrac{1}{2}\tilde{\mathbf{x}}(\mathbf{A}+\tilde{\mathbf{A}})\mathbf{x}=\tilde{\mathbf{x}}\mathbf{S}\mathbf{x}, \tag{7}$$

where

$$\mathbf{S}=\tfrac{1}{2}(\mathbf{A}+\tilde{\mathbf{A}}), \tag{8}$$

which is evidently a symmetric matrix.

EXAMPLE 2: Express the quadratic form $\tilde{\mathbf{x}}\mathbf{A}\mathbf{x}$, where

$$\mathbf{x}=[x_1\ x_2\ x_3],\ \mathbf{A}=\begin{bmatrix} 3 & -1 & 5 \\ 4 & 2 & -6 \\ 9 & 2 & 4 \end{bmatrix},$$

in the form $\tilde{\mathbf{x}}\mathbf{S}\mathbf{x}$, where $\mathbf{S}$ is a symmetric matrix.

Solution: The quadratic form is explicitly given by

$$\tilde{\mathbf{x}}\mathbf{A}\mathbf{x}=3x_1^2+2x_2^2+4x_2^3+3x_1x_2-4x_2x_3+14x_3x_1. \tag{9}$$

Let us construct a matrix $\mathbf{S}$ by writing

$$\mathbf{S}=\tfrac{1}{2}(\mathbf{A}+\tilde{\mathbf{A}})=\begin{bmatrix} 3 & 3/2 & 7 \\ 3/2 & 2 & -2 \\ 7 & -2 & 4 \end{bmatrix}, \tag{10}$$

so that

$$\tilde{\mathbf{x}}\mathbf{S}\mathbf{x}=3x_1^2+2x_2^2+4x_3^2+3x_1x_2-4x_2x_3+14x_3x_1, \tag{11}$$

which is the same as in Eq. (9).

Hermitian form

If $\mathbf{H}$ is a hermitian matrix, the quadratic form $\mathbf{x}^\dagger\mathbf{H}\mathbf{x}$ ($\mathbf{x}$ real or complex) is called a *hermitian form*. A useful theorem regarding hermitian forms is proved in the following example.

EXAMPLE 3: Show that for any vector $\mathbf{x}$, real or complex, the value of a hermitian form is always real.

Solution: Let $\mathbf{H}$ be a hermitian matrix and let

$$Q=\mathbf{x}^\dagger\,\mathbf{H}\mathbf{x}. \tag{12}$$

Since Q is a scalar,

$$Q^\dagger=Q^*. \tag{13}$$

Now, from Eq. (12), we have

$$Q^{\dagger}=(\mathbf{x}^{\dagger}\,\mathbf{H}\mathbf{x})^{\dagger}=\mathbf{x}^{\dagger}\,\mathbf{H}^{\dagger}\,\mathbf{x}=\mathbf{x}^{\dagger}\,\mathbf{H}\mathbf{x}=Q. \tag{14}$$

Together with Eq. (13), this shows that $Q^*=Q$, which proves the result that Q is real.

EXAMPLE 4: Evaluate the hermitian form $\mathbf{x}^{\dagger}\,\mathbf{H}\mathbf{x}$, where

$$\mathbf{x}=\begin{bmatrix} x_1 \\ x_2 \\ x_3 \end{bmatrix}, \qquad \mathbf{H}=\begin{bmatrix} a & p-iq & r-is \\ p+iq & b & u-iv \\ r+is & u+iv & c \end{bmatrix}, \tag{15}$$

where $a, b, c, p, q, r, s, u, v$ are real numbers.

Solution: The desired hermitian form is given by

$$\mathbf{x}^{\dagger}\,\mathbf{H}\mathbf{x}=[x_1^*\; x_3^*\; x_3^*]\begin{bmatrix} a & p-iq & r-is \\ p+iq & b & u-iv \\ r+is & u+iv & c \end{bmatrix}\begin{bmatrix} x_1 \\ x_2 \\ x_3 \end{bmatrix}$$

$$=a|\,x_1\,|^2+b|\,x_2\,|^2+c|\,x_3\,|^2+2\mathrm{Re}\,[(p-iq)x_1^*x_2 +(u-iv)x_2^*x_3+(r-is)x_1^*x_3], \tag{16}$$

which is a real number for any real or complex values of x_1, x_2, x_3.

Principal axis transformation

Let $\mathbf{x}$ and $\mathbf{A}$ be given by

$$\mathbf{x}=\begin{bmatrix} x_1 \\ x_2 \end{bmatrix}, \qquad \mathbf{A}=\begin{bmatrix} a & h \\ h & b \end{bmatrix}, \tag{17}$$

where a, b, h are real constants and x_1 and x_2 are real variables. The equation

$$\tilde{\mathbf{x}}\mathbf{A}\mathbf{x}=c, \tag{18}$$

written explicitly as the quadratic equation

$$ax_1^2+2hx_1x_2+bx_2^2=c, \tag{19}$$

then denotes a conic section in the two-dimensional (x_1, x_2) space. To see what this conic section is, let us make the following coordinate transformation.

Since $\mathbf{A}$ is a real symmetric matrix, it can be diagonalized by an orthogonal matrix, say $\mathbf{P}$, so that

$$\mathbf{P}^{-1}\mathbf{A}\mathbf{P}=\begin{bmatrix} \lambda_1 & 0 \\ 0 & \lambda_2 \end{bmatrix}\equiv\mathbf{\Lambda}, \tag{20}$$

where λ_1 and λ_2 are the eigenvalues of $\mathbf{A}$. Eq. (18) then becomes

$$\tilde{\mathbf{x}}\mathbf{P}\mathbf{P}^{-1}\mathbf{A}\mathbf{P}\mathbf{P}^{-1}\mathbf{x}=c \Rightarrow (\mathbf{P}^{-1}\mathbf{x})^T\mathbf{\Lambda}(\mathbf{P}^{-1}\mathbf{x})=c. \tag{21}$$

If we define new coordinates by

$$\mathbf{y} \equiv \begin{bmatrix} y_1 \\ y_2 \end{bmatrix} = \mathbf{P}^{-1} \begin{bmatrix} x_1 \\ x_2 \end{bmatrix} = \mathbf{P}^{-1}\mathbf{x}, \tag{22}$$

then Eq. (21) takes the form

$$\tilde{\mathbf{y}}\mathbf{\Lambda}\mathbf{y} = c \Rightarrow \lambda_1 y_1^2 + \lambda_2 y_2^2 = c. \tag{23a}$$

When this is written in the familiar form

$$\frac{y_1^2}{c/\lambda_1} + \frac{y_2^2}{c/\lambda_2} = 1, \tag{23b}$$

it is evident that the equation represents (a) an ellipse if c, λ_1 and λ_2 are of the same sign, (b) a hyperbola if λ_1 and λ_2 have opposite signs, (c) an imaginary curve if λ_1 and λ_2 are of the same sign but c has the opposite sign, and (d) a pair of straight lines, a single straight line, a point or no real curve if one or more of λ_1, λ_2 and c are zero.

The cases of interest among those mentioned above are the first two. When Eq. (23b) denotes an ellipse, it is evident that the semi-major and the semi-minor axes are given by $(c/\lambda_1)^{1/2}$ and $(c/\lambda_2)^{1/2}$, respectively. Moreover, the principal axes of the conic section are along the new coordinate axes y_1 and y_2. The presence of the cross term x_1x_2 in Eq. (19) shows that the coordinate axes are not parallel to the principal axes of the conic section. By making an orthogonal transformation from (x_1, x_2) to (y_1, y_2), the equation of the conic section assumes the standard form and allows us to determine the lengths and the directions of the principal axes. The process is therefore called the *principal axis transformation*, and is illustrated in Fig. 8.

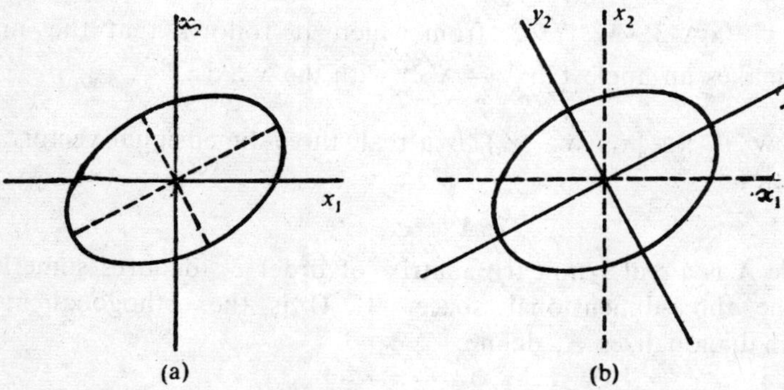

Figure 8. (a) An ellipse whose principal axes do not coincide with the coordinate axes. (b) An ellipse whose principal axes coincide with the coordinate axes.

EXAMPLE 5: Show that the equation

$$8x^2+2\sqrt{2}\,xy+7y^2=3 \tag{24}$$

represents an ellipse. Find the lengths of the principal axes and their directions.

Solution: The given equation can be written as

$$\tilde{\mathbf{x}}\mathbf{A}\mathbf{x}=3, \tag{25}$$

where

$$\mathbf{x}=\begin{bmatrix} x \\ y \end{bmatrix}, \quad \mathbf{A}=\begin{bmatrix} 8 & \sqrt{2} \\ \sqrt{2} & 7 \end{bmatrix}. \tag{26}$$

The eigenvalues and the normalized eigenvectors of **A** are found to be

$$\text{(i)}\ 9, \{\sqrt{2/3}\quad 1/\sqrt{3}\}, \qquad \text{(ii)}\ 6, \{1/\sqrt{3}\quad -\sqrt{2/3}\}. \tag{27}$$

The diagonalizing matrix is therefore

$$\mathbf{P}=\begin{bmatrix} \sqrt{2/3} & 1/\sqrt{3} \\ 1/\sqrt{3} & -\sqrt{2/3} \end{bmatrix}, \tag{28}$$

and if we define new coordinates by

$$\mathbf{y}\equiv\begin{bmatrix} x' \\ y' \end{bmatrix}=\mathbf{P}^{-1}\mathbf{x}=\begin{bmatrix} \sqrt{2/3} & 1/\sqrt{3} \\ 1/\sqrt{3} & -\sqrt{2/3} \end{bmatrix}\begin{bmatrix} x \\ y \end{bmatrix}, \tag{29}$$

the given equation becomes

$$9x'^2+6y'^2=3 \Rightarrow 3x'^2+2y'^2=1. \tag{30}$$

It is then evident that the equation represents an ellipse and that the length of the semi-major axis is $1/\sqrt{2}$, that of the semi-minor axis is $1/\sqrt{3}$ and the major axis is along y'. From Eq. (29), we have $y'=x/\sqrt{3}-\sqrt{2}y/\sqrt{3}$, from which it follows that the major axis makes an angle $\tan^{-1}(-\sqrt{2})$ with the x axis.

Now if $\mathbf{x}\equiv\{x_1\quad x_2\quad x_3\}$ is a real three-dimensional vector, the equation

$$\tilde{\mathbf{x}}\mathbf{A}\mathbf{x}=c, \tag{31}$$

where **A** is a real symmetric matrix of order 3, denotes some locus in the three-dimensional space. If **U** is the orthogonal matrix which diagonalizes **A**, define

$$\mathbf{y}=\mathbf{U}^{-1}\mathbf{x}\equiv\{y_1\quad y_2\quad y_3\}, \tag{32}$$

so that Eq. (31) takes the form

$$\tilde{\mathbf{y}}\mathbf{\Lambda}\mathbf{y}=c, \tag{33}$$

where Λ is a diagonal matrix containing the eigenvalues of **A**. If these are denoted by $\lambda_1, \lambda_2, \lambda_3$, Eq. (33) becomes

$$\frac{y_1^2}{c/\lambda_1}+\frac{y_2^2}{c/\lambda_2}+\frac{y_3^2}{c/\lambda_3}=1. \tag{34}$$

If all the three denominators are positive, Eq. (34) represents an ellipsoid; if one is negative, it denotes a hyperboloid of one sheet; if two are negative, it denotes a hyperboloid of two sheets. In all other cases, Eq. (34) denotes a surface, a curve, a point, or no real locus. In the first three cases of interest, the principal axes are along y_1, y_2 and y_3. When the equation represents an ellipsoid, the lengths of the principal axes are $2(c/\lambda_i)^{1/2}$, $i=1, 2, 3$.

EXAMPLE 6: Show that the equation

$$3x^2+3y^2-4z^2-18xy+12=0 \tag{35}$$

represents a hyperboloid of one sheet and determine the direction cosines of its principal axes.

Solution: The given equation can be written as

$$\tilde{\mathbf{x}}\mathbf{A}\mathbf{x}=12, \tag{36}$$

where

$$\mathbf{x}=\begin{bmatrix} x \\ y \\ z \end{bmatrix}, \mathbf{A}=\begin{bmatrix} -3 & 9 & 0 \\ 9 & -3 & 0 \\ 0 & 0 & 4 \end{bmatrix}. \tag{37}$$

The eigenvalues and the normalized eigenvectors of **A** are found to be

$$\text{(i)}\ 6, \{1/\sqrt{2}\quad 1/\sqrt{2}\quad 0\}; \qquad \text{(ii)}\ 4, \{0\quad 0\quad 1\};$$
$$\text{(iii)}\ -12, \{1/\sqrt{2}\quad -1/\sqrt{2}\quad 0\}. \tag{38}$$

The diagonalizing matrix is therefore

$$\mathbf{P}=\begin{bmatrix} 1/\sqrt{2} & 0 & 1/\sqrt{2} \\ 1/\sqrt{2} & 0 & -1/\sqrt{2} \\ 0 & 1 & 0 \end{bmatrix}. \tag{39}$$

If we define new coordinates by

$$\mathbf{y}\equiv\{x'\quad y'\quad z'\}=\mathbf{P}^{-1}\{x\quad y\quad z\}, \tag{40}$$

Eq. (35) becomes

$$\frac{x'^2}{2}+\frac{y'^2}{3}-\frac{z'^2}{1}=1. \tag{41}$$

This clearly indicates that the equation represents a hyperboloid of one sheet. Its principal axes are along x', y' and z' whose direction cosines are just the components of the eigenvectors of **A** as given below:

$$\begin{aligned}&\text{direction cosines of } x'=(1/\sqrt{2},\ \ 1/\sqrt{2},\ \ 0);\\ &\text{direction cosines of } y'=(0,\ \ 0,\ \ 1);\\ &\text{direction cosines of } z'=(1/\sqrt{2},\ \ -1/\sqrt{2},\ \ 0).\end{aligned}\tag{42}$$

Exercise 11

1. For any matrix **A** and vector **x**, show that $\widetilde{\mathbf{xAx}}=\tilde{\mathbf{x}}\tilde{\mathbf{A}}\mathbf{x}$.

*2. Write the quadratic form $(a_1x_1+a_2x_2+\ldots+a_nx_n)^2$ in the form $\tilde{\mathbf{x}}\mathbf{Ax}$ with **A** symmetric. What is the rank of **A**?

3. Let $\mathbf{u}_1, \mathbf{u}_2, \ldots, \mathbf{u}_k$ and **x** be n-dimensional vectors. Show that the quadratic form $(\tilde{\mathbf{u}}_1\mathbf{x})^2+(\tilde{\mathbf{u}}_2\mathbf{x})^2+\ldots+(\tilde{\mathbf{u}}_k\mathbf{x})^2$ can be expressed as $\tilde{\mathbf{x}}\mathbf{U}\tilde{\mathbf{U}}\mathbf{x}$, where **U** is a matrix of order $n\times k$ whose i-th column is the vector $\mathbf{u}_i$.

4. If **A** is a real symmetric matrix and $\mathbf{Ax}=\lambda\mathbf{x}$, $\mathbf{Ay}=\mu\mathbf{y}$, where $\lambda\neq\mu$, show that $\tilde{\mathbf{x}}\mathbf{Ay}=0$ and that **x** and **y** are orthogonal.

5. If $\mathbf{x}^\dagger\mathbf{Hx}$ is real for every choice of the vector **x**, show that **H** is hermitian. [This is the converse of what was proved in Example 3.]

*6. Show that a quadratic form $\mathbf{x}^\dagger\mathbf{Ax}$ is positive for every choice of the vector **x** if every eigenvalue of **A** is positive. Show that the converse is not necessarily true.

7. If **A** is an antihermitian matrix, show that the value of $\mathbf{x}^\dagger\mathbf{Ax}$ is purely imaginary or zero for every choice of **x**.

8. Interpret geometrically the equation $\tilde{\mathbf{x}}\mathbf{Ax}=c$, where $\mathbf{x}=\{x\ \ y\ \ z\}$, and

(a) $\mathbf{A}=\begin{bmatrix}0&1&1\\1&0&1\\1&1&0\end{bmatrix}$, $c=4$; (b) $\mathbf{A}=\begin{bmatrix}2.6&-0.2&0.2\\0.2&2.1&0.4\\0.4&0.2&2.8\end{bmatrix}$, $c=15$.

*9. The potential experienced by a particle in a certain mechanical system is $V(\mathbf{r})=(x-y)^2+(y-z)^2+(z-x)^2$. Show that there is a line along which the potential energy of the particle is constant. Also show that the potential energy of the particle increases as it moves away from this line.

12. Functions of a matrix

As we define and study various functions of a variable in algebra, it is possible to define and evaluate functions of a matrix. We shall study the following functions of a matrix in this section: integral powers (positive and negative), fractional powers (roots), exponential, logarithm, trigonometric and hyperbolic functions.

There are two methods by which a function of a matrix can be evaluated. The first is a rather straightforward method based on the diagonalization of a matrix and is therefore applicable to diagonalizable matrices only. The second method is based on the existence of a minimal polynomial and can be used to evaluate functions of any matrix.

Functions of a diagonalizable matrix

Let $\mathbf{A}$ be a diagonalizable matrix and let $\mathbf{P}$ be a diagonalizing matrix for $\mathbf{A}$, so that

$$\mathbf{P}^{-1}\mathbf{A}\mathbf{P}=\mathbf{\Lambda}, \ \mathbf{A}=\mathbf{P}\mathbf{\Lambda}\mathbf{P}^{-1}, \tag{1}$$

where $\mathbf{\Lambda}$ is a diagonal matrix containing the eigenvalues of $\mathbf{A}$. Now, if f is any function of a matrix, then we have

$$f(\mathbf{A})=\mathbf{P}f(\mathbf{\Lambda})\mathbf{P}^{-1}. \tag{2}$$

Thus, if we can define a function of a diagonal matrix, we can define and evaluate the function of any diagonalizable matrix. The discussion of this section evidently applies to square matrices only.

Powers of a matrix

We have, in fact, had many occasions so far in this book of using the powers of a matrix. Thus, we define the square of a matrix by $\mathbf{A}^2=\mathbf{AA}$, the cube by $\mathbf{A}^3=\mathbf{AAA}$, etc. In general, if k is a positive integer, we define the k-th power of $\mathbf{A}$ as a matrix obtained by multiplying $\mathbf{A}$ with itself k times, that is,

$$\mathbf{A}^k=\mathbf{AAA}\ldots\mathbf{A} \ (k \text{ times}). \tag{3}$$

If $\mathbf{A}$ is nonsingular, we have defined its inverse $\mathbf{A}^{-1}$ as a matrix whose product with $\mathbf{A}$ gives the unit matrix. The negative powers of $\mathbf{A}$ are then similarly defined. If m is a negative integer, let $k=-m$, so that

$$\mathbf{A}^m=(\mathbf{A}^{-1})^k=\mathbf{A}^{-1}\mathbf{A}^{-1}\mathbf{A}^{-1}\ldots\mathbf{A}^{-1} \ (k \text{ times}). \tag{4}$$

Finally, in analogy with the functions of a variable, we define

$$\mathbf{A}^0 = \mathbf{I}. \tag{5}$$

Although all the integral powers of **A** have thus been defined in a straightforward manner, the actual evaluation may be tedious for large values of k. The calculation is considerably simplified by using the diagonalizability of **A**. For, taking the k-th power of **A** and using the second of Eqs. (1), we have

$$\begin{aligned} \mathbf{A}^k &= (\mathbf{P}\Lambda\mathbf{P}^{-1})(\mathbf{P}\Lambda\mathbf{P}^{-1})\ldots(\mathbf{P}\Lambda\mathbf{P}^{-1}) \ (k \text{ times}) \\ &= \mathbf{P}\Lambda^k\mathbf{P}^{-1}. \end{aligned} \tag{6}$$

Similarly, if $m = -k$ is a negative integer and **A** is nonsingular, then

$$\mathbf{A}^m = \mathbf{P}\Lambda^m\mathbf{P}^{-1} = \mathbf{P}(\Lambda^{-1})^k\mathbf{P}^{-1}. \tag{7}$$

EXAMPLE 1: Find $\mathbf{A}^k$, where k is any integer, positive or negative, and

$$\mathbf{A} = \begin{bmatrix} 4/3 & \sqrt{2}/3 \\ \sqrt{2}/3 & 5/3 \end{bmatrix} \tag{8}$$

Solution: The eigenvalues and the eigenvectors of **A** are found to be

$$\text{(i)} \quad 1, \{\sqrt{2} \quad -1\} \ ; \quad \text{(ii)} \quad 2, \{1 \quad \sqrt{2}\}. \tag{9}$$

We therefore have

$$\mathbf{P} = \begin{bmatrix} \sqrt{2} & 1 \\ -1 & \sqrt{2} \end{bmatrix}, \quad \mathbf{P}^{-1}\mathbf{A}\mathbf{P} = \begin{bmatrix} 1 & 0 \\ 0 & 2 \end{bmatrix} \equiv \Lambda. \tag{10}$$

The matrix **A** is seen to be nonsingular. For any integral k, therefore, we have

$$\begin{aligned} \mathbf{A}^k = \mathbf{P}\Lambda^k\mathbf{P}^{-1} &= \begin{bmatrix} \sqrt{2} & 1 \\ -1 & \sqrt{2} \end{bmatrix} \begin{bmatrix} 1 & 0 \\ 0 & 2^k \end{bmatrix} \begin{bmatrix} \sqrt{2}/3 & -1/3 \\ 1/3 & \sqrt{2}/3 \end{bmatrix} \\ &= \tfrac{1}{3} \begin{bmatrix} 2^k + 2 & (2^k - 1)\sqrt{2} \\ (2^k - 1)\sqrt{2} & 2^{k+1} + 1 \end{bmatrix}. \end{aligned} \tag{11}$$

Note that, in particular, $\mathbf{A}^0 = \mathbf{I}$. Also,

$$\mathbf{A}^{50} = \tfrac{1}{3} \begin{bmatrix} 2^{50} + 2 & (2^{50} - 1)\sqrt{2} \\ (2^{50} - 1)\sqrt{2} & 2^{51} + 1 \end{bmatrix}, \tag{12}$$

$$\mathbf{A}^{-10} = \tfrac{1}{3} \begin{bmatrix} 2^{-10} + 2 & (2^{-10} - 1)\sqrt{2} \\ (2^{-10} - 1)\sqrt{2} & 2^{-9} + 1 \end{bmatrix}. \tag{13}$$

Roots of a matrix

In elementary algebra, we say that y is a k-th root of x if $y^k = x$. Similarly, we shall say that *a matrix* **B** *is a k-th root of a matrix* **A** if $\mathbf{B}^k=\mathbf{A}$. The object is to find all the matrices $\mathbf{B}$ which satisfy this relation for a given matrix $\mathbf{A}$.

To begin with, consider a diagonal matrix $\mathbf{\Lambda}$ whose elements are given by $(\mathbf{\Lambda})_{ij}=d_i\delta_{ij}$. It is evident that $\mathbf{\Lambda}^k$ is again a diagonal matrix whose diagonal elements are d_i^k, i.e., $(\mathbf{\Lambda}^k)_{ij}=d_i^k\delta_{ij}$. Now let $p=1/k$ and consider a diagonal matrix $\mathbf{D}$ whose elements are given by $(\mathbf{D})_{ij}=d_i^p\delta_{ij}$. Clearly, the k-th power of $\mathbf{D}$ will equal $\mathbf{\Lambda}$, that is $\mathbf{D}^k=\mathbf{\Lambda}$. Then consider the matrix $\mathbf{B}=\mathbf{PDP}^{-1}=\mathbf{P\Lambda}^p\mathbf{P}^{-1}$. Taking the k-th power of $\mathbf{B}$, we find

$$\mathbf{B}^k=(\mathbf{P\Lambda}^p\mathbf{P}^{-1})(\mathbf{P\Lambda}^p\mathbf{P}^{-1})\ldots(\mathbf{P\Lambda}^p\mathbf{P}^{-1})\ (k \text{ times})=\mathbf{P\Lambda P}^{-1}=\mathbf{A}. \quad (14)$$

Thus, $\mathbf{B}=\mathbf{P\Lambda}^p\mathbf{P}^{-1}$ is a k-th root of $\mathbf{A}$. The same result holds good for any fractional power. Thus, if q is any fraction, we have

$$\mathbf{A}^q=\mathbf{P\Lambda}^q\mathbf{P}^{-1}. \quad (15)$$

EXAMPLE 2: Find $\mathbf{A}^{3/7}$ where $\mathbf{A}$ is the matrix of Eq. (8).

Solution: In this case, the diagonal matrix $\mathbf{\Lambda}$ is given by Eq. (10). Hence we have

$$\mathbf{\Lambda}^{3/7}=\begin{bmatrix} 1 & 0 \\ 0 & 2^{3/7} \end{bmatrix}. \quad (16)$$

This gives

$$\mathbf{A}^{3/7}=\mathbf{P\Lambda}^{3/7}\mathbf{P}^{-1}=\tfrac{1}{3}\begin{bmatrix} 2^{3/7}+2 & (2^{3/7}-1)\sqrt{2} \\ (2^{3/7}-1)\sqrt{2} & 2^{10/7}+1 \end{bmatrix}. \quad (17)$$

It should be realized that the k-th root of a number is not unique. In fact, there are exactly k k-th roots of any number except zero. Similarly, the k-th root of a matrix will not be unique. If a matrix $\mathbf{A}$ has m nonzero eigenvalues, there will be k^m matrices whose k-th power equals $\mathbf{A}$.

EXAMPLE 3: Find all the square roots of the matrix

$$\mathbf{A}=\begin{bmatrix} 3/2 & 1/2 \\ 1/2 & 3/2 \end{bmatrix}. \quad (18)$$

Solution: The eigenvalues and the eigenvectors of the given matrix are found to be

$$\text{(i) } 2, \{1 \quad 1\}\ ; \quad \text{(ii) } 1, \{1 \ -1\}. \quad (19)$$

We therefore have

$$\mathbf{P}=\begin{bmatrix} 1 & 1 \\ 1 & -1 \end{bmatrix}, \quad \mathbf{P}^{-1}=\tfrac{1}{2}\begin{bmatrix} 1 & 1 \\ 1 & -1 \end{bmatrix}, \quad \mathbf{\Lambda}=\begin{bmatrix} 2 & 0 \\ 0 & 1 \end{bmatrix}. \quad (20)$$

The square roots of $\mathbf{A}$ are

$$\Lambda^{1/2}=\begin{bmatrix} \pm\sqrt{2} & 0 \\ 0 & \pm 1 \end{bmatrix}. \tag{21}$$

We can choose any of the four matrices $\Lambda^{1/2}$ to obtain $\mathbf{A}^{1/2}$. Using the relation

$$\mathbf{A}^{1/2}=\mathbf{P}\Lambda^{1/2}\mathbf{P}^{-1}, \tag{22}$$

we find that $\mathbf{A}$ has four square roots given by $\pm\mathbf{B}$ and $\pm\mathbf{C}$, where

$$\mathbf{B}=\begin{bmatrix} (\sqrt{2}+1)/2 & (\sqrt{2}-1)/2 \\ (\sqrt{2}-1)/2 & (\sqrt{2}+1)/2 \end{bmatrix},$$

$$\mathbf{C}=\begin{bmatrix} (\sqrt{2}-1)/2 & (\sqrt{2}+1)/2 \\ (\sqrt{2}+1)/2 & (\sqrt{2}-1)/2 \end{bmatrix}. \tag{23}$$

Series

A series such as

$$\mathbf{S}=\sum_{k=0}^{\infty} a_k\mathbf{A}^k \tag{24}$$

in a matrix $\mathbf{A}$, where a_k are scalar coefficients, is said to converge if and only if every element of the right hand side converges. In that case, the series of Eq. (24) is equal to the matrix $f(\mathbf{A})$, of the same order as $\mathbf{A}$, whose elements are given by

$$[f(\mathbf{A})]_{ij}=\sum_{k=0}^{\infty} a_k(\mathbf{A}^k)_{ij}. \tag{25}$$

We shall state without proof the result that *a series $f(\mathbf{A})$ in a matrix $\mathbf{A}$ is convergent if and only if the corresponding algebraic series $f(\lambda)$ is convergent for every eigenvalue λ_i of* $\mathbf{A}$. Thus, if

$$f(\lambda)=\sum_{k=0}^{\infty} a_k\lambda^k \tag{26}$$

exists for $|\lambda| < R$, then

$$f(\mathbf{A})=\sum_{k=0}^{\infty} a_k\mathbf{A}^k \tag{27}$$

exists if and only if every eigenvalue λ_i of $\mathbf{A}$ satisfies $|\lambda_i| < R$. R is called the radius of convergence of the series.

Exponential of a matrix

The exponential series is defined by

$$e^{\lambda}=\sum_{k=0}^{\infty} \lambda^k/k!, \tag{28}$$

and is convergent for every finite value of λ. Similarly, we shall defiine the exponential of a matrix **A** by

$$e^{\mathbf{A}}=\sum_{k=0}^{\infty} \mathbf{A}^k/k!, \tag{29}$$

which will be a matrix of the same order as **A** and will exist provided all the elements (and hence eigenvalues) of **A** are finite.

To begin with, let us obtain the exponential of a diagonal matrix. Let $\mathbf{\Lambda}$ be a diagonal matrix with elements $(\mathbf{\Lambda})_{ij}=\lambda_i\delta_{ij}$. Then

$$\exp(\mathbf{\Lambda})=\sum_{k=0}^{\infty} \mathbf{\Lambda}^k/k!, \tag{30}$$

The ij-th element of exp ($\mathbf{\Lambda}$) will be given by

$$[\exp(\mathbf{\Lambda})]_{ij}=\sum_{k=0}^{\infty} (\mathbf{\Lambda}^k)_{ij}/k! =\sum_{k=0}^{\infty} \lambda_i^k\delta_{ij}/k!=\exp(\lambda_i)\,\delta_{ij}. \tag{31}$$

It is therefore evident that if

$$\mathbf{\Lambda}=\begin{bmatrix} \lambda_1 & & & & \\ & \lambda_2 & & \mathbf{0} & \\ & & \cdot & & \\ & & & \cdot & \\ & \mathbf{0} & & & \cdot \\ & & & & \lambda_n \end{bmatrix}, \tag{32}$$

then

$$\exp(\mathbf{\Lambda})=\begin{bmatrix} \exp(\lambda_1) & & & & \\ & \exp(\lambda_2) & & \mathbf{0} & \\ & & \cdot & & \\ & & & \cdot & \\ & \mathbf{0} & & & \cdot \\ & & & & \exp(\lambda_n) \end{bmatrix}. \tag{33}$$

Now consider the series

$$\exp\mathbf{A}=\mathbf{I}+\mathbf{A}+\frac{\mathbf{A}^2}{2!}+\frac{\mathbf{A}^3}{3!}+\ldots+\frac{\mathbf{A}^k}{k!}+\ldots. \tag{34}$$

Let **P** be a matrix which brings **A** to the diagonal form $\mathbf{\Lambda}$. Multiplying Eq. (34) from the left by $\mathbf{P}^{-1}$ and from the right by **P** and remembering that $\mathbf{P}^{-1}\mathbf{A}^k\mathbf{P}=\mathbf{\Lambda}^k$, we have

$$\mathbf{P}^{-1}(\exp\mathbf{A})\,\mathbf{P}=\mathbf{I}+\mathbf{\Lambda}+\frac{\mathbf{\Lambda}^2}{2!}+\frac{\mathbf{\Lambda}^3}{3!}+\ldots+\frac{\mathbf{\Lambda}^k}{k!}+\ldots=\exp\mathbf{\Lambda}. \tag{35}$$

It follows immediately that

$$\exp\mathbf{A}=\mathbf{P}(\exp\mathbf{\Lambda})\mathbf{P}^{-1}. \tag{36}$$

Having defined the exponential function, it is possible to define the matrix exponent of any number. In elementary algebra, we have $a^x = \exp(x \ln a)$, where ln denotes the natural logarithm. Similarly, we define

$$a^{\mathbf{A}} = \exp(\mathbf{A} \ln a) = \mathbf{P} \exp(\mathbf{\Lambda} \ln a) \mathbf{P}^{-1}, \tag{37}$$

where

$$\exp(\mathbf{\Lambda} \ln a) \equiv a^{\mathbf{\Lambda}} = \begin{bmatrix} a^{\lambda_1} & & & & \\ & a^{\lambda_2} & & \mathbf{0} & \\ & & \cdot & & \\ & & & \cdot & \\ & \mathbf{0} & & & \cdot \\ & & & & a^{\lambda_n} \end{bmatrix}. \tag{38}$$

EXAMPLE 4: Find $e^{\mathbf{A}}$ and $4^{\mathbf{A}}$ if **A** is the matrix of Eq. (18).

Solution: The matrices **P** and **Λ** are given in Eqs. (20). We have

$$e^{\mathbf{\Lambda}} = \begin{bmatrix} e^2 & 0 \\ 0 & e \end{bmatrix}, \quad 4^{\mathbf{\Lambda}} = \begin{bmatrix} 4^2 & 0 \\ 0 & 4 \end{bmatrix} = \begin{bmatrix} 16 & 0 \\ 0 & 4 \end{bmatrix}. \tag{39}$$

Therefore,

$$e^{\mathbf{A}} = \mathbf{P} e^{\mathbf{\Lambda}} \mathbf{P}^{-1} = \tfrac{1}{2} \begin{bmatrix} e^2+e & e^2-e \\ e^2-e & e^2+e \end{bmatrix}, \tag{40a}$$

$$4^{\mathbf{A}} = \mathbf{P}\, 4^{\mathbf{\Lambda}} \mathbf{P}^{-1} = \begin{bmatrix} 10 & 6 \\ 6 & 10 \end{bmatrix}. \tag{40b}$$

Logarithm of a matrix

We say that x is the natural logarithm of y if $e^x = y$, and write it as $x = \ln y$. Similarly, given a matrix **A**, we shall say that a matrix **B** is the natural logarithm of **A** if $e^{\mathbf{B}} = \mathbf{A}$. Therefore, by definition,

$$\mathbf{B} = \ln \mathbf{A} \Leftrightarrow \exp \mathbf{B} = \mathbf{A}. \tag{41}$$

We shall first find the logarithm of a diagonal matrix. If $\mathbf{\Lambda} \equiv [\lambda_i \delta_{ij}]$ is a diagonal matrix, its natural logarithm is a diagonal matrix of the same order given by

$$\mathbf{D} \equiv \ln \mathbf{\Lambda} = \begin{bmatrix} \ln \lambda_1 & & & & \\ & \ln \lambda_2 & & 0 & \\ & & \cdot & & \\ & & & \cdot & \\ & 0 & & & \cdot \\ & & & & \ln \lambda_n \end{bmatrix}. \tag{42}$$

To prove this, consider exp(**D**). Remembering that $\exp(\ln x)=x$ and using Eq. (33), it is easy to see that

$$\exp(\mathbf{D})=\exp(\ln \mathbf{\Lambda})=\mathbf{\Lambda}. \tag{43}$$

Therefore, by definition, it follows that **D** is the natural logarithm of **Λ**.

Let **A** be a diagonalizable matrix of Eqs. (1). We now state that

$$\mathbf{B}\equiv\ln \mathbf{A}=\mathbf{P}(\ln \mathbf{\Lambda})\mathbf{P}^{-1}=\mathbf{PDP}^{-1}. \tag{44}$$

To prove this, again, we must show that $\exp(\mathbf{B})=\mathbf{A}$, We have

$$\begin{aligned} e^{\mathbf{B}} &= \mathbf{I}+\mathbf{B}+\frac{\mathbf{B}^2}{2\,!}+\ldots+\frac{\mathbf{B}^k}{k\,!}+\ldots \\ &= \mathbf{I}+\mathbf{PDP}^{-1}+\frac{(\mathbf{PDP}^{-1})^2}{2\,!}+\ldots+\frac{(\mathbf{PDP}^{-1})^k}{k\,!}+\ldots \\ &= \mathbf{P}\left[\mathbf{I}+\mathbf{D}+\frac{\mathbf{D}^2}{2\,!}+\ldots+\frac{\mathbf{D}^k}{k\,!}+\ldots\right]\mathbf{P}^{-1} \\ &= \mathbf{P}e^{\mathbf{D}}\mathbf{P}^{-1}=\mathbf{P\Lambda P}^{-1}=\mathbf{A}. \end{aligned} \tag{45}$$

EXAMPLE 5: Find the logarithm of the matrix

$$\mathbf{A}=\begin{bmatrix} 39 & -50 & -20 \\ 15 & -16 & -10 \\ 30 & -50 & -11 \end{bmatrix}. \tag{46}$$

Solution: For the given matrix **A**, the matrices **P**, **Λ** and $\mathbf{P}^{-1}$ are found to be

$$\mathbf{P}=\begin{bmatrix} 3 & 1 & 2 \\ 1 & 1 & 1 \\ 2 & -1 & 2 \end{bmatrix}, \quad \mathbf{\Lambda}=\begin{bmatrix} 9 & 0 & 0 \\ 0 & 9 & 0 \\ 0 & 0 & -6 \end{bmatrix},$$

$$\mathbf{P}^{-1}=\tfrac{1}{3}\begin{bmatrix} 3 & -4 & -1 \\ 0 & 2 & -1 \\ -3 & 5 & 2 \end{bmatrix}. \tag{47}$$

We have

$$\ln \mathbf{\Lambda}=\begin{bmatrix} \ln 9 & 0 & 0 \\ 0 & \ln 9 & 0 \\ 0 & 0 & \ln 6+i\pi \end{bmatrix}. \tag{48}$$

Therefore ln **A** is given by

$$\ln \mathbf{A}=\mathbf{P}(\ln \mathbf{\Lambda})\mathbf{P}^{-1}=\tfrac{1}{3}\begin{bmatrix} 9a-6b & 10(b-a) & 4(b-a) \\ 3(a-b) & 5b-2a & 2(b-a) \\ 6(a-b) & 10(b-a) & 4b-a \end{bmatrix}, \tag{49a}$$

where

$$a=\ln 9,\ b=\ln(-6)=\ln 6+i\pi. \tag{49b}$$

Hyperbolic and trigonometric functions

The exponential function also leads to the hyperbolic and the trigonometric functions of a matrix. Thus, for any square matrix **A**, we define the hyperbolic functions by

$$\sinh \mathbf{A}=\tfrac{1}{2}(e^{\mathbf{A}}-e^{-\mathbf{A}})=\sum_{k=0}^{\infty} \mathbf{A}^{2k+1}/(2k+1)!, \tag{50a}$$

$$\cosh \mathbf{A}=\tfrac{1}{2}(e^{\mathbf{A}}+e^{-\mathbf{A}})=\sum_{k=0}^{\infty} \mathbf{A}^{2k}/(2k)!, \tag{50b}$$

which exist for any matrix **A** with finite elements. Similarly, the trigonometric functions are defined by

$$\sin \mathbf{A}=\frac{1}{2i}(e^{i\mathbf{A}}-e^{-i\mathbf{A}})=\sum_{k=0}^{\infty} (-1)^k\mathbf{A}^{2k+1}/(2k+1)!, \tag{51a}$$

$$\cos \mathbf{A}=\tfrac{1}{2}(e^{i\mathbf{A}}+e^{-i\mathbf{A}})=\sum_{k=0}^{\infty} (-1)^k\mathbf{A}^{2k}/(2k)!, \tag{51b}$$

which also exist for any matrix **A** with finite elements.

EXAMPLE 6 : Find $\sin \pi\mathbf{A}$ and $\cos \pi\mathbf{A}$, where

$$\mathbf{A}=\begin{bmatrix} -47/2 & 53 & 30 \\ -12 & 53/2 & 15 \\ 2 & -7/2 & -2 \end{bmatrix}. \tag{52}$$

Solution : The matrices **P**, $\mathbf{\Lambda}$ and $\mathbf{P}^{-1}$ associated with the given matrix **A** are found to be

$$\mathbf{P}=\begin{bmatrix} 5 & 8 & 1 \\ 0 & 2 & 1 \\ 4 & 3 & -1 \end{bmatrix}, \quad \mathbf{\Lambda}=\begin{bmatrix} \tfrac{1}{2} & 0 & 0 \\ 0 & 1 & 0 \\ 0 & 0 & -\tfrac{1}{2} \end{bmatrix},$$

$$\mathbf{P}^{-1}=\begin{bmatrix} 5 & -11 & -6 \\ -4 & 9 & 5 \\ 8 & -17 & -10 \end{bmatrix}. \tag{53}$$

We therefore have

$$\sin \pi\mathbf{\Lambda}=\sum_{k=0}^{\infty} (-1)^k(\pi\mathbf{\Lambda})^{2k+1}/(2k+1)!$$

$$=\begin{bmatrix} \sin\frac{\pi}{2} & 0 & 0 \\ 0 & \sin\pi & 0 \\ 0 & 0 & -\sin\frac{\pi}{2} \end{bmatrix}=\begin{bmatrix} 1 & 0 & 0 \\ 0 & 0 & 0 \\ 0 & 0 & -1 \end{bmatrix}. \tag{54}$$

Therefore

$$\sin \pi\mathbf{A}=\mathbf{P} \sin (\pi\Lambda)\, \mathbf{P}^{-1}=\begin{bmatrix} 17 & -38 & -20 \\ -8 & 17 & 10 \\ 28 & -61 & -34 \end{bmatrix}. \tag{55}$$

Similarly,

$$\cos \pi\Lambda=\begin{bmatrix} 0 & 0 & 0 \\ 0 & -1 & 0 \\ 0 & 0 & 0 \end{bmatrix}, \tag{56}$$

so that

$$\cos \pi\mathbf{A}=\mathbf{P} \cos (\pi\Lambda)\, \mathbf{P}^{-1}=\begin{bmatrix} 32 & -72 & -40 \\ 8 & -18 & -10 \\ 12 & -27 & -15 \end{bmatrix}. \tag{57}$$

Evaluation of functions using Cayley-Hamilton theorem

The method discussed so far for evaluating the various functions of a matrix is based on the principle that if $\mathbf{A}=\mathbf{P}\Lambda\mathbf{P}^{-1}$, where Λ is diagonal, then $f(\mathbf{A})=\mathbf{P}f(\Lambda)\mathbf{P}^{-1}$. It has, however, the major drawback that it is applicable to diagonalizable matrices only. There is an alternative method[1] for evaluating the functions of a matrix which is based on the use of the Cayley-Hamilton theorem and which is applicable to any matrix.

In Section 10, we have shown that any polynomial of whatever degree in a matrix is equal to a polynomial of degree $\leqslant m-1$, where m is the degree of the minimal polynomial. The result, in fact, holds good not only for polynomials but also for any arbitrary function of a matrix provided the function is sufficiently differentiable. Thus if the degree of the minimal polynomial of a matrix $\mathbf{A}$ is m, any function $f(\mathbf{A})$ can be expressed as a linear combination of the m linearly independent matrices $\mathbf{I}, \mathbf{A}, \mathbf{A}^2, \ldots, \mathbf{A}^{m-1}$, i.e.,

$$f(\mathbf{A})=r(\mathbf{A}), \tag{58}$$

where

$$r(\mathbf{A})=\alpha_{m-1}\mathbf{A}^{m-1}+\alpha_{m-2}\mathbf{A}^{m-2}+\ldots+\alpha_1\mathbf{A}+\alpha_0\mathbf{I}. \tag{59}$$

The scalars α_i are determined as follows. If λ_i is a k-fold degenerate eigenvalue of $\mathbf{A}$, the algebraic functions $f(\lambda)$ and $r(\lambda)$ satisfy the k equations

[1]The remaining part of this section, except Examples 12(a) and 13, is not for the reader who has skipped Section 10.

$$f(\lambda_i)=r(\lambda_i),$$
$$\frac{df(\lambda_i)}{d\lambda}=\frac{dr(\lambda_i)}{d\lambda},$$
$$\frac{d^2 f(\lambda_i)}{d\lambda^2}=\frac{d^2 r(\lambda_i)}{d\lambda^2}, \tag{60}$$
$$\vdots$$
$$\frac{d^{k-1} f(\lambda_i)}{d\lambda^{k-1}}=\frac{d^{k-1} r(\lambda_i)}{d\lambda^{k-1}}.$$

Here the notation $d^l f(\lambda_i)/d\lambda^l$ denotes the l-th derivative of $f(\lambda)$ evaluated at $\lambda=\lambda_i$. We shall now use this method to solve some of the problems discussed earlier in this section and also apply it to nondiagonalizable matrices.

EXAMPLE 7: Find $\mathbf{A}^p$, where p is any number and $\mathbf{A}$ is the matrix of Eq. (8).

Solution: The matrix $\mathbf{A}$ is of order 2 and has distinct eigenvalues $\lambda_1=1$, $\lambda_2=2$. The degree of the minimal polynomial is therefore $m=2$. We have $f(\mathbf{A})=\mathbf{A}^p$, so that $f(\lambda)=\lambda^p$. Let $r(\mathbf{A})=\alpha_1\mathbf{A}+\alpha_0\mathbf{I}$, so that $r(\lambda)=\alpha_1\lambda+\alpha_0$. Since both the eigenvalues are nondegenerate, we have the two conditions

$$f(\lambda_1)=r(\lambda_1),\ f(\lambda_2)=r(\lambda_2), \tag{61}$$

which give

$$1=\alpha_1+\alpha_0,\ 2^p=2\alpha_1+\alpha_0. \tag{62}$$

The solution is found to be

$$\alpha_1=2^p-1,\quad \alpha_0=2-2^p. \tag{63}$$

Putting these back in $r(\mathbf{A})$, we have

$$f(\mathbf{A})=r(\mathbf{A})=\alpha_1\mathbf{A}+\alpha_0\mathbf{I},$$

or

$$\mathbf{A}^p=(2^p-1)\begin{bmatrix} 4/3 & 2/3 \\ 2/3 & 5/3 \end{bmatrix}+(2-2^p)\begin{bmatrix} 1 & 0 \\ 0 & 1 \end{bmatrix}$$
$$=\tfrac{1}{3}\begin{bmatrix} 2^p+2 & (2^p-1)\sqrt{2} \\ (2^p-1)\sqrt{2} & 2^{p+1}+1 \end{bmatrix}, \tag{64}$$

which agrees with Eq. (11). *If p is nonintegral, this will not be the only matrix equal to* $\mathbf{A}^p$.

EXAMPLE 8: Find all the square roots of the matrix of Eq. (18).

Solution: The matrix $\mathbf{A}$ is of order 2 with distinct eigen values $\lambda_1=2$, $\lambda_2=1$; hence $m=2$. We have $f(\mathbf{A})=\mathbf{A}^{1/2}$, $f(\lambda)=\lambda^{1/2}$; let

$r(\mathbf{A})=\alpha_1\mathbf{A}+\alpha_0\mathbf{I}$, so that $r(\lambda)=\alpha_1\lambda+\alpha_0$. The coefficients α_1 and α_0 satisfy the conditions

$$f(\lambda_1)=r(\lambda_1),\ f(\lambda_2)=r(\lambda_2), \tag{65}$$

or

$$\pm\sqrt{2}=2\alpha_1+\alpha_0,\ \pm 1=\alpha_1+\alpha_0. \tag{66}$$

These give the four sets of solutions

$$\alpha_1=\pm(\sqrt{2}-1),\ \alpha_0=\pm(2-\sqrt{2}); \tag{67a}$$

$$\alpha_1=\pm(\sqrt{2}+1),\ \alpha_0=\mp(2+\sqrt{2}). \tag{67b}$$

Using these in the equation $\mathbf{A}^{1/2}=\alpha_1\mathbf{A}+\alpha_0\mathbf{I}$, we get the four square roots $\pm\mathbf{B}$ and $\pm\mathbf{C}$, where $\mathbf{B}$ and $\mathbf{C}$ are given in Eqs. (23).

EXAMPLE 9: Determine $4^{\mathbf{A}}$ and ln $\mathbf{A}$, where $\mathbf{A}$ is the matrix of Eq. (18).

Solution: We have $\lambda_1=2$, $\lambda_2=1$, $m=2$.

(a) Here $f(\mathbf{A})=4^{\mathbf{A}}$, so that $f(\lambda)=4^{\lambda}$. Let $r(\mathbf{A})=\alpha_1\mathbf{A}+\alpha_0\mathbf{I}$, so that $r(\lambda)=\alpha_1\lambda+\alpha_0$. We have the conditions

$$f(1)=r(1),\ f(2)=r(2) \Rightarrow 4=\alpha_1+\alpha_0,\ 16=2\alpha_1+\alpha_0.$$

The solution is $\alpha_1=12$, $\alpha_0=-8$. Hence

$$4^{\mathbf{A}}=12\begin{bmatrix} 3/2 & 1/2 \\ 1/2 & 3/2 \end{bmatrix}-8\begin{bmatrix} 1 & 0 \\ 0 & 1 \end{bmatrix}=\begin{bmatrix} 10 & 6 \\ 6 & 10 \end{bmatrix}, \tag{68}$$

which agrees with Eq. (40b).

(b) Here $f(\mathbf{A})=\ln \mathbf{A}$, $f(\lambda)=\ln\lambda$. Let $r(\mathbf{A})=\alpha_1\mathbf{A}+\alpha_0\mathbf{I}$, so that $r(\lambda)=\alpha_1\lambda+\alpha_0$. The conditions on α_1 and α_0 are

$$\ln 1\equiv 0=\alpha_1+\alpha_0,\ \ln 2=2\alpha_1+\alpha_0 \Rightarrow \alpha_1=\ln 2,\ \alpha_0=-\ln 2.$$

Hence

$$\ln \mathbf{A}=\ln 2\begin{bmatrix} 3/2 & 1/2 \\ 1/2 & 3/2 \end{bmatrix}-\ln 2\begin{bmatrix} 1 & 0 \\ 0 & 1 \end{bmatrix}$$

$$=(\ln\sqrt{2})\begin{bmatrix} 1 & 1 \\ 1 & 1 \end{bmatrix}. \tag{69}$$

EXAMPLE 10: Find $e^{3\mathbf{A}}$, $e^{-\mathbf{A}}$ and cosh $(\mathbf{A}t)$, where $\mathbf{A}$ is the matrix of Eq. (10.19).

Solution: The given matrix has eigenvalues $\lambda_1=1$ with multiplicity 1 and $\lambda_2=2$ with multiplicity 2. It would be convenient to find the function $f(\mathbf{A})=\exp(\mathbf{A}t)$, where t is a parameter. We therefore take $f(\lambda)=\exp(\lambda t)$. Also, let

$$r(\mathbf{A})=\alpha_2\mathbf{A}^2+\alpha_1\mathbf{A}+\alpha_0\mathbf{I}, \tag{70a}$$

$$r(\lambda)=\alpha_2\lambda^2+\alpha_1\lambda+\alpha_0. \tag{70b}$$

The coefficients α_0, α_1 and α_2 are determined by the conditions

$$f(1)=r(1),\ f(2)=r(2),\ f'(\lambda)\Big|_{\lambda=2}=r'(\lambda)\Big|_{\lambda=2}. \tag{71}$$

These give

$$e^t=\alpha_2+\alpha_1+\alpha_0,$$
$$e^{2t}=4\alpha_2+2\alpha_1+\alpha_0,$$
$$te^{2t}=4\alpha_2+\alpha_1,$$

which have the solution

$$\alpha_2=(t-1)e^{2t}+e^t,\ \alpha_1=(4-3t)e^{2t}-4e^t,\ \alpha_0=(2t-3)e^{2t}+4e^t. \tag{72}$$

The matrix exp $(\mathbf{A}t)$ is therefore given by

$$e^{\mathbf{A}t}=\alpha_2\begin{bmatrix}9 & 3 & -4\\ 8 & 4 & -4\\ 10 & 6 & -4\end{bmatrix}+\alpha_1\begin{bmatrix}3 & 1 & -1\\ 2 & 2 & -1\\ 2 & 2 & 0\end{bmatrix}+\alpha_0\begin{bmatrix}1 & 0 & 0\\ 0 & 1 & 0\\ 0 & 0 & 1\end{bmatrix}$$

$$=\begin{bmatrix}2te^{2t}+e^t & e^{2t}-e^t & -te^{2t}\\ 2te^{2t} & e^{2t} & -te^{2t}\\ (4t-2)e^{2t}+2e^t & 2e^{2t}-2e^t & (1-2t)e^{2t}\end{bmatrix}. \tag{73}$$

From this, we have

$$e^{3\mathbf{A}}=\begin{bmatrix}6e^6+e^3 & e^6-e^3 & -3e^6\\ 6e^6 & e^6 & -3e^6\\ 10e^6+2e^3 & 2e^6-2e^3 & -5e^6\end{bmatrix}. \tag{74a}$$

$$e^{-\mathbf{A}}=\begin{bmatrix}-2e^{-2}+e^{-1} & e^{-2}-e^{-1} & e^{-2}\\ -2e^{-2} & e^{-2} & e^{-2}\\ -6e^{-2}+2e^{-1} & 2e^{-2}-2e^{-1} & 3e^{-2}\end{bmatrix}. \tag{74b}$$

The function cosh $\mathbf{A}t$ can be found out either directly by taking $f(\lambda)=\cosh(\lambda t)$ and following the same procedure, or by finding $e^{-\mathbf{A}t}$ [replacing t by $-t$ in Eq. (73)] and using $\cosh \mathbf{A}t=(e^{\mathbf{A}t}+e^{-\mathbf{A}t})/2$. The result is

$$\cosh \mathbf{A}t=\begin{bmatrix}2t\sinh 2t+\cosh t & \cosh 2t-\cosh t & -t\sinh 2t\\ 2t\sinh 2t & \cosh 2t & -t\sinh 2t\\ 2t\sinh 2t+2\cosh t-2\cosh 2t & 2\cosh 2t-2\cosh t & \cosh 2t-2t\sinh 2t\end{bmatrix}. \tag{75}$$

EXAMPLE 11: Find ln $\mathbf{A}$, where

$$\mathbf{A}=\begin{bmatrix} 2 & -2 & 0 \\ 0 & 2 & 0 \\ 1 & -3 & 2 \end{bmatrix}. \tag{76}$$

Solution: The given matrix has only one distinct eigenvalue, $\lambda=2$, with multiplicity 3. We have

$$f(\mathbf{A})=\ln \mathbf{A},\ f(\lambda)=\ln \lambda;$$

$$r(\mathbf{A})=\alpha_2\mathbf{A}^2+\alpha_1\mathbf{A}+\alpha_0\mathbf{I},\ r(\lambda)=\alpha_2\lambda^2+\alpha_1\lambda+\alpha_0. \tag{77}$$

The coefficients are determined by the conditions

$$f(2)=r(2),\ f'(2)=r'(2),\ f''(2)=r''(2), \tag{78}$$

which give the system of equations

$$\ln 2=4\alpha_2+2\alpha_1+\alpha_0,\ \tfrac{1}{2}=4\alpha_2+\alpha_1,\ -\tfrac{1}{4}=2\alpha_2, \tag{79}$$

having the solution

$$\alpha_0=\ln 2-3/2,\ \alpha_1=1,\ \alpha_2=-1/8. \tag{80}$$

Therefore

$$\ln \mathbf{A}=-\tfrac{1}{8}\begin{bmatrix} 4 & -8 & 0 \\ 0 & 4 & 0 \\ 4 & -14 & 4 \end{bmatrix}+\begin{bmatrix} 2 & -2 & 0 \\ 0 & 2 & 9 \\ 1 & -3 & 2 \end{bmatrix}$$
$$+(\ln 2-3/2)\begin{bmatrix} 1 & 0 & 0 \\ 0 & 1 & 0 \\ 0 & 0 & 1 \end{bmatrix}$$
$$=\begin{bmatrix} \ln 2 & -1 & 0 \\ 0 & \ln 2 & 0 \\ 1/2 & -5/4 & \ln 2 \end{bmatrix}. \tag{81}$$

EXAMPLE 12: For any square matrix **A**, show that

$$\det(\exp \mathbf{A})=\exp(\mathrm{Tr}\,\mathbf{A}). \tag{82}$$

Solution: We shall prove this result by two methods. The first is applicable to diagonalizable matrices only while the second is general.

(a) If **A** is a diagonalizable matrix, let

$$\mathbf{P}^{-1}\mathbf{A}\mathbf{P}=\mathbf{\Lambda},\ \mathbf{A}=\mathbf{P}\mathbf{\Lambda}\mathbf{P}^{-1}, \tag{83}$$

where $\mathbf{\Lambda}$ is a diagonal matrix containing the eigenvalues λ_i of **A**. From the second of Eqs. (83), it follows that

$$\exp \mathbf{A}=\mathbf{P}(\exp \mathbf{\Lambda})\,\mathbf{P}^{-1}. \tag{84}$$

Therefore,

$$\det(\exp \mathbf{A})=(\det \mathbf{P})\det(\exp \mathbf{\Lambda})\det(\mathbf{P}^{-1})=\det(\exp \mathbf{\Lambda}). \tag{85}$$

Now, exp Λ is a diagonal matrix with diagonal elements exp λ_i. Hence it is evident that

$$\det(\exp \mathbf{A})=(\exp \lambda_1)(\exp \lambda_2)\ldots(\exp \lambda_n)$$
$$=\exp(\lambda_1+\lambda_2+\ldots+\lambda_n)=\exp(\text{Tr }\mathbf{A}), \tag{86}$$

by using Eq. (9.10a). The result is thus proved.

(b) Now we shall give a proof of the theorem without assuming that $\mathbf{A}$ is diagonalizable. Any square matrix can be brought to a triangular form by a similarity transformation. Let, therefore,

$$\mathbf{P}^{-1}\mathbf{A}\mathbf{P}=\mathbf{T},\ \mathbf{A}=\mathbf{P}\mathbf{T}\mathbf{P}^{-1}, \tag{87}$$

where $\mathbf{T}$ is a triangular matrix whose diagonal elements are $(\mathbf{T})_{ii}=\lambda_i$, which are the eigenvalues of $\mathbf{A}$. From Eq. (87), we have

$$\exp \mathbf{A}=\sum_{k=0}^{\infty}\mathbf{A}^k/k!=\mathbf{P}\left[\sum_{k=0}^{\infty}\mathbf{T}^k/k!\right]\mathbf{P}^{-1}=\mathbf{P}(\exp \mathbf{T})\,\mathbf{P}^{-1}. \tag{88}$$

Since $\mathbf{T}$ is triangular, the diagonal elements of the k-th power of $\mathbf{T}$ are λ_i^k, where k is a positive integer. Therefore, it is evident from the exponential series that the diagonal elements of exp $\mathbf{T}$ are exp λ_i. Since the determinant of a triangular matrix equals the product of its diagonal elements, we have

$$\det(\exp \mathbf{T})=\exp(\lambda_1+\lambda_2+\ldots+\lambda_n)=\exp(\text{Tr }\mathbf{A}). \tag{89}$$

From Eq. (88), we have

$$\det(\exp \mathbf{A})=\det(\exp \mathbf{T}), \tag{90}$$

which, together with eq. (89), proves the result.

EXAMPLE 13: Let $\mathbf{A}$ and $\mathbf{B}$ be two square matrices of the same order and let $\mathbf{C}=[\mathbf{A},\mathbf{B}]$. If $[\mathbf{C},\mathbf{A}]=[\mathbf{C},\mathbf{B}]=\mathbf{0}$, prove that

$$e^{\mathbf{A}}e^{\mathbf{B}}=e^{\mathbf{A}+\mathbf{B}+\frac{1}{2}\mathbf{C}}. \tag{91}$$

Solution: Let λ be a parameter and let

$$f(\lambda)=e^{\lambda\mathbf{A}}\,e^{\lambda\mathbf{B}}\,e^{-\lambda(\mathbf{A}+\mathbf{B})}. \tag{92}$$

Differentiating with respect to λ, we have

$$\frac{df}{d\lambda}=\left[e^{\lambda\mathbf{A}}\,\mathbf{A}e^{\lambda\mathbf{B}}+e^{\lambda\mathbf{A}}\,e^{\lambda\mathbf{B}}\,\mathbf{B}-e^{\lambda\mathbf{A}}\,e^{\lambda\mathbf{B}}\,(\mathbf{A}+\mathbf{B})\right]\,e^{-\lambda(\mathbf{A}+\mathbf{B})}. \tag{93}$$

By virtue of the fact that $[[\mathbf{A},\mathbf{B}],\mathbf{A}]=\mathbf{0}$, the result of Exercise 12 reduces in this particular case to

$$e^{\lambda\mathbf{A}}\,\mathbf{B}e^{-\lambda\mathbf{A}}=\mathbf{B}+\lambda\,[\mathbf{A},\mathbf{B}]. \tag{94}$$

Multiplying from the left by $e^{-\mathbf{A}\lambda}$, we have

$$\mathbf{B}\,e^{-\lambda\mathbf{A}}=e^{-\lambda\mathbf{A}}\,\mathbf{B}+\lambda\,[\mathbf{A},\mathbf{B}]\,e^{-\lambda\mathbf{A}}. \tag{95}$$

Interchanging $\mathbf{A}$ and $\mathbf{B}$ and replacing λ by $-\lambda$ in Eq. (95), we get

$$\mathbf{A}e^{\lambda\mathbf{B}}=e^{\lambda\mathbf{B}}\ \mathbf{A}+\lambda\ [\mathbf{A},\mathbf{B}]\ e^{\lambda\mathbf{B}}=e^{\lambda\mathbf{B}}\ \mathbf{A}+\lambda\mathbf{C}\ e^{\lambda\mathbf{B}}. \tag{96}$$

Substituting this in the first term on the right hand side of Eq. (93), we find

$$\frac{df}{d\lambda}=\left[e^{\lambda\mathbf{A}}e^{\lambda\mathbf{B}}\mathbf{A}+\lambda\mathbf{C}e^{\lambda\mathbf{A}}e^{\lambda\mathbf{B}}+e^{\lambda\mathbf{A}}e^{\lambda\mathbf{B}}\mathbf{B}-e^{\lambda\mathbf{A}}e^{\lambda\mathbf{B}}(\mathbf{A}+\mathbf{B})\right]e^{-\lambda(\mathbf{A}+\mathbf{B})}$$

$$=\lambda\mathbf{C}e^{\lambda\mathbf{A}}e^{\lambda\mathbf{B}}e^{-\lambda(\mathbf{A}+\mathbf{B})}=\lambda\mathbf{C}f(\lambda), \tag{97}$$

or

$$\frac{df}{f}=\mathbf{C}\lambda d\lambda. \tag{98}$$

Noting that **A** and **B** are independent of λ, we can integrate Eq. (98), which, on using the initial value $f(0)=\mathbf{I}$, gives

$$f(\lambda)=\exp(\mathbf{C}\lambda^2/2), \tag{99a}$$

or

$$e^{\lambda\mathbf{A}}e^{\lambda\mathbf{B}}e^{-\lambda(\mathbf{A}+\mathbf{B})}=e^{\mathbf{C}\lambda^2/2}. \tag{99b}$$

Since **C** commutes with **A** and **B**, $e^{\mathbf{C}}$ commutes with $e^{\mathbf{A}}$, $e^{\mathbf{B}}$ and $e^{(\mathbf{A}+\mathbf{B})}$. Multiplying Eq. (99b) from the right by $e^{\lambda(\mathbf{A}+\mathbf{B})}$, we have

$$e^{\lambda\mathbf{A}}e^{\lambda\mathbf{B}}=e^{\lambda(\mathbf{A}+\mathbf{B})}e^{\mathbf{C}\lambda^2/2}=e^{\lambda(\mathbf{A}+\mathbf{B})+\mathbf{C}\lambda^2/2}, \tag{100}$$

because of the commutability mentioned above. Finally, putting $\lambda=1$, we get the desired result of Eq. (91).

This shows that the *law of indices* $e^a e^b=e^{a+b}$, which holds good for numbers, does not hold good in case of noncommuting matrices.

Exercise 12

1. Show that if **A** and **B** commute with each other, then $f(\mathbf{A})$ and $g(\mathbf{B})$, where f and g are arbitrary functions expressible as power series, also commute with each other.

*2. Prove that $e^{\mathbf{A}}$ is nonsingular even if **A** is singular. Also show that the inverse of $e^{\mathbf{A}}$ is $e^{-\mathbf{A}}$.

3. Show that $a^{\mathbf{0}}=\mathbf{I}$, where a is any nonzero scalar and **0** is the null matrix.

4. Let **A** be the matrix of Eq. (10.19). Obtain approximations to exp **A** and cosh **A** by retaining terms upto $\mathbf{A}^5$ in their series expansions. Compare these with the exact results obtained in Example 10.

*5. Obtain $\ln(\mathbf{I}+\mathbf{A})$, where **A** is the matrix

$$\mathbf{A}=\begin{bmatrix} 3/8 & 1/8 & 1/8 \\ 0 & 1/3 & 0 \\ 1/8 & 1/24 & 3/8 \end{bmatrix},$$

by any of the methods outlined in this section. Evaluate the expression

$$\mathbf{A}-\frac{\mathbf{A}^2}{2}+\frac{\mathbf{A}^3}{3}-\frac{\mathbf{A}^4}{4}$$

and compare this with $\ln(\mathbf{I}+\mathbf{A})$. [This is a generalization of the logarithmic series

$$\ln(1+z)=\sum_{n=1}^{\infty}(-1)^{n+1}z^n/n, \quad |z|\leqslant 1, \quad z\neq -1,$$

to matrices. You will notice that all the eigenvalues of **A** have magnitude less than unity.]

6. Obtain $\exp(-\mathbf{A}t)$ for the matrix of Eq. (10.19). [See Eq. (73).] Show explicitly that $\exp(-\mathbf{A}t)\exp(\mathbf{A}t)=\mathbf{I}$.

7. Let **B** be the matrix of Eq. (81). Show that $\exp \mathbf{B}=\mathbf{A}$, where **A** is the matrix of Eq. (76), thus providing a check on the result of Example 11.

*8. Find $\mathbf{A}^k, e^{\mathbf{A}}$ and $\ln \mathbf{A}$, where

$$\mathbf{A}=\begin{bmatrix} 1 & 1 & 1 \\ 0 & 1 & 1 \\ 0 & 0 & 1 \end{bmatrix}.$$

Sum the series

$$\sum_{k=0}^{\infty}\mathbf{A}^k/k!$$

exactly and verify that the result agrees with the direct calculation of $e^{\mathbf{A}}$.

9. Write down the matrix $\mathbf{A}^{-k}$, where **A** is the matrix of Eq. (8). Show explicitly that $\mathbf{A}^k\mathbf{A}^{-k}=\mathbf{I}$.

*10. If **A** and **B** are any square matrices of the same order and λ is a parameter, show that

$$e^{\lambda\mathbf{A}}\,\mathbf{B}_e{}^{-\lambda\mathbf{A}}=\mathbf{B}+\lambda[\mathbf{A},\mathbf{B}]+\frac{\lambda^2}{2!}[\mathbf{A},[\mathbf{A},\mathbf{B}]]+\frac{\lambda^3}{3!}[\mathbf{A},[\mathbf{A},[\mathbf{A},\mathbf{B}]]]+\ldots.$$

[If **B** is the Hamiltonian, this transformation is known as the renormalization of the Hamiltonian in quantum mechanics.]

13. Kronecker sum and product of matrices

In various problems in quantum mechanics, one often comes across matrices which are in the form of a *Kronecker sum* or a *Kronecker product* (also known as *direct sum* and *direct product*, respectively) of two or more matrices. These two operations are discussed in this section.

Kronecker sum of matrices

The *Kronecker*, or *direct*, *sum of two square matrices* $\mathbf{A}\equiv[a_{ij}]$ of order m and $\mathbf{B}\equiv[b_{ij}]$ of order n is a square matrix $\mathbf{C}$ of order $m+n$ defined by

$$\mathbf{C}=\mathbf{A}\oplus\mathbf{B}\equiv\left[\begin{array}{c|c}\mathbf{A} & \mathbf{0}_1\\ \hline \mathbf{0}_2 & \mathbf{B}\end{array}\right]$$

$$=\left[\begin{array}{ccc|ccc} a_{11} & \cdots & a_{1m} & & & \\ \vdots & & & & \mathbf{0}_1 & \\ a_{m1} & \cdots & a_{mm} & & & \\ \hline & & & b_{11} & \cdots & b_{1n} \\ & \mathbf{0}_2 & & \vdots & & \\ & & & b_{n1} & \cdots & b_{nn}\end{array}\right], \tag{1}$$

where $\mathbf{0}_1$ and $\mathbf{0}_2$ are null matrices of order $m\times n$ and $n\times m$, respectively. Here the symbol $\oplus$ stands for the Kronecker sum. This idea can be easily extended to more than two matrices.

EXAMPLE 1: Find the Kronecker sum of the three matrices

$$\mathbf{A}=a,\ \mathbf{B}=\begin{bmatrix} b & c\\ d & e\end{bmatrix},\ \mathbf{C}=\begin{bmatrix} f & g & h\\ i & j & k\\ x & y & z\end{bmatrix}. \tag{2}$$

Solution: The direct sum of the given matrices is a matrix of order $1+2+3=6$, given by

$$\mathbf{D}=\mathbf{A}\oplus\mathbf{B}\oplus\mathbf{C}=\left[\begin{array}{c|cc|ccc} a & 0 & 0 & 0 & 0 & 0\\ \hline 0 & b & c & 0 & 0 & 0\\ 0 & d & e & 0 & 0 & 0\\ \hline 0 & 0 & 0 & f & g & h\\ 0 & 0 & 0 & i & j & k\\ 0 & 0 & 0 & x & y & z\end{array}\right]. \tag{3}$$

Such a matrix, which has nonvanishing elements in square blocks along the main diagonal and zeros elsewhere, is said to be in the *block diagonalized form.* It is easy to see that if a matrix $\mathbf{M}$ is the direct sum of a number of matrices such as

$$\mathbf{M}=\mathbf{A}_1\oplus\mathbf{A}_2\oplus\ldots\oplus\mathbf{A}_k, \tag{4}$$

then the following equations are satisfied:

$$\det \mathbf{M}=(\det \mathbf{A}_1)\ (\det \mathbf{A}_2)\ldots(\det \mathbf{A}_k), \tag{5a}$$

$$\mathrm{Tr}\ \mathbf{M}=\mathrm{Tr}\ \mathbf{A}_1+\mathrm{Tr}\ \mathbf{A}_2+\ldots+\mathrm{Tr}\ \mathbf{A}_k, \tag{5b}$$

$$\mathbf{M}^{-1}=\mathbf{A}_1^{-1}\oplus\mathbf{A}_2^{-1}\oplus\ldots\oplus\mathbf{A}_k^{-1}. \tag{5c}$$

Thus, (a) *the determinant of a direct-sum matrix is the product of the determinants of the constituent matrices,* (b) *the trace of a direct-sum matrix is the sum of the traces of the constituent matrices,* and (c) *the inverse of a direct-sum matrix is the direct sum of the inverses of the constituent matrices, assuming that the matrix is nonsingular*

EXAMPLE 2: If $\mathbf{A}_1$ and $\mathbf{A}_2$ are two square matrices of the same order, say m, and $\mathbf{B}_1$ and $\mathbf{B}_2$ are two square matrices of the same order, say n, then prove that

$$(\mathbf{A}_1\oplus\mathbf{B}_1)(\mathbf{A}_2\oplus\mathbf{B}_2)=(\mathbf{A}_1\mathbf{A}_2)\oplus(\mathbf{B}_1\mathbf{B}_2). \tag{6}$$

Solution: To prove this, let us write the left hand side of Eq. (6) by using the definition of Eq. (1) as

$$(\mathbf{A}_1\oplus\mathbf{B}_1)(\mathbf{A}_2\oplus\mathbf{B}_2)=\begin{bmatrix}\mathbf{A}_1 & \mathbf{0}\\ \mathbf{0} & \mathbf{B}_1\end{bmatrix}\begin{bmatrix}\mathbf{A}_2 & \mathbf{0}\\ \mathbf{0} & \mathbf{B}_2\end{bmatrix}. \tag{7}$$

Notice that the two partitioned matrices on the right hand side have dimensions suitable for multiplication by the methods of Section 6. Using Eq. (6.5) in the right hand side of the above equation, we have

$$(\mathbf{A}_1\oplus\mathbf{B}_1)(\mathbf{A}_2\oplus\mathbf{B}_2)=\begin{bmatrix}\mathbf{A}_1\mathbf{A}_2 & \mathbf{0}\\ \mathbf{0} & \mathbf{B}_1\mathbf{B}_2\end{bmatrix}=(\mathbf{A}_1\mathbf{A}_2)\oplus(\mathbf{B}_1\mathbf{B}_2), \tag{8}$$

which proves the desired result.

Kronecker product of matrices

The *Kronecker*, or *direct*, *product* (also known as *outer product*) of two matrices $\mathbf{A}\equiv[a_{lm}]$ of order $L\times M$ and $\mathbf{B}\equiv[b_{pq}]$ of order $P\times Q$ is a matrix of order $I\times J$, where $I=LP$ and $J=MQ$, given by

$$\mathbf{C}=\mathbf{A}\otimes\mathbf{B}\equiv\begin{bmatrix}a_{11}\mathbf{B} & a_{12}\mathbf{B} & \ldots & a_{1M}\mathbf{B}\\ a_{21}\mathbf{B} & a_{22}\mathbf{B} & \ldots & a_{2M}\mathbf{B}\\ \vdots & & & \\ a_{L1}\mathbf{B} & a_{L2}\mathbf{B} & \ldots & a_{LM}\mathbf{B}\end{bmatrix}, \tag{9}$$

where an 'element' $a_{lm}\mathbf{B}$ stands for a block matrix of order $P\times Q$ given by

$$a_{lm}\mathbf{B}=\begin{bmatrix} a_{lm}b_{11} & a_{lm}b_{12} & \cdots & a_{lm}b_{1Q} \\ a_{lm}b_{21} & a_{lm}b_{22} & \cdots & a_{lm}b_{2Q} \\ \cdot & & & \\ \cdot & & & \\ \cdot & & & \\ a_{lm}b_{P1} & a_{lm}b_{P2} & \cdots & a_{lm}b_{PQ} \end{bmatrix} \tag{10}$$

To obtain an element of $\mathbf{C}$ in terms of the elements of $\mathbf{A}$ and $\mathbf{B}$, we use the notation $\mathbf{C}\equiv[c_{lp,mq}]$, where a row of $\mathbf{C}$ is denoted by a dual symbol (lp) and a column of $\mathbf{C}$ by a dual symbol (mq), such that

$$c_{lp,mq}=a_{lm}b_{pq}. \tag{11}$$

The rows and the columns of $\mathbf{C}$ may now be relabeled by two new indices i and j $(1\leqslant i\leqslant I,\ 1\leqslant j\leqslant J)$ so that

$$\mathbf{C}\equiv[c_{ij}]\equiv[c_{lp,mq}]. \tag{12}$$

Note that multiplication of a matrix by a scalar as defined in Eq. (2.36) is a special case of the direct product. When one of the two matrices is of order 1×1, the direct product reduces to multiplication by a scalar. The rather complicated notation developed above may be made clear by an example.

EXAMPLE 3: Obtain the direct product of the two matrices

$$\mathbf{A}=\begin{matrix}(1)\\(2)\\(3)\end{matrix}\overset{\begin{matrix}(1) & (2)\end{matrix}}{\begin{bmatrix} a & b \\ c & d \\ e & f \end{bmatrix}},\quad \mathbf{B}=\begin{matrix}(1)\\(2)\end{matrix}\overset{\begin{matrix}(1) & (2) & (3)\end{matrix}}{\begin{bmatrix} r & s & t \\ x & y & z \end{bmatrix}}, \tag{13}$$

where the numbering of the rows and the columns of the two matrices has been explicitly shown.

Solution: The matrix $\mathbf{A}$ is of order 3×2 and $\mathbf{B}$ of order 2×3. According to the definition of Eq. (9), the direct product of these two matrices is a matrix of order 6×6 given by

$$\mathbf{C}=\mathbf{A}\otimes\mathbf{B}=\begin{bmatrix} a\begin{bmatrix} r & s & t \\ x & y & z \end{bmatrix} & b\begin{bmatrix} r & s & t \\ x & y & z \end{bmatrix} \\ c\begin{bmatrix} r & s & t \\ x & y & z \end{bmatrix} & d\begin{bmatrix} r & s & t \\ x & y & z \end{bmatrix} \\ e\begin{bmatrix} r & s & t \\ x & y & z \end{bmatrix} & f\begin{bmatrix} r & s & t \\ x & y & z \end{bmatrix} \end{bmatrix}$$

$$
\begin{array}{c} \\ =(11) \\ (12) \\ (21) \\ (22) \\ (31) \\ (32) \end{array}
\begin{array}{c} \begin{array}{cccccc} (11) & (12) & (13) & (21) & (22) & (23) \end{array} \\
\left[\begin{array}{cccccc}
ar & as & at & br & bs & bt \\
ax & ay & az & bx & by & bz \\
cr & cs & ct & dr & ds & dt \\
cx & cy & cz & dx & dy & dz \\
er & es & et & fr & fs & ft \\
ex & ey & ez & fx & fy & fz
\end{array}\right]. \end{array} \tag{14}
$$

Then an element of **C** is, for example,

$$(\mathbf{C})_{31,\,22}=fs=(\mathbf{A})_{32}(\mathbf{B})_{12}, \tag{15}$$

which is consistent with Eq. (12). It is important to note that *the rows and the columns of the direct-product matrix are labeled by different schemes* in the dual symbol notation. For example, the third row of the matrix **C** above is labeled as the (21)-row, whereas the third column is labeled as the (13)-column. The elements of the (21)-row of **C** are obtained by taking the product of each element of the second row of **A** with every element of the first row of **B**. Similarly, the elements of the (13)-column of **C** are obtained by taking the product of each element of the first column of **A** with every element of the third column of **B**.

The rows and the columns of the direct-product matrix may be relabeled according to the usual familiar scheme by identifying each dual symbol with one index. This can be done by choosing

$$i=(l-1)P+p,\; j=(m-1)Q+q. \tag{16}$$

Then

$$(\mathbf{C})_{lp,mq}\equiv(\mathbf{C})_{ij}=(\mathbf{C})_{(l-1)P+p,(m-1)Q+q}. \tag{17}$$

For example, referring to the element of Eq. (15), we have $l=3$, $p=1$, $m=q=2$, so that $i=5, j=5$. It is evident that fs is indeed the (5, 5)-element of the direct-product matrix **C** of Eq. (14).

The concept can be easily extended to the direct product of any number of matrices. There is no restriction on the order of the matrices whose direct product is to be taken. In the case of more than two matrices, *the operation of direct product is associative*, that is,

$$(\mathbf{A}\otimes\mathbf{B})\otimes\mathbf{C}=\mathbf{A}\otimes(\mathbf{B}\otimes\mathbf{C})\equiv\mathbf{A}\otimes\mathbf{B}\otimes\mathbf{C}. \tag{18}$$

The operation is also distributive with respect to matrix addition. Thus,

$$\mathbf{A}\otimes(\mathbf{B}+\mathbf{C})=\mathbf{A}\otimes\mathbf{B}+\mathbf{A}\otimes\mathbf{C}. \tag{19}$$

EXAMPLE 4: If **A**, **B**, **C** and **D** are any matrices whose dimensions are such that the ordinary matrix products **AC** and **BD** are defined, then show that

$$(\mathbf{A}\otimes\mathbf{B})(\mathbf{C}\otimes\mathbf{D})=(\mathbf{AC})\otimes(\mathbf{BD}). \tag{20}$$

Solution: Let the matrices **A**, **B**, **C** and **D** be given by

$$\mathbf{A}\equiv[a_{ij}]_{L\times M},\ \mathbf{B}\equiv[b_{ij}]_{P\times Q},\ \mathbf{C}\equiv[c_{ij}]_{M\times N},\ \mathbf{D}\equiv[d_{ij}]_{Q\times R}, \tag{21}$$

where the orders of the matrices are chosen in such a way that the ordinary matrix products **AC** and **BD** are defined ; otherwise they are arbitrary. The direct products $\mathbf{A}\otimes\mathbf{B}$ and $\mathbf{C}\otimes\mathbf{D}$ can be expressed in the partitioned form as

$$\mathbf{A}\otimes\mathbf{B}=\begin{bmatrix} a_{11}\mathbf{B} & a_{12}\mathbf{B}\ldots a_{1M}\mathbf{B} \\ \vdots & \\ a_{i_1}\mathbf{B} & a_{i_2}\mathbf{B}\ldots a_{iM}\mathbf{B} \\ \vdots & \\ a_{L_1}\mathbf{B} & a_{L_2}\mathbf{B}\ldots a_{LM}\mathbf{B} \end{bmatrix},$$

$$\mathbf{C}\otimes\mathbf{D}=\begin{bmatrix} c_{1_1}\mathbf{D}\ldots c_{1_j}\mathbf{D}\ldots c_{1N}\mathbf{D} \\ c_{21}\mathbf{D}\ldots c_{2_j}\mathbf{D}\ldots c_{2N}\mathbf{D} \\ \vdots \\ c_{M_1}\mathbf{D}\ldots c_{M_j}\mathbf{D}\ldots c_{MN}\mathbf{D} \end{bmatrix}. \tag{22}$$

The partitions of the matrices $\mathbf{A}\otimes\mathbf{B}$ and $\mathbf{C}\otimes\mathbf{D}$ are seen to be such that their ordinary matrix product can be taken. The *ij*-th element of the product $(\mathbf{A}\otimes\mathbf{B})(\mathbf{C}\otimes\mathbf{D})$ is then seen to be

$$[(\mathbf{A}\otimes\mathbf{B})(\mathbf{C}\otimes\mathbf{D})]_{ij}=a_{i_1}c_{1_j}\mathbf{BD}+a_{i_2}c_{2_j}\mathbf{BD}+\ldots+a_{iM}c_{M_j}\mathbf{BD}$$
$$=(\mathbf{AC})_{ij}\mathbf{BD}. \tag{23}$$

The complete matrix can therefore be written as

$$(\mathbf{A}\otimes\mathbf{B})(\mathbf{C}\otimes\mathbf{D})=\begin{bmatrix} (\mathbf{AC})_{11}\mathbf{BD} & (\mathbf{AC})_{12}\mathbf{BD}\ldots(\mathbf{AC})_{1N}\mathbf{BD} \\ (\mathbf{AC})_{21}\mathbf{BD} & (\mathbf{AC})_{22}\mathbf{BD}\ldots(\mathbf{AC})_{2N}\mathbf{BD} \\ \vdots & \\ (\mathbf{AC})_{L_1}\mathbf{BD} & (\mathbf{AC})_{L_2}\mathbf{BD}\ldots(\mathbf{AC})_{LN}\mathbf{BD} \end{bmatrix}$$
$$=(\mathbf{AC})\otimes(\mathbf{BD}), \tag{24}$$

by the definition of Eq. (9) for the direct product. The matrix $(\mathbf{AC})\otimes(\mathbf{BD})$ is seen to be of order $LP\times NR$.

EXAMPLE 5: Let the notation $\mathbf{A}^{[k]}$ denote the direct product of the matrix **A** with itself k times, that is,

$$\mathbf{A}^{[k]}=\mathbf{A}\otimes\mathbf{A}\otimes\mathbf{A}\otimes\ldots\otimes\mathbf{A}\ (k \text{ times}), \tag{25}$$

which will be a matrix of order $m^k\times n^k$ if **A** is of order $m\times n$. Let **A** and **B** be two matrices whose ordinary matrix product can be taken. Then show that

$$(\mathbf{AB})^{[k]}=(\mathbf{A})^{[k]}(\mathbf{B})^{[k]}. \tag{26}$$

Solution: Replacing $\mathbf{B}$ by $\mathbf{A}$ and $\mathbf{C}$ and $\mathbf{D}$ by $\mathbf{B}$ in Eq. (20), we have

$$(\mathbf{A}\otimes\mathbf{A})(\mathbf{B}\otimes\mathbf{B})=(\mathbf{AB})\otimes(\mathbf{AB}),$$

or

$$\mathbf{A}^{[2]}\mathbf{B}^{[2]}=(\mathbf{AB})^{[2]}. \tag{27}$$

Consider now the triple direct product

$$\begin{aligned}(\mathbf{AB})^{[3]}&=(\mathbf{AB})\otimes(\mathbf{AB})\otimes(\mathbf{AB})\\&=[(\mathbf{A}\otimes\mathbf{A})(\mathbf{B}\otimes\mathbf{B})]\otimes(\mathbf{AB})\\&=(\mathbf{A}\otimes\mathbf{A}\otimes\mathbf{A})(\mathbf{B}\otimes\mathbf{B}\otimes\mathbf{B})=\mathbf{A}^{[3]}\mathbf{B}^{[3]}.\end{aligned} \tag{28}$$

Thus we see that Eq. (26) holds good for $k=2$ and $k=3$. Let us assume that it holds good for the index $k-1$, that is,

$$(\mathbf{AB})^{[k-1]}=(\mathbf{A})^{[k-1]}(\mathbf{B})^{[k-1]}. \tag{29}$$

Taking the direct product of both sides of Eq. (29) with $\mathbf{AB}$ and using Eq. (20) on the right hand side, we find

$$(\mathbf{AB})^{[k-1]}\otimes(\mathbf{AB})=[(\mathbf{A})^{[k-1]}(\mathbf{B})^{[k-1]}]\otimes(\mathbf{AB}),$$

or

$$(\mathbf{AB})^{[k]}=[(\mathbf{A})^{[k-1]}\otimes\mathbf{A}]\,[(\mathbf{B})^{[k-1]}\otimes\mathbf{B}]=(\mathbf{A})^{[k]}(\mathbf{B})^{[k]},$$

which shows that the result holds good for the index k if it holds good for the index $k-1$. Since we have already shown that it holds for $k=2$, it follows that Eq. (26) is true for every positive integral value of k.

An important result concerning the direct product of matrices from the point of view of the eigenvalue problem is discussed in the following example.

EXAMPLE 6: Let $\mathbf{A}$ be a square matrix of order M with eigenvalues $\lambda_1, \lambda_2, \ldots, \lambda_M$ and the corresponding eigenvectors $\mathbf{x}_1, \mathbf{x}_2, \ldots, \mathbf{x}_M$. Let $\mathbf{B}$ be a square matrix of order N with eigenvalues $\mu_1, \mu_2, \ldots, \mu_N$ and the corresponding eigenvectors $\mathbf{y}_1, \mathbf{y}_2, \ldots, \mathbf{y}_N$. Then show that the direct-product matrix $\mathbf{A}\otimes\mathbf{B}$ has the MN eigenvalues $\lambda_i\mu_j$ with the corresponding eigenvectors $\mathbf{x}_i\otimes\mathbf{y}_j$, $1\leqslant i\leqslant M$, $1\leqslant j\leqslant N$.

Solution: It is given that

$$\begin{aligned}\mathbf{A}\mathbf{x}_i&=\lambda_i\mathbf{x}_i,\ 1\leqslant i\leqslant M;\\ \mathbf{B}\mathbf{y}_j&=\mu_j\mathbf{y}_j,\ 1\leqslant j\leqslant N.\end{aligned} \tag{30}$$

Taking the direct product of the respective sides of the above two equations, we get

$$(\mathbf{A}\mathbf{x}_i)\otimes(\mathbf{B}\mathbf{y}_j)=\lambda_i\mu_j(\mathbf{x}_i\otimes\mathbf{y}_j), \tag{31}$$

where λ_i and μ_j can be taken out of the direct product because they are scalars.

Using Eq. (20), we have

$$(\mathbf{A}\mathbf{x}_i)\otimes(\mathbf{B}\mathbf{y}_j)=(\mathbf{A}\otimes\mathbf{B})(\mathbf{x}_i\otimes\mathbf{y}_j). \tag{32}$$

This, together with Eq. (31), gives

$$(\mathbf{A}\otimes\mathbf{B})(\mathbf{x}_i\otimes\mathbf{y}_j)=\lambda_i\mu_j\,(\mathbf{x}_i\otimes\mathbf{y}_j), \tag{33}$$

showing that $\mathbf{x}_i\otimes\mathbf{y}_j$ is an eigenvector of $\mathbf{A}\otimes\mathbf{B}$ with the eigenvalue $\lambda_i\mu_j$ for $1\leqslant i\leqslant M$, $1\leqslant j\leqslant N$.

Exercise 13

1. Obtain the direct sum of the following matrices:

(a) $\begin{bmatrix} 2 & -3 \\ 5 & 1 \end{bmatrix}$, $\begin{bmatrix} 9 & 3 & 1 \\ 0 & -5 & 4 \\ 2 & 4 & 6 \end{bmatrix}$ and $\begin{bmatrix} 1 & 2 \\ -3 & 4 \end{bmatrix}$;

*(b) $\begin{bmatrix} a^2 & ab \\ bc & c^2 \end{bmatrix}$ and $\begin{bmatrix} c-a & b^2 \\ b^2 & a-c \end{bmatrix}$;

(c) $\begin{bmatrix} 2 & 3 & 1 \\ -1 & 0 & 5 \\ 4 & 9 & 6 \end{bmatrix}$ and $\begin{bmatrix} 1 & 2 & 3 & -4 \\ 5 & 6 & -7 & 8 \\ 9 & -10 & 11 & 12 \\ -13 & 14 & 15 & 16 \end{bmatrix}$.

2. Prove Eqs. (5).

3. Let

$$\mathbf{A}_1=\begin{bmatrix} 1 & 3 \\ 2 & 4 \end{bmatrix}, \qquad \mathbf{A}_2=\begin{bmatrix} -2 & 0 \\ 5 & 2 \end{bmatrix},$$

$$\mathbf{B}_1=\begin{bmatrix} 1 & 5 & -6 \\ 2 & 0 & -1 \\ 3 & 0 & -4 \end{bmatrix}, \qquad \mathbf{B}_2=\begin{bmatrix} 4 & -2 & 0 \\ 0 & 1 & 5 \\ -3 & 0 & 2 \end{bmatrix}.$$

Verify that these matrices satisfy Eq. (6).

4. Obtain the direct products of the following matrices :

(a) $\begin{bmatrix} 2 & 3 & 1 \\ -1 & 0 & 4 \end{bmatrix}$ with $\begin{bmatrix} 1 & -1 & 1 \\ 2 & 0 & 2 \\ 2 & -1 & 0 \end{bmatrix}$;

(b) $\begin{bmatrix} 1 & -1 & 1 \\ 2 & 0 & 2 \\ 2 & -1 & 0 \end{bmatrix}$ with $\begin{bmatrix} 2 & 3 & 1 \\ -1 & 0 & 4 \end{bmatrix}$;

*(c) $\begin{bmatrix} a^2 & ab \\ bc & c^2 \end{bmatrix}$ with $\begin{bmatrix} b^2 & bd \\ cd & c^2 \end{bmatrix}$.

5. Obtain the eigenvalues and the eigenvectors of the matrix

$$\mathbf{A}=\begin{bmatrix} 0 & 3/2 & 0 & 1/2 \\ 3/2 & 0 & 1/2 & 0 \\ 0 & 1/2 & 0 & 3/2 \\ 1/2 & 0 & 3/2 & 0 \end{bmatrix}.$$

Then show that **A** is the direct product of

$$\mathbf{B}=\begin{bmatrix} 3/2 & 1/2 \\ 1/2 & 3/2 \end{bmatrix} \text{ and } \mathbf{C}=\begin{bmatrix} 0 & 1 \\ 1 & 0 \end{bmatrix}.$$

Find the eigenvalues and the eigenvectors of **B** and **C** and verify the result of Example 6 for these matrices.

6. For the matrices **B** and **C** of Exercise 5 above, obtain the matrix $\mathbf{C}\otimes\mathbf{B}$. Compare it with the matrix $\mathbf{A}=\mathbf{B}\otimes\mathbf{C}$.

14. Matrices in classical and quantum mechanics

Finally in this section, we shall discuss some applications of matrix algebra in physics. It is hardly possible to enumerate the vast number of occasions where matrices arise in physics and therefore only a few typical cases are considered.

The rotation matrix

Let x, y, z be a cartesian coordinate system in a three-dimensional space and let $\mathbf{u}=\{u_1\ u_2\}$ be a vector in the xy-plane. Consider a rotation of the coordinate system about the z-axis through an angle θ in the anticlockwise sense and let us denote the new coordinate axes by x', y', z. If the same vector $\mathbf{u}$ has components u_1', u_2' relative to the new system, it is evident that

$$\begin{aligned} u_1' &= u_1 \cos\theta + u_2 \sin\theta, \\ u_2' &= -u_1 \sin\theta + u_2 \cos\theta. \end{aligned} \tag{1}$$

This could be written in the matrix notation as

$$\begin{bmatrix} u_1' \\ u_2' \end{bmatrix} = \begin{bmatrix} \cos\theta & \sin\theta \\ -\sin\theta & \cos\theta \end{bmatrix} \begin{bmatrix} u_1 \\ u_2 \end{bmatrix}. \tag{2}$$

If we denote the rotation operator by $R_z(\theta)$, we could write the transformation of the vector $\mathbf{u}$ to $\mathbf{u}'=\{u_1'\ u_2'\}$ symbolically by

$$\mathbf{u} \to \mathbf{u}' = R_z(\theta)\,\mathbf{u}. \tag{3}$$

Comparing Eqs. (2) and (3), we see that the effect of the operator $R_z(\theta)$ is to multiply the column vector $\{u_1\ u_2\}$ by the 2×2 matrix appearing on the right hand side of Eq. (2). We say that this matrix *represents* the operator $R_z(\theta)$ in the two-dimensional subspace (x, y). It is often expressed by giving an equality sign and writing

$$R_z(\theta) = \begin{bmatrix} \cos\theta & \sin\theta \\ -\sin\theta & \cos\theta \end{bmatrix}, \tag{4}$$

which is an orthogonal matrix.

It should be realized that the representation of an operator depends on the choice of the coordinate system. Let us now consider a vector $\mathbf{u}=\{u_1\quad u_2\quad u_3\}$ and suppose the same rotation $R_z(\theta)$ is applied on the system so that the vector changes to $\mathbf{u}'=\{u_1'\ u_2'\ u_3'\}$. Since the rotation is about the z axis, the z-component of the vector will remain unchanged. The new components are given in terms of the old components by the equations

$$\begin{aligned} u_1' &= u_1 \cos\theta + u_2 \sin\theta, \\ u_2' &= -u_1 \sin\theta + u_2 \cos\theta, \\ u_3' &= u_3. \end{aligned} \tag{5}$$

This shows that the operator $R_z(\theta)$ is represented by the matrix

$$R_z(\theta) = \begin{bmatrix} \cos\theta & \sin\theta & 0 \\ -\sin\theta & \cos\theta & 0 \\ 0 & 0 & 1 \end{bmatrix} \tag{6}$$

in the three-dimensional space (x, y, z).

Let us now consider a rotation about any arbitrary axis passing through the origin of the coordinate system. It is known that any general rotation can be obtained by three successive rotations, each performed about one of the coordinate axes. In the Euler's method, a general rotation is the result of the following three successive rotations: (1) a rotation through α about the z axis, followed by (2) a rotation through β about the new y axis, followed by (3) a rotation through γ about the new z axis. The angles α, β, γ are called the *Euler angles* and the rotation may be denoted by the operator $R(\alpha, \beta, \gamma)$. If $\mathbf{u} = \{u_1 \; u_2 \; u_3\}$ is the original vector and $\mathbf{u}' = \{u_1' \; u_2' \; u_3'\}$ the transformed vector, then $\mathbf{u}' = R(\alpha, \beta, \gamma)\mathbf{u}$. The matrix representing $R(\alpha, \beta, \gamma)$ can be found as follows.

According to the above definition, the operator $R_z(\alpha)$ acts on $\mathbf{u}$ first, giving rise to the vector $R_z(\alpha)\mathbf{u}$, where $R_z(\alpha)$ has a matrix similar to that in Eq. (6). On this vector acts $R_y(\beta)$, giving the vector $R_y(\beta)R_z(\alpha)\mathbf{u}$. Since $R_y(\beta)$ is a rotation about the y axis, its matrix representation can be obtained in a similar way as above and is found to be

$$R_y(\beta) = \begin{bmatrix} \cos\beta & 0 & \sin\beta \\ 0 & 1 & 0 \\ -\sin\beta & 0 & \cos\beta \end{bmatrix}. \tag{7}$$

Finally, $R_z(\gamma)$ acts on the new vector giving the final vector

$$\mathbf{u}' = R(\alpha, \beta, \gamma)\,\mathbf{u} = R_z(\gamma)\, R_y(\beta)\, R_z(\alpha)\mathbf{u}; \tag{8}$$

$$\begin{bmatrix} u_1' \\ u_2' \\ u_3' \end{bmatrix} = \begin{bmatrix} \cos\gamma & \sin\gamma & 0 \\ -\sin\gamma & \cos\gamma & 0 \\ 0 & 0 & 1 \end{bmatrix} \begin{bmatrix} \cos\beta & 0 & \sin\beta \\ 0 & 1 & 0 \\ -\sin\beta & 0 & \cos\beta \end{bmatrix} \times \begin{bmatrix} \cos\alpha & \sin\alpha & 0 \\ -\sin\alpha & \cos\alpha & 0 \\ 0 & 0 & 1 \end{bmatrix} \begin{bmatrix} u_1 \\ u_2 \\ u_3 \end{bmatrix}. \tag{9}$$

Working out the matrix product, we find[1]

$$R(\alpha, \beta, \gamma) = \begin{bmatrix} \cos\gamma\cos\alpha - \sin\gamma\cos\beta\sin\alpha & \cos\gamma\sin\alpha + \sin\gamma\cos\beta\cos\alpha & \sin\gamma\sin\beta \\ -\sin\gamma\cos\alpha - \cos\gamma\cos\beta\sin\alpha & -\sin\gamma\sin\alpha + \cos\gamma\cos\beta\cos\alpha & \cos\gamma\sin\beta \\ \sin\beta\sin\alpha & -\sin\beta\cos\alpha & \cos\beta \end{bmatrix}. \tag{10}$$

This is known as the *three-dimensional rotation matrix.*

Pauli spin matrices

In the problem of the intrinsic spin angular momentum of an electron in quantum mechanics, we come across the spin angular momentum operator **s** which can be expressed as

$$s_x = \tfrac{1}{2}\hbar\,\sigma_x,\; s_y = \tfrac{1}{2}\hbar\,\sigma_y,\; s_z = \tfrac{1}{2}\hbar\sigma_z, \tag{11}$$

where $\hbar = h/2\pi$, h being the Planck's constant, and σ_x, σ_y. σ_z are certain entities which are to be determined. s_x, s_y, s_z and σ_x, σ_y, σ_z are hermitian. It is known that they satisfy the relations

$$\sigma_x^2 = \sigma_y^2 = \sigma_z^2 = 1; \tag{12a}$$

$$\sigma_x\sigma_y + \sigma_y\sigma_x = 0,\; \sigma_y\sigma_z + \sigma_z\sigma_y = 0,\; \sigma_z\sigma_x + \sigma_x\sigma_z = 0. \tag{12b}$$

We wish to solve these equations for σ_x, σ_y and σ_z. Eqs. (12b) show that the σ's anticommute with each other. They are therefore not ordinary numbers. It is known that matrices may anticommute with each other. Hence let us try 2×2 matrices for the σ's. Then the right hand side of Eqs. (12a) is of course to be understood as the 2×2 unit matrix **I**.

Since σ_x, σ_y, σ_z are hermitian, it is possible to choose a representation in which some of them, though not all, are diagonal. In fact not more than one of them can be diagonalized at a time because they do not commute with each other, while diagonal matrices must commute with each other. Let us choose σ_z to be represented by a diagonal matrix.

Now since σ_z is a diagonal matrix and $\sigma_z^2 = \mathbf{I}$, it follows that all the diagonal elements of σ_z are $+1$ or -1. Not both the elements can have the same sign, for then σ_z would be a constant matrix and

[1]See also H. Goldstein, *Classical Mechanics* (Addison-Wesley, Reading, Mass., 1969), Section 4.4; A. Messiah, *Quantum Mechanics*, Vol. II (North-Holland, Amsterdam, 1965 Section C.10.

would not anticommute with σ_x and σ_y; in fact, a constant matrix commutes with every other matrix. Therefore σ_z can be either of the form $\begin{bmatrix} 1 & 0 \\ 0 & -1 \end{bmatrix}$ or $\begin{bmatrix} -1 & 0 \\ 0 & 1 \end{bmatrix}$. Choosing one or the other of these matrices is immaterial because as we have seen in Section 9, the order of the diagonal elements can be changed at will by a suitable rearrangement of the basis vectors [see discussion on page 96]. Hence, choosing the first of these matrices, we finally have

$$\sigma_z = \begin{bmatrix} 1 & 0 \\ 0 & -1 \end{bmatrix}. \tag{13}$$

Now we wish to obtain σ_x and σ_y. Let $\begin{bmatrix} a & b \\ c & d \end{bmatrix}$ be a matrix which anticommutes with σ_z, so that

$$\begin{bmatrix} 1 & 0 \\ 0 & -1 \end{bmatrix}\begin{bmatrix} a & b \\ c & d \end{bmatrix} + \begin{bmatrix} a & b \\ c & d \end{bmatrix}\begin{bmatrix} 1 & 0 \\ 0 & -1 \end{bmatrix} = \begin{bmatrix} 2a & 0 \\ 0 & -2d \end{bmatrix} \equiv \begin{bmatrix} 0 & 0 \\ 0 & 0 \end{bmatrix}, \tag{14}$$

giving $a=d=0$. The matrix which anticommutes with σ_z therefore reduces to the form $\begin{bmatrix} 0 & b \\ c & 0 \end{bmatrix}$. If, further, its square is to be equal to the unit matrix, we must have

$$\begin{bmatrix} 0 & b \\ c & 0 \end{bmatrix}\begin{bmatrix} 0 & b \\ c & 0 \end{bmatrix} = \begin{bmatrix} bc & 0 \\ 0 & bc \end{bmatrix} \equiv \begin{bmatrix} 1 & 0 \\ 0 & 1 \end{bmatrix} \Rightarrow bc = 1. \tag{15}$$

σ_x and σ_y must therefore be of the form

$$\sigma_x = \begin{bmatrix} 0 & p \\ q & 0 \end{bmatrix}, \quad \sigma_y = \begin{bmatrix} 0 & r \\ s & 0 \end{bmatrix} \tag{16}$$

with

$$pq = 1, \; rs = 1. \tag{17}$$

But σ_x and σ_y must also anticommute with each other:

$$\sigma_x\sigma_y + \sigma_y\sigma_x = 0 \Rightarrow ps + qr = 0. \tag{18}$$

Thus we have three equations to determine p, q, r, s. Choosing $p=1$ arbitrarily, Eqs. (17) and (18) give $q=1$, $r=-s$ and $r^2=s^2=-1$. Hence we can take $s=i$, $r=-i$. We thus have the *Pauli spin matrices*, in a representation in which σ_z is diagonal, given by

$$\sigma_x = \begin{bmatrix} 0 & 1 \\ 1 & 0 \end{bmatrix}, \quad \sigma_y = \begin{bmatrix} 0 & -i \\ i & 0 \end{bmatrix}, \quad \sigma_z = \begin{bmatrix} 1 & 0 \\ 0 & -1 \end{bmatrix}. \tag{19}$$

Dirac matrices

In the Dirac's relativistic formulation of the wave equation of a free particle, the Hamiltonian is written as

$$H=c(\alpha_x p_x+\alpha_y p_y+\alpha_z p_z)+\beta mc^2, \tag{20}$$

where c is the velocity of light, (p_x, p_y, p_z) the linear momentum vector of the particle of mass m, and $\alpha_x, \alpha_y, \alpha_z, \beta$ are quantities to be determined. It is found that these entities satisfy the relations

$$\alpha_x^2=\alpha_y^2=\alpha_z^2=\beta^2=1; \tag{21a}$$

$$\{\alpha_x, \alpha_y\}=\{\alpha_y, \alpha_z\}=\{\alpha_z, \alpha_x\}=\{\alpha_x, \beta\}=\{\alpha_y, \beta\}=\{\alpha_z, \beta\}=0. \tag{21b}$$

Eqs. (21b) show that each of the four entities $\alpha_x, \alpha_y, \alpha_z, \beta$ anticommutes with the other three. Therefore again they are not ordinary numbers but must be matrices. Exercise 1 further shows that four mutually anticommuting matrices of order 2 cannot be found; hence α_i and β must be matrices of order greater than 2.

Also, as before, not more than one of the four matrices can be made diagonal at a time. Let us choose β to be the diagonal matrix whose diagonal elements must be ± 1, but not all of the same sign. Suppose β in a partitioned form has the structure

$$\beta=\begin{bmatrix} \mathbf{I}_m & \mathbf{0} \\ \mathbf{0} & -\mathbf{I}_n \end{bmatrix}, \tag{22}$$

where $m+n>2$ as just shown above. Let α_i, identically partitioned as β, be given by

$$\alpha_i=\begin{bmatrix} \mathbf{A} & \mathbf{B} \\ \mathbf{C} & \mathbf{D} \end{bmatrix}. \tag{23}$$

The anticommutation of α_i with β then shows that $\mathbf{A}=\mathbf{D}=\mathbf{0}$, while the relation $\alpha_i^2=\mathbf{I}$ shows that

$$\mathbf{BC}=\mathbf{I},\ \mathbf{CB}=\mathbf{I}. \tag{24}$$

Note that $\mathbf{B}$ is of order $m\times n$ and $\mathbf{C}$ of order $n\times m$. But the discussion of Appendix II shows that $\mathbf{B}$ and $\mathbf{C}$ must be square in order to satisfy Eqs. (24). Hence $n=m$, and we choose $m=2$ as the value giving matrices of the smallest possible order satisfying the desired properties. The matrix β then becomes

$$\beta=\begin{bmatrix} \mathbf{I}_2 & \mathbf{0} \\ \mathbf{0} & -\mathbf{I}_2 \end{bmatrix}=\begin{bmatrix} 1 & 0 & 0 & 0 \\ 0 & 1 & 0 & 0 \\ 0 & 0 & -1 & 0 \\ 0 & 0 & 0 & -1 \end{bmatrix}. \tag{25}$$

Our experience with the Pauli spin matrices then suggests that we may choose $\mathbf{B}=\mathbf{C}=\sigma_x$ in α_x and similarly construct α_y and α_z to give

$$\alpha_x = \begin{bmatrix} \mathbf{0} & \sigma_x \\ \sigma_x & \mathbf{0} \end{bmatrix} = \begin{bmatrix} 0 & 0 & 0 & 1 \\ 0 & 0 & 1 & 0 \\ 0 & 1 & 0 & 0 \\ 1 & 0 & 0 & 0 \end{bmatrix}; \tag{26a}$$

$$\alpha_y = \begin{bmatrix} \mathbf{0} & \sigma_y \\ \sigma_y & \mathbf{0} \end{bmatrix} = \begin{bmatrix} 0 & 0 & 0 & -i \\ 0 & 0 & i & 0 \\ 0 & -i & 0 & 0 \\ i & 0 & 0 & 0 \end{bmatrix}; \tag{26b}$$

$$\alpha_z = \begin{bmatrix} \mathbf{0} & \sigma_z \\ \sigma_z & \mathbf{0} \end{bmatrix} = \begin{bmatrix} 0 & 0 & 1 & 0 \\ 0 & 0 & 0 & -1 \\ 1 & 0 & 0 & 0 \\ 0 & -1 & 0 & 0 \end{bmatrix}. \tag{26c}$$

These four are known as *Dirac matrices*.

Infinite matrices

If the number of rows or the number of columns of a matrix is infinite, the matrix is said to be of infinite order, or an *infinite matrix*.[2] The number of rows or columns of an infinite matrix may be denumerably infinite or nondenumerably infinite. Further the number of rows may be denumerably infinite and the number of columns nondenumerably infinite, or vice versa. In the remainder of this section, we shall use the letters i, j, k to denote integers $(1 \leqslant i, j, k < \infty)$, and x, y, z to denote all real numbers.

For example, a matrix $\mathbf{A} \equiv [a_{ij}]$ with $1 \leqslant i, j < \infty$ will be an infinite matrix for which the number of rows and the number of columns are both denumerably infinite. A matrix $\mathbf{A} \equiv [a_{ix}]$, where $1 \leqslant i < \infty$ and x varies either over a finite interval $[c, d]$ or over an infinite interval, is an infinite matrix for which the number of rows is denumerably infinite and the number of columns is nondenumerably infinite. Finally, $\mathbf{A} \equiv [a_{xy}]$, where x and y vary over the same or different, finite or infinite intervals, is an infinite matrix for which the number of rows and the number of columns are both nondenumerably infinite.

If $\mathbf{r}$ stands for a triplet of components x, y, z, one could have an infinite matrix with elements $a_{i\mathbf{r}}$ or $a_{\mathbf{r}\mathbf{r}'}$, where $\mathbf{r}'$ denotes the triplet x', y', z', etc. One comes across such matrices almost at every step in quantum mechanics.

[2]An infinite matrix does not mean a matrix whose elements are infinite. If any element of a matrix is infinite, we say that the matrix does not exist.

An infinite matrix is said to be square provided its rows and columns are labeled by the same scheme; otherwise it is rectangular, Thus, a matrix with elements a_{ij} $(1 \leqslant i,\ j < \infty)$ is square; a matrix with elements a_{xy} is square provided x and y vary over the same interval; a matrix with elements $a_{\mathbf{rr}'}$ is square provided x and x' vary over the same interval and so do y, y' and z, z'. On the other hand, matrices with elements a_{ix}, or a_{xy} where x and y vary over different intervals, are rectangular.

The addition of two infinite matrices is defined provided they are both of the same order, i.e., their rows are labeled by the same scheme and their columns are labeled by the same scheme. For the product of two infinite matrices to be defined, it is necessary that all the sums or integrals occurring in the elements of the product matrix be finite.

Thus, let $\mathbf{A} \equiv [a_{ij}]$ and $\mathbf{B} \equiv [b_{ij}]$ be discrete infinite matrices. Their product matrix $\mathbf{C} = \mathbf{AB}$ has elements

$$c_{ij} = \sum_{k=1}^{\infty} a_{ik} b_{kj},\ 1 \leqslant i,\ j < \infty, \tag{27}$$

provided that the series on the right hand side of Eq. (27) is finite for every value of i and j. In case the rows or columns are denoted by a continuous variable, the summation must be replaced by an integration. Thus, let $\mathbf{A} \equiv [a_{ix}]$ $(1 \leqslant i < \infty,\ x \in [c, d])$ and $\mathbf{B} \equiv [b_{xy}]$ $(x \in [c, d],\ y \in [p, q])$ be two infinite matrices. Their product is the matrix $\mathbf{C}$ with elements

$$c_{iy} = \int_{c}^{d} a_{ix} b_{xy}\, dx,\ 1 \leqslant i < \infty,\ y \in [p, q], \tag{28}$$

provided the integral in Eq. (28) is finite for every value of i and y on their respective intervals.

While discussing a general infinite matrix, in which case we do not know whether the rows and columns are denumerably or non-denumerably infinite, we shall use Greek indices to label the elements. A Greek index may take discrete values or continuous values or partly discrete and partly continuous values. The notation $\int_{\mu}$ will be used to denote a summation over the discrete values and an integration over the continuous values of μ.

It can be easily shown that an infinite hermitian matrix must be square. For if $\mathbf{A} \equiv [a_{\mu\nu}]$ is a hermitian matrix, it follows that $a_{\mu\nu}^* = a_{\nu\mu} \forall \mu, \nu$. This requires that both μ and ν run over the same set of values, implying that $\mathbf{A}$ is square.

We define a discrete infinite unit matrix $\mathbf{I}$ as having elements

$(\mathbf{I})_{\mu\nu}=\delta_{\mu\nu}$, where μ and ν are discrete variables. On the other hand, a continuous unit matrix is defined as having elements $(\mathbf{I})_{\mu\nu}=\delta(\mu-\nu)$, where μ and ν are continuous variables and $\delta(\mu-\nu)$ is the Dirac's delta function. Thus, *it is not true that the diagonal elements of a continuous unit matrix are equal to unity*, though its off-diagonal elements are all zero.

It is shown in Appendix II that a finite nonsingular matrix must be square. However, an infinite nonsingular matrix may be rectangular.[3]

Matrix representation of an operator

The basic problem in quantum mechanics is to solve the eigenvalue equation

$$P\psi=\lambda\psi, \tag{29}$$

where P is an operator corresponding to some observable of the system under consideration and hence is a hermitian operator, and λ and ψ are respectively the eigenvalue and the eigenvector of P. Eq. (29) may have many solutions and can therefore be written in the form

$$P\psi_\mu=\lambda_\mu\psi_\mu. \tag{30}$$

In fact, for an actual physical system, Eq. (30) has an infinite number of solutions, corresponding to an infinite number of eigenvalues and eigenfunctions of the operator. The variable μ in Eq. (30), therefore, takes an infinite number of values and these may be discrete or continuous or partly discrete and partly continuous. The set of all the linear combinations of the infinite number of eigenfunctions ψ_μ is the Hilbert space on which P acts, and the eigenfunctions ψ_μ serve as a basis in this space, which may be chosen to be orthonormal since P is hermitian.

If ϕ is any (infinite-dimensional) vector in the Hilbert space of P, it may be expressed as a linear combination of the basis functions in the form

$$\phi=\int_\mu a_\mu\psi_\mu. \tag{31}$$

The vector ϕ should be regarded as a column vector.

The crucial point is that the eigenvalues and the eigenfunctions of the operator P are not known beforehand. However, its Hilbert

[3]For more detail, see J. von Neumann, *Mathematical Foundations of Quantum Mechanics* (Princeton University Press, Princeton, N. J., 1955).

space is known. So we choose any suitable set $\{\phi_\mu\}$ of basis functions in the Hilbert space and work out the scalar products

$$P_{\mu\nu}=(\phi_\mu,\ P\phi_\nu). \tag{32}$$

If we arrange these in the form of a matrix, we get the *matrix representation of the operator* or the *operator matrix*

$$\mathbf{P}\equiv[P_{\mu\nu}], \tag{33}$$

which is of infinite order. The element $P_{\mu\nu}$ is known as *the matrix element of the operator between the μ-th and the ν-th basis functions.* If we are able to diagonalize this matrix by means of a suitable unitary transformation, we can get the desired eigenvalues as well as the eigenfunctions.

If the eigenfunctions are themselves chosen as the basis functions, the representation of P turns out to be a diagonal matrix. This follows from the fact that

$$(\psi_\mu,\ P\psi_\nu)=(\psi_\mu,\ \lambda_\nu\psi_\nu)=\lambda_\nu\ (\psi_\mu,\ \psi_\nu)=\begin{cases}\lambda_\mu\delta_{\mu\nu}\\ \lambda_\mu\ \delta(\mu-\nu),\end{cases} \tag{34}$$

where we have used the orthonormality of the eigenfunctions. In other words, *functions of that basis which makes an operator diagonal are the eigenfunctions of that operator.*

Exercise 14

1. Show that there is no 2×2 matrix which anticommutes with all the three Pauli spin matrices.

2. Verify that the Pauli spin matrices of Eqs. (19) satisfy Eqs. (12). Also show that

   (a) $[\sigma_x,\ \sigma_y]=2i\sigma_z,\ [\sigma_y,\ \sigma_z]=2i\sigma_x,\ [\sigma_z,\ \sigma_x]=2i\sigma_y$;

   (b) $\sigma^2\equiv\sigma_x{}^2+\sigma_y{}^2+\sigma_z{}^2=3\mathbf{I}$;

   (c) $[\sigma^2,\ \sigma_i]=0,\ i=x,\ y,\ z$;

   (d) $[\sigma_z,\ \sigma^\pm]=\ \pm2\sigma^\pm$, where $\sigma^\pm=\sigma_x\pm i\sigma_y$;

   (e) $[\sigma^+,\ \sigma^-]=4\sigma_z$.

*3. Find a similarity transformation matrix which reduces σ_x to a diagonal form. In this new representation, find the matrices for $\sigma_x,\ \sigma_y,\ \sigma_z$.

4. Do the same as above and obtain $\sigma_x,\ \sigma_y,\ \sigma_z$ in a representation in which σ_y is diagonal.

5. Show that the matrices $\alpha_x,\ \alpha_y,\ \alpha_z$ and β satisfy Eqs. (27).

Bibliography for chapter 1

ARFKEN, G., *Mathematical Methods for Physicists*, Chapter 4, New York: Academic Press (1968).

BELLMAN, R., *Introduction to Matrix Analysis*, 2nd ed., New York: McGraw-Hill (1970).

BRONSON, R., *Matrix Methods*, New York: Academic Press (1969).

EISELE, J. A., and R. M. MASON, *Applied Matrix and Tensor Analysis*, New York: Wiley (1970).

FINKBEINER, D. T., *Introduction to Matrices and Linear Transformations*, 2nd ed., San Francisco: W. H. Freeman (1966).

GANTMACHER, F. R., *Application of the Theory of Matrices*, New York: Wiley (1959).

GERE, J. M., and W. WEAVER, *Matrix Algebra for Engineers*, New York: Van Nostrand (1965).

HOFFMAN, K., and R. KUNZE, *Linear Algebra*, Englewood Cliffs: Prentice-Hall (1961).

HOHN, F. E., *Elementary Matrix Algebra*, New York: Macmillan (1964).

KAPUR, J. N., and M. K. SINGAL, *Matrices*, New Delhi: R. Chand (1969).

KREYSZIG, E., *Advanced Engineering Mathematics*, 3rd ed., New York: Wiley (1972).

LANCASTER, P., *Theory of Matrices*, New York: Academic Press (1969).

MARCUS, M., and H. MINC, *Elementary Linear Algebra*, New York: Macmillan (1968).

MARGENAU, H., and G. M. MURPHY, *Mathematics for Physics and Chemistry*, New Delhi: Affiliated East-West Press (1966).

MILLS, G., *Introduction to Linear Algebra for Social Scientists*, London: George Allen and Unwin (1969).

PETTOFREZZO, A. J., *Matrices and Transformations*, Englewood Cliffs: Prentice-Hall (1966).

PIPES, L. A., *Applied Mathematics for Engineers and Physicists*, Chapter 4, New York: McGraw-Hill (1958).

RAGHAVARAO, D., *Matrix Theory*, New Delhi: Oxford & IBH (1972).

REINER, I., *Introduction to Matrix Theory and Linear Algebra*, New York: Holt, Rinehart and Winston (1971).

SCHNEIDER, H., and G. P. BARKER, *Matrices and Linear Algebra*, New York: Holt, Rinehart and Winston (1968).

SEARLE, S. R., *Matrix Algebra for the Biological Sciences*, New York: Wiley (1966).

SHANTI NARAYAN, *A Textbook of Matrices*, 5th ed., Delhi: S. Chand (1967).

YAARI, M. E., *Linear Algebra for Social Sciences*, Englewood Cliffs: Prentice-Hall (1971).

II Tensor Analysis

Tensors are a natural and logical generalization of vectors. It is well-known that the use of vectors is absolutely essential in the mathematical study of a large number of physical phenomena. In a similar way, tensor analysis has also found applications in various branches of physics. These applications can be broadly divided into two categories: (a) applications in non-relativistic physics, and (b) applications in the theories of relativity. In this chapter, the general definition of a tensor is given, followed by a discussion of the algebra of tensors, calculus involving tensors, and a number of physical situations where the use of tensors is expedient. Only general tensors are considered.

15. Introduction

In this section, we shall discuss in a very elementary manner a few occasions where the methods of tensor algebra come in handy. Then we shall give the definition of a tensor and the transformation law of tensors.

Occurrence of tensors in physics

We are familiar with elementary physical laws such as that *the acceleration of a body is proportional to the force acting on it* or that *the electric current in a medium is proportional to the applied electric field*, which can be stated mathematically as

$$\mathbf{a}=\mathbf{F}/m, \tag{1a}$$

$$\mathbf{j}=\sigma\mathbf{E}, \tag{1b}$$

where **a** is the acceleration, **F** the force, m the mass of the body, **j** the electric current, **E** the electric field and σ the electrical conductivity. It should be realized, however, that these laws are a special case and apply strictly only to isotropic[1] media or to media possessing high symmetry. In reality, many media are anisotropic, and as a result, the acceleration is not necessarily parallel to the applied force, or the current flows in a direction different from that of the electric field, etc. In such a situation, Eq. (1b), for example, can be written in a generalized form as

$$\begin{aligned} j_x&=\sigma_{xx}E_x+\sigma_{xy}E_y+\sigma_{xz}E_z,\\ j_y&=\sigma_{yx}E_x+\sigma_{yy}E_y+\sigma_{yz}E_z,\\ j_z&=\sigma_{zx}E_x+\sigma_{zy}E_y+\sigma_{zz}E_z, \end{aligned} \tag{2}$$

where j_x, j_y, j_z and E_x, E_y, E_z are respectively the cartesian components of **j** and **E** and σ_{ij} $(i, j=x, y, z)$ are said to be the components of the *conductivity tensor* of the medium. Eq. (1a) can be similarly generalized with $(1/m)_{ij}$ denoting the components of the *mass tensor* (actually, the reciprocal mass tensor) of the particle in the medium. Such a situation arises in the study of the motion of an electron in a crystalline solid, where we speak of the *effective mass tensor* of the electron. Some other tensors of physical interest are discussed in Section 18.

In relativistic physics, again, tensors are widely used in the

[1]A medium whose properties are the same in all directions is said to be isotropic.

special theory and much more so in the general theory. These applications are discussed in Sections 21 and 22.

Notation and conventions

Consider an N-dimensional space and let $x^1, x^2, \ldots, x^N$ be any set of coordinates in this space. It is important to note that in writing x^i, i is merely a superscript on x and does *not* denote the i-th power of x. When occasion arises for using superscripts and powers together, we shall write these as $(x^i)^2$, etc. The N-dimensional space under consideration will be denoted by V_N.

A notation such as $f \equiv f(t)$ will mean that f is a function of t.

Let $\bar{x}^\alpha$ $(1 \leqslant \alpha \leqslant N)$ be another set[2] of coordinates in the same space V_N. Clearly, each of the coordinates x^i will be a function of the N coordinates $\bar{x}^\alpha$, and vice versa. Thus, we can write

$$x^i \equiv x^i(\bar{x}^1, \bar{x}^2, \ldots, \bar{x}^N), \ 1 \leqslant i \leqslant N; \tag{3a}$$

$$\bar{x}^\alpha \equiv \bar{x}^\alpha(x^1, x^2, \ldots, x^N), \ 1 \leqslant \alpha \leqslant N. \tag{3b}$$

EXAMPLE 1: Express the cartesian and the spherical polar coordinates as functions of each other.

Solution: Let x, y, z denote the cartesian coordinates and r, θ, ϕ the spherical polar coordinates in a three-dimensional space. The coordinates x, y, z are independent of each other; similarly, r, θ, ϕ are independent of each other. But the coordinates of one set are functions of those of the other set. The cartesian coordinates are related to the spherical polar coordinates by

$$x = r \sin\theta \cos\phi, \ y = r \sin\theta \sin\phi, \ z = r \cos\theta. \tag{4}$$

The inverse transformation is

$$r = (x^2 + y^2 + z^2)^{\frac{1}{2}}, \ \theta = \tan^{-1}[(x^2 + y^2)^{\frac{1}{2}}/z], \ \phi = \tan^{-1}(y/x). \tag{5}$$

Differentiating Eqs. (3), we find that

$$dx^i = \sum_{\alpha=1}^{N} \frac{\partial x^i}{\partial \bar{x}^\alpha} d\bar{x}^\alpha, \ 1 \leqslant i \leqslant N; \tag{6a}$$

$$d\bar{x}^\alpha = \sum_{i=1}^{N} \frac{\partial \bar{x}^\alpha}{\partial x^i} dx^i, \ 1 \leqslant \alpha \leqslant N. \tag{6b}$$

A summation convention due to Einstein simplifies the notation for the right hand sides of the above equations. According to this

[2]Latin letters will be used to denote the unbarred coordinates, while Greek letters will be used to denote the barred coordinates. This convention will be used throughout this chapter.

convention, *if an index* (except N) *is repeated in a term, summation over it from* 1 *to* N *is implied.* Thus, Eqs. (6) can be simply written as

$$dx^i = \frac{\partial x^i}{\partial \bar{x}^\alpha} d\bar{x}^\alpha, \quad 1 \leqslant i \leqslant N; \tag{7a}$$

$$d\bar{x}^\alpha = \frac{\partial \bar{x}^\alpha}{\partial x^i} dx^i, \quad 1 \leqslant \alpha \leqslant N. \tag{7b}$$

The index α occurs twice on the right hand side of Eq. (7a) while i occurs twice on the right hand side of Eq. (7b). Summation over these from 1 to N is therefore implied in the respective equations. This convention will be used throughout this chapter.

If an index appears only once in any term, it has a definite value—any value between 1 *and* N. Such an index is called a *free index.* Thus, in Eq. (3), i and α are free indices. In Eqs. (6a) and (7a), i is a free index, while in Eqs. (6b) and (7b), α is a free index. Hereafter, the specification such as '$1 \leqslant i \leqslant N$' will be dropped and, for example, Eq. (7a) can be further abbreviated to

$$dx^i = \frac{\partial x^i}{d\bar{x}^\alpha} d\bar{x}^\alpha. \tag{8}$$

Eq. (8), in fact, stands for a set of N equations, one for each value of i between 1 and N.

An index which is repeated and over which summation is implied is called a *dummy index.* Clearly, *a dummy index can be replaced by any other index which does not appear in the same term.* These conventions are clarified by means of the following example.

EXAMPLE 2: Let a_i, b_i, c_i, d_i, $1 \leqslant i \leqslant N$, be four sets of N quantities each. Then, according to the Einstein's summation convention, we have

$$a_i b_i \equiv a_1 b_1 + a_2 b_2 + \ldots + a_N b_N. \tag{9}$$

Similarly,

$$a_i b_j c_j \equiv a_i b_1 c_1 + a_i b_2 c_2 + \ldots + a_i b_N c_N, \tag{10}$$

where i is a free index and has a fixed value between 1 and N. Moreover, it is obvious that

$$a_i b_i = a_j b_j = a_k b_k, \text{ etc.} \tag{11}$$

Also,

$$a_i b_j c_j = a_i b_k c_k = a_i b_l c_l, \text{ etc.} \tag{12}$$

But in Eqs. (12), the dummy index j cannot be replaced by i because i already appears in the same term. Thus,

$$a_i b_j c_j \neq a_i b_i c_i \ . \tag{13}$$

This can be easily verified, because whereas the left hand side of Eq. (13) is explicitly given by Eq. (10), its right hand side is seen to be

$$a_i b_i c_i \equiv a_1 b_1 c_1 + a_2 b_2 c_2 + \ldots + a_N b_N c_N, \tag{14}$$

which is evidently different from $a_i b_j c_j$.

Finally, consider the expressions $a_i b_i c_i d_i$ and $a_i b_i c_j d_j$. These are explicitly given by

$$a_i b_i c_i d_i \equiv a_1 b_1 c_1 d_1 + a_2 b_2 c_2 d_2 + \ldots + a_N b_N c_N d_N, \tag{15a}$$

$$a_i b_i c_j d_j \equiv \Big(\sum_{i=1}^{N} a_i\, b_i\Big)\Big(\sum_{j=1}^{N} c_j d_j\Big)$$
$$=(a_1 b_1 + a_2 b_2 + \ldots + a_N b_N)(c_1 d_1 + c_2 d_2 + \ldots + c_N d_N). \tag{15b}$$

Eq. (15a) thus does not contain the cross terms which appear in Eq. (15b). Evidently, therefore,

$$a_i b_i c_i d_i \neq a_i b_i c_j d_j. \tag{16}$$

On the other hand,

$$a_i b_i c_j d_j = a_i b_i c_k d_k = a_l b_l c_i d_i, \text{ etc.} \tag{17}$$

Since the coordinates x^i are independent of each other, it is evident that

$$\frac{dx^i}{dx^j} = \begin{cases} 1 \text{ if } i=j; \\ 0 \text{ if } i \neq j. \end{cases} \tag{18}$$

We define the Kronecker delta symbol by

$$\delta^i_j = \begin{cases} 1 \text{ if } i=j; \\ 0 \text{ if } i \neq j. \end{cases} \tag{19}$$

Eq. (18) can then be written as

$$\frac{dx^i}{dx^j} = \delta^i_j. \tag{20}$$

Similarly, the coordinates $\bar{x}^\alpha$ are also independent of each other, so that

$$\frac{d\bar{x}^\alpha}{d\bar{x}^\beta} = \delta^\alpha_\beta. \tag{21}$$

If x^i are functions of $\bar{x}^\alpha$, as in Eq. (3a), then by partial differentiation, it follows that

$$\frac{dx^i}{dx^j} = \frac{\partial x^i}{\partial \bar{x}^\alpha} \ \frac{\partial \bar{x}^\alpha}{\partial x^j}, \tag{22}$$

where, as per convention, summation over α is implied. Comparing Eq. (22) with Eq. (20), we have

$$\frac{\partial x^i}{\partial \bar{x}^\alpha}\frac{\partial \bar{x}^\alpha}{\partial x^j}=\delta^i_j. \tag{23a}$$

Similarly,

$$\frac{\partial \bar{x}^\alpha}{\partial x^k}\frac{\partial x^k}{\partial \bar{x}^\beta}=\delta^\alpha_\beta. \tag{23b}$$

Contravariant vector

Tensors are defined by means of their properties of transformation under coordinate transformations. Vectors are a special case of tensors.

Let a physical entity be characterized by the N functions A^i when expressed in the x^i coordinate system. When measured in the coordinate system $\bar{x}^\alpha$, let the same entity be characterized by the components $\bar{A}^\alpha$. Then A^i are said to be the components of a *contravariant vector* if they transform under coordinate transformations according to

$$\bar{A}^\alpha=\frac{\partial \bar{x}^\alpha}{\partial x^j}A^j. \tag{24}$$

This relation can be easily inverted to express A^i in terms of $\bar{A}^\alpha$. Multiplying Eq. (24) by $\partial x^k/\partial \bar{x}^\alpha$ and summing over α, we have

$$\frac{\partial x^k}{\partial \bar{x}^\alpha}\bar{A}^\alpha=\frac{\partial x^k}{\partial \bar{x}^\alpha}\frac{\partial \bar{x}^\alpha}{\partial x^j}A^j. \tag{25}$$

On using Eq. (23a), this becomes

$$\frac{\partial x^k}{\partial \bar{x}^\alpha}\bar{A}^\alpha=\delta^k_j A^j=A^k. \tag{26}$$

Replacing k by i, this becomes

$$A^i=\frac{\partial x^i}{\partial \bar{x}^\alpha}\bar{A}^\alpha \tag{27}$$

Covariant vector

A set of N quantities A_i which are functions of the N coordinates x^i are said to be the components of a *covariant vector* if they transform according to

$$\bar{A}_\alpha=\frac{\partial x^j}{\partial \bar{x}^\alpha}A_j \tag{28}$$

under a change of coordinates from x^i to $\bar{x}^\alpha$, where $\bar{A}_\alpha$ are the components of the vector in the barred coordinate system. The inverse transformation is

$$A_i = \frac{\partial \bar{x}^\alpha}{\partial x^i} \bar{A}_\alpha. \tag{29}$$

EXAMPLE 3: Show that velocity and acceleration are contravariant vectors and that the gradient of a scalar field is a covariant vector.

Solution: (a) If t denotes time, dividing Eq. (7b) by dt, we get

$$\frac{d\bar{x}^\alpha}{dt} = \frac{\partial \bar{x}^\alpha}{\partial x^i} \frac{dx^i}{dt}. \tag{30}$$

If we define the components of velocity in the barred and the unbarred coordinate systems respectively by

$$\bar{v}^\alpha = d\bar{x}^\alpha/dt, \; v^i = dx^i/dt, \tag{31}$$

Eq. (30) becomes

$$\bar{v}^\alpha = \frac{\partial \bar{x}^\alpha}{\partial x^i} v^i. \tag{32}$$

From the definition of Eq. (24), it is evident that *the velocity* v^i *is a contravariant vector*. Taking time derivatives once again, it follows that acceleration is also a contravariant vector,[3] proving the desired result.

(b) Let $\phi \equiv \phi(x^i)$ be a scalar field. Being a scalar field, its functional form remains unaltered under coordinate transformations, so that

$$\phi(x^i) = \bar{\phi}(\bar{x}^\alpha) = \phi(\bar{x}^\alpha). \tag{33}$$

The gradient of the scalar field will be a vector whose components can be defined by

$$A_i = \partial\phi/\partial x^i, \quad \bar{A}_\alpha = \partial\bar{\phi}/\partial\bar{x}^\alpha = \partial\phi/\partial\bar{x}^\alpha. \tag{34}$$

On using partial derivatives, it is evident that we can write

$$\frac{\partial\phi}{\partial x^i} = \frac{\partial\phi}{\partial\bar{x}^\alpha} \frac{\partial\bar{x}^\alpha}{\partial x^i}, \tag{35}$$

or

$$A_i = \frac{\partial \bar{x}^\alpha}{\partial x^i} \bar{A}_\alpha. \tag{36}$$

By the definition of Eq. (29), it is evident that *the gradient of a scalar field is a covariant vector.*

[3]The coefficients $d\bar{x}^\alpha/dx^j$ are independent of time for fixed coordinate systems. There is a subtle distinction to be made here. The coordinates x^j in dx^j/dt are the coordinates of, say, a particle in motion, while the coefficients $d\bar{x}^\alpha/dx^j$ only denote a relation between two coordinate systems, which is independent of time.

Tensors of second rank

A set of N^2 functions A^{ij} are said to be the components of a *contravariant tensor of rank two* if they transform according to

$$\bar{A}^{\alpha\beta}=\frac{\partial\bar{x}^{\alpha}}{\partial x^{i}}\frac{\partial\bar{x}^{\beta}}{\partial x^{j}}A^{ij} \tag{37}$$

under coordinate transformations, where $\bar{A}^{\alpha\beta}$ are the components of the tensor in the barred coordinate system.

A set of N^2 functions A_{ij} are said to be the components of a *covariant tensor of the second rank* if they transform according to

$$\bar{A}_{\alpha\beta}=\frac{\partial x^{i}}{\partial\bar{x}^{\alpha}}\frac{\partial x^{j}}{\partial\bar{x}^{\beta}}A_{ij} \tag{38}$$

under coordinate transformations.

Finally, a set of N^2 functions $A^{i}{}_{j}$ are said to be the components of a *tensor of contravariant rank one and covariant rank one* (or simply a *mixed tensor of the second rank*) if they transform according to

$$\bar{A}^{\alpha}{}_{\beta}=\frac{\partial\bar{x}^{\alpha}}{\partial x^{i}}\frac{\partial x^{j}}{\partial\bar{x}^{\beta}}A^{i}{}_{j} \tag{39}$$

under coordinate transformations.

A tensor of the second rank can be written as a square matrix of order N, just as a tensor of the first rank can be treated as an N-component vector. Thus, a covariant tensor of rank two can be written as

$$A_{ij}=\begin{bmatrix} A_{11} & A_{12} & \cdots & A_{1N} \\ A_{21} & A_{22} & \cdots & A_{2N} \\ \vdots & & & \\ A_{N1} & A_{N2} & \cdots & A_{NN} \end{bmatrix}. \tag{40}$$

General definition

The above definitions can be easily generalized for a tensor of arbitrary rank. Thus, a set of N^{p+q} functions $A^{i_1 i_2 \ldots i_p}_{j_1 j_2 \ldots j_q}$ are said to be the components of a *tensor of contravariant rank p and covariant rank q* (of total rank $p+q$) if they transform under coordinate transformations according to

$$\bar{A}^{\alpha_1\alpha_2\ldots\alpha_p}_{\beta_1\beta_2\ldots\beta_q}=\frac{\partial\bar{x}^{\alpha_1}}{\partial x^{i_1}}\cdots\frac{\partial\bar{x}^{\alpha_p}}{\partial x^{i_p}}\frac{\partial x^{l_1}}{\partial\bar{x}^{\beta_1}}\cdots\frac{\partial x^{l_q}}{\partial\bar{x}^{\beta_q}}A^{i_1 i_2\ldots i_p}_{l_1 l_2\ldots l_q}. \tag{41}$$

Here α_r $(1\leqslant r\leqslant p)$ and β_s $(1\leqslant s\leqslant q)$ are free indices each having any value between 1 and N, and i_r and l_s are dummy indices with summation over each from 1 to N being implicit. Since each of the indices α_r and β_s can take N values, the tensor has N^{p+q} components.

It is clear from this that a scalar and a vector are particular cases of a tensor. *A vector is a tensor of rank one* and hence has N components in an N-dimensional space. *A scalar is a tensor of rank zero* and thus has $N^0=1$ component. If A is a scalar, Eq. (41) with $p=q=0$ reduces to $\bar{A}=A$, showing that *a scalar is invariant under any coordinate transformation.* Obvious physical examples of a scalar are length, mass, energy, volume, etc., which are independent of the choice of the coordinate system.

It is important to note the essential difference between a contravariant and a covariant tensor. In the case of a contravariant tensor, the tensor is represented by components in the *directions of coordinate increase*, whereas in the case of a covariant tensor, the tensor is represented by components in the *directions orthogonal to constant coordinate surfaces.* The physical examples discussed in Example 3 will help in clarifying this point. The velocity and the acceleration vectors are represented in terms of components in the directions of coordinate increase, while the gradient vector is represented in terms of components in the directions orthogonal to the constant coordinate surfaces. In the case of a cartesian coordinate system, the coordinate direction x^i coincides with the direction orthogonal to the constant-x^i surface, so that the distinction between the covariant and the contravariant tensors vanishes (see Example 18.7).

It should be evident from Eqs. (6) that the coordinate differentials dx^i are the components of a contravariant tensor of rank one—the infinitesimal displacement vector. It is important to note that the coordinates x^i, in spite of the notation, are *not* the components of a tensor.

EXAMPLE 4: The cartesian components of the acceleration vector are

$$a_x=\frac{d^2x}{dt^2},\quad a_y=\frac{d^2y}{dt^2},\quad a_z=\frac{d^2z}{dt^2}. \tag{42}$$

Find the components of the acceleration vector in the spherical polar coordinates.

Solution: Since acceleration is a contravariant vector, its components will transform according to Eq. (24), which can be written as

$$\bar{a}^\alpha=\frac{\partial\bar{x}^\alpha}{\partial x^i}\,a^i. \tag{43}$$

We take x^1, x^2, x^3 to be the cartesian coordinates x, y, z and $\bar{x}^1$, $\bar{x}^2$, $\bar{x}^3$ to be the spherical polar coordinates r, θ, ϕ. Similarly, we identify a^i with the cartesian components of acceleration, i.e., $a^1 \equiv a_x$, $a^2 \equiv a_y$, $a^3 \equiv a_z$, which are given in Eqs. (42), and $\bar{a}^\alpha$ with the polar components, i.e., $\bar{a}^1 \equiv a_r$, $\bar{a}^2 \equiv a_\theta$, $\bar{a}^3 \equiv a_\phi$, which are to be found out. Eq. (43), written explicitly with the above identification in mind, becomes

$$\begin{aligned} a_r &= \frac{\partial r}{\partial x}\frac{d^2x}{dt^2} + \frac{\partial r}{\partial y}\frac{d^2y}{dt^2} + \frac{\partial r}{\partial z}\frac{d^2z}{dt^2}, \\ a_\theta &= \frac{\partial \theta}{\partial x}\frac{d^2x}{dt^2} + \frac{\partial \theta}{dy}\frac{d^2y}{dt^2} + \frac{\partial \theta}{dz}\frac{d^2z}{dt^2}, \\ a_\phi &= \frac{\partial \phi}{\partial x}\frac{d^2x}{dt^2} + \frac{\partial \phi}{\partial y}\frac{d^2y}{dt^2} + \frac{\partial \phi}{\partial z}\frac{d^2z}{dt^2}. \end{aligned} \tag{44}$$

We thus need the nine partial derivatives of r, θ, ϕ with respect to x, y, z. These are found from Eqs. (5) and after converting all the cartesian coordinates to spherical polar coordinates by using Eqs. (4), they are given by[4]

$$\begin{aligned} &\frac{\partial r}{\partial x} = \sin\theta\cos\phi, \quad &&\frac{\partial r}{\partial y} = \sin\theta\sin\phi, \quad &&\frac{\partial r}{\partial z} = \cos\theta, \\ &\frac{\partial \theta}{\partial x} = \frac{\cos\theta\cos\phi}{r}, &&\frac{\partial \theta}{\partial y} = \frac{\cos\theta\sin\phi}{r}, &&\frac{\partial \theta}{\partial z} = -\frac{\sin\theta}{r}, \\ &\frac{\partial \phi}{dx} = -\frac{\sin\phi}{r\sin\theta}, &&\frac{\partial \phi}{\partial y} = \frac{\cos\phi}{r\sin\theta}, &&\frac{\partial \phi}{\partial z} = 0. \end{aligned} \tag{45}$$

The first of Eqs. (44), for example, then becomes

$$\begin{aligned} a_r = {} & \sin\theta\cos\phi\, d^2(r\sin\theta\cos\phi)/dt^2 \\ & + \sin\theta\sin\phi\, d^2(r\sin\theta\sin\phi)/dt^2 + \cos\theta\, d^2(r\cos\theta)/dt^2. \end{aligned} \tag{46}$$

The first term on the right hand side of the above equation is worked out as follows. We have

$$\frac{d}{dt}(r\sin\theta\cos\phi) = \dot{r}\sin\theta\cos\phi + r\cos\theta\cos\phi\,\dot{\theta} - r\sin\theta\sin\phi\,\dot{\phi}, \tag{47}$$

and hence

$$\begin{aligned} \frac{d^2}{dt^2}(r\sin\theta\cos\phi) = {} & \ddot{r}\sin\theta\cos\phi + 2\dot{r}\,\dot{\theta}\cos\theta\cos\phi \\ & -2\sin\theta\sin\phi\,\dot{r}\,\dot{\phi} - r\sin\theta\cos\phi\,\dot{\theta}^2 - 2r\cos\theta\sin\phi\,\dot{\theta}\,\dot{\phi} \\ & + r\cos\theta\cos\phi\,\ddot{\theta} - r\sin\theta\cos\phi\,\dot{\phi}^2 - r\sin\theta\sin\phi\,\ddot{\phi}. \end{aligned} \tag{48}$$

Working out other terms similarly, we finally find

$$a_r = \ddot{r} - r\dot{\theta}^2 - r\sin^2\theta\,\dot{\phi}^2. \tag{49}$$

[4]It should be remembered that in the case of partial derivatives, for example, $\partial r/\partial x \neq 1/(\partial x/\partial r)$.

The other components can be determined in the same way and the result is found to be

$$\begin{aligned}\bar{a}^1 &\equiv a_r = \ddot{r} - r\dot{\theta}^2 - r\sin^2\theta\,\dot{\phi}^2,\\ \bar{a}^2 &\equiv a_\theta = \ddot{\theta} - \sin\theta\cos\phi\,\dot{\phi}^2,\\ \bar{a}^3 &\equiv a_\phi = \ddot{\phi} + \frac{2}{r}\,\dot{r}\dot{\phi} + 2\cot\theta\,\dot{\theta}\dot{\phi}.\end{aligned} \tag{50}$$

It should be noted that only the component a_r has the dimensions of acceleration, whereas a_θ and a_ϕ have the dimensions $(\text{time})^{-2}$. This is not surprising, because the spherical polar coordinates themselves do not all have the same dimensions.

As a special case, the above equations can be reduced to the two-dimensional (x, y)-plane by taking $\theta = \pi/2$ and replacing r by ρ. For the polar components of acceleration in the polar coordinates ρ, ϕ, this gives

$$a_\rho = \ddot{\rho} - \rho\dot{\phi}^2, \quad a_\phi = \ddot{\phi} + \frac{2}{\rho}\dot{\rho}\dot{\phi}. \tag{51}$$

EXAMPLE 5: Determine divA^i, divA_i and $\nabla^2 I$ in cylindrical polar coordinates, where A^i and A_i are vectors and I is a scalar.

Solution: (a) In cartesian coordinates, we have

$$\operatorname{div} A^i = \frac{\partial A^x}{\partial x} + \frac{\partial A^y}{\partial y} + \frac{\partial A^z}{\partial z}\,. \tag{52}$$

The cylindrical coordinates ρ, ϕ, z are related to the cartesian coordinates by

$$x = \rho\cos\phi, \quad y = \rho\sin\phi, \quad z = z. \tag{53}$$

If the components of A^i in cylindrical coordinates are denoted by A^ρ, A^ϕ, A^z, we have

$$A^x = \frac{\partial x}{\partial \rho}A^\rho + \frac{\partial x}{\partial \phi}A^\phi + \frac{\partial x}{\partial z}A^z = \cos\phi\,A^\rho - \rho\sin\phi\,A^\phi, \tag{54a}$$

$$A^y = \frac{\partial y}{\partial \rho}A^\rho + \frac{\partial y}{\partial \phi}A^\phi + \frac{\partial y}{\partial z}A^z = \sin\phi\,A^\rho + \rho\cos\phi\,A^\phi, \tag{54b}$$

$$A^z = A^z. \tag{54c}$$

Therefore

$$\frac{\partial A^x}{\partial x} = \frac{\partial A^x}{\partial \rho}\frac{\partial \rho}{\partial x} + \frac{\partial A^x}{\partial \phi}\frac{\partial \phi}{\partial x} + \frac{\partial A^x}{\partial z}\frac{\partial z}{\partial x} = \cos\phi\,\frac{\partial A^x}{\partial \rho} - \frac{\sin\phi}{\rho}\frac{\partial A^x}{\partial \phi}\,, \tag{55a}$$

$$\frac{\partial A^y}{\partial y} = \frac{\partial A^y}{\partial \rho}\frac{\partial \rho}{\partial y} + \frac{\partial A^y}{\partial \phi}\frac{\partial \phi}{\partial y} + \frac{\partial A^y}{\partial z}\frac{\partial z}{\partial y} = \sin\phi\,\frac{\partial A^y}{\partial \rho} + \frac{\cos\phi}{\rho}\frac{\partial A^y}{\partial \phi}\,, \tag{55b}$$

while $\partial A^z/\partial z$ remains the same. Using Eqs. (54) in Eqs. (55) gives

$$\frac{\partial A^x}{\partial x}=\cos^2\phi\,\frac{\partial A^\rho}{\partial \rho}-\frac{\sin\phi\cos\phi}{\rho}\frac{\partial A^\rho}{\partial \phi}-\rho\sin\phi\cos\phi\frac{\partial A^\phi}{\partial \rho}$$

$$+\sin^2\phi\,\frac{\partial A^\phi}{\partial \phi}+\frac{\sin^2\phi}{\rho}A^\rho, \tag{56a}$$

$$\frac{\partial A^y}{\partial y}=\sin^2\phi\,\frac{\partial A^\rho}{\partial \rho}+\frac{\sin\phi\cos\phi}{\rho}\frac{\partial A^\rho}{\partial \phi}+\rho\sin\phi\cos\phi\frac{\partial A^\phi}{\partial \rho}$$

$$+\cos^2\phi\,\frac{\partial A^\phi}{\partial \phi}+\frac{\cos^2\phi}{\rho}A^\rho. \tag{56b}$$

Adding these and $\partial A^z/\partial z$, we have

$$\text{div } A^i=\frac{\partial A^\rho}{\partial \rho}+\frac{\partial A^\phi}{\partial \phi}+\frac{\partial A^z}{\partial z}+\frac{A^\rho}{\rho}. \tag{57}$$

(b) For the covariant vector A_i, we have[5]

$$\text{div } A_i=\frac{\partial A_x}{\partial x}+\frac{\partial A_y}{\partial y}+\frac{\partial A_z}{\partial z}\;. \tag{58}$$

In terms of the cylindrical components A_ρ , A_ϕ, A_z, these are given by

$$A_x=\frac{\partial \rho}{\partial x}A_\rho+\frac{\partial \phi}{\partial x}A_\phi=\cos\phi\,A_\rho-\frac{\sin\phi}{\rho}A_\phi,$$

$$A_y=\frac{\partial \rho}{dy}A_\rho+\frac{\partial \phi}{\partial y}A_\phi=\sin\phi\,A_\rho+\frac{\cos\phi}{\rho}A_\phi, \tag{59}$$

$$A_z=A_z.$$

Following a procedure similar to that in part (a) above, we finally arrive at the result

$$\text{div } A_i=\frac{\partial A_\rho}{\partial \rho}+\frac{1}{\rho^2}\frac{\partial A_\phi}{\partial \phi}+\frac{\partial A_z}{\partial z}+\frac{A_\rho}{\rho}. \tag{60}$$

(c) Since I is a scalar field, it remains invariant under any coordinate transformation. We have

$$\nabla^2 I=\frac{\partial^2 I}{\partial x^2}+\frac{\partial^2 I}{\partial y^2}+\frac{\partial^2 I}{\partial z^2}\;. \tag{61}$$

Now,

$$\frac{\partial I}{\partial x}=\frac{\partial I}{\partial \rho}\frac{\partial \rho}{\partial x}+\frac{\partial I}{\partial \phi}\frac{\partial \phi}{\partial x}+\frac{\partial I}{\partial z}\frac{\partial z}{\partial x}=\cos\phi\,\frac{\partial I}{\partial \rho}-\frac{\sin\phi}{\rho}\frac{\partial I}{\partial \phi}. \tag{62}$$

Hence

$$\frac{\partial^2 I}{\partial x^2}=\frac{\partial}{\partial x}\left(\frac{\partial I}{\partial x}\right)=\frac{\partial}{\partial \rho}\left(\frac{\partial I}{\partial x}\right)\frac{\partial \rho}{\partial x}+\frac{\partial}{\partial \phi}\left(\frac{\partial I}{\partial x}\right)\frac{\partial \phi}{\partial x}. \tag{63}$$

[5]We shall see later, in Example 18.7, that in a cartesian coordinate system, $A^i=A_i$.

Substituting Eq. (62) in this and simplifying, we get

$$\frac{\partial^2 I}{\partial x^2}=\cos^2\phi\frac{\partial^2 I}{\partial\rho^2}+\frac{\sin 2\phi}{\rho}\left(\frac{\partial I}{\rho\partial\phi}-\frac{\partial^2 I}{\partial\rho\partial\phi}\right)+\frac{\sin^2\phi}{\rho}\left(\frac{\partial I}{\partial\rho}+\frac{\partial^2 I}{\rho\partial\phi^2}\right). \tag{64a}$$

In a similar way, we have

$$\frac{\partial^2 I}{\partial y^2}=\sin^2\phi\frac{\partial^2 I}{\partial\rho^2}-\frac{\sin 2\phi}{\rho}\left(\frac{\partial I}{\rho\partial\phi}-\frac{\partial^2 I}{\partial\rho\partial\phi}\right)+\frac{\cos^2\phi}{\rho}\left(\frac{\partial I}{\partial\rho}+\frac{\partial^2 I}{\rho\partial\phi^2}\right). \tag{64b}$$

Adding Eqs. (64) and $\partial^2 I/\partial z^2$, we finally get

$$\nabla^2 I=\frac{\partial^2 I}{\partial\rho^2}+\frac{1}{\rho}\frac{\partial I}{\partial\rho}+\frac{1}{\rho^2}\frac{\partial^2 I}{\partial\phi^2}+\frac{\partial^2 I}{\partial z^2}. \tag{65}$$

Exercise 15

1. Obtain the inverse transformations of those given in Eqs. (37), (38), (39) and (41).

*2. What is the value of δ^i_i in an N-dimensional space? Also evaluate $\delta^i_j\delta^j_i$ and $\delta^i_j\delta^j_k\delta^k_i$.

3. Show that (a) $\delta^i_jA_{ik}=A_{jk}$; (b) $\delta^p_i\delta^j_qA^i_j=A^p_q$.

4. The cartesian components of velocity are $dx/dt, dy/dt, dz/dt$. Prove that its components in spherical polar coordinates are $dr/dt, d\theta/dt, d\phi/dt$.

*5. Find the components of acceleration in cylindrical coordinates ρ, z, ϕ which are related to the cartesian coordinates by $x=\rho\cos\phi$, $y=\rho\sin\phi$, $z=z$. Compare this result with that obtained for polar coordinates in a two-dimensional plane at the end of Example 4.

*6. Obtain the components of the gradient of a scalar field in terms of polar coordinates in a two-dimensional space. Also obtain the components of the gradient of a scalar field in terms of spherical polar coordinates in a three-dimensional space.

*7. Show that the second derivatives of a scalar field f, that is, $A_{ij}=\partial^2 f/\partial x^i\partial x^j$, are not the components of a second rank tensor.

8. Show that in spherical polar coordinates r, θ, ϕ,

(a) $\operatorname{div} A^i=\dfrac{\partial A^r}{\partial r}+\dfrac{\partial A^\theta}{\partial\theta}+\dfrac{\partial A^\phi}{\partial\phi}+\dfrac{2A^r}{r}+\cot\theta\, A^\theta$;

(b) $\operatorname{div} A_i=\dfrac{\partial A_r}{\partial r}+\dfrac{1}{r^2}\dfrac{\partial A_\theta}{\partial\theta}+\dfrac{1}{r^2\sin^2\theta}\dfrac{\partial A_\phi}{\partial\phi}+\dfrac{2A_r}{r}+\dfrac{\cot\theta}{r^2}A_\theta$;

(c) $\nabla^2 I=\dfrac{\partial^2 I}{\partial r^2}+\dfrac{1}{r^2}\dfrac{\partial^2 I}{\partial\theta^2}+\dfrac{1}{r^2\sin^2\theta}\dfrac{\partial^2 I}{\partial\phi^2}+\dfrac{2}{r}\dfrac{\partial I}{\partial r}+\dfrac{\cot\theta}{r^2}\dfrac{\partial I}{\partial\theta}$.

*9. The cartesian components of the velocity vector of a fluid in motion in a two-dimensional plane are $v_x=x^2$, $v_y=y^2$. Find the polar components of the velocity vector in terms of polar coordinates ρ, ϕ.

16. The algebra of tensors

After defining a tensor, the logical next step is to define the equality of tensors, the null tensor, addition, subtraction and multiplication of tensors, and the various rules connected with these operations.

Equality and null tensor

Two tensors $A^{i_1\ldots i_p}_{j_1\ldots j_q}$ and $B^{i_1\ldots i_p}_{j_1\ldots j_q}$ are said to be equal if and only if they have the same contravariant rank and the same covariant rank and every component of one is equal to the corresponding component of the other, i.e.,

$$A^{i_1 i_2\ldots i_p}_{j_1 j_2\ldots j_q}=B^{i_1 i_2\ldots i_p}_{j_1 j_2\ldots j_q}. \tag{1}$$

If all the N^r components of a tensor of total rank r identically vanish, it is said to be a *null tensor.* It is evident from Eq. (15.41) that if all the components of a tensor vanish in any coordinate system, they vanish in all coordinate systems.

If two tensors have the same contravariant rank and the same covariant rank, we shall say that they are of the *same type.*

Addition and subtraction

Two tensors can be added or subtracted provided they are of the same type. Thus, let $A^{i_1\ldots i_p}_{j_1\ldots j_q}$ and $B^{i_1\ldots i_p}_{j_1\ldots j_q}$ be two tensors each having contravariant rank p and covariant rank q. The sum of these two tensors is defined as a tensor $C^{i_1\ldots i_p}_{j_1\ldots j_q}$ of the same rank as the original tensors and whose components are respectively equal to the sum of the corresponding components of the two tensors, i.e.,

$$C^{i_1 i_2\ldots i_p}_{j_1 j_2\ldots j_q}=A^{i_1 i_2\ldots i_p}_{j_1 j_2\ldots j_q}+B^{i_1 i_2\ldots i_p}_{j_1 j_2\ldots j_q}. \tag{2}$$

To prove that the sum of two tensors of the same type is also a tensor, we write the transformation of the components of $B^{i_1\ldots i_p}_{j_1\ldots j_q}$ in a form similar to Eq. (15.41). This gives

$$\bar{B}^{\alpha_1\ldots\alpha_p}_{\beta_1\ldots\beta_q}=\frac{\partial\bar{x}^{\alpha_1}}{\partial x^{i_1}}\cdots\frac{\partial\bar{x}^{\alpha_p}}{\partial x^{i_p}}\frac{\partial x^{l_1}}{\partial\bar{x}^{\beta_1}}\cdots\frac{\partial x^{l_q}}{\partial\bar{x}^{\beta_q}}B^{i_1\ldots i_p}_{l_1\ldots l_q}. \tag{3}$$

Adding the respective sides of Eq. (3) and Eq. (15.41), we have

$$\bar{A}^{\alpha_1\ldots\alpha_p}_{\beta_1\ldots\beta_q}+\bar{B}^{\alpha_1\ldots\alpha_p}_{\beta_1\ldots\beta_q}=\frac{\partial\bar{x}^{\alpha_1}}{\partial x^{i_1}}\cdots\frac{\partial\bar{x}^{\alpha_p}}{\partial x^{i_p}}\frac{\partial x^{l_1}}{\partial\bar{x}^{\beta_1}}\cdots\frac{\partial x^{l_q}}{\partial\bar{x}^{\beta_q}}(A^{i_1\ldots i_p}_{l_1\ldots l_q}+B^{i_1\ldots i_p}_{l_1\ldots l_q}). \tag{4}$$

If we write the sum of the components of the two tensors in the barred coordinate system as

$$\bar{C}^{\alpha_1\ldots\alpha_p}_{\beta_1\ldots\beta_q}=\bar{A}^{\alpha_1\ldots\alpha_p}_{\beta_1\ldots\beta_q}+\bar{B}^{\alpha_1\ldots\alpha_p}_{\beta_1\ldots\beta_q}, \tag{5}$$

Eq. (4) becomes

$$\bar{C}^{\alpha_1\ldots\alpha_p}_{\beta_1\ldots\beta_q}=\frac{\partial\bar{x}^{\alpha_1}}{\partial x^{i_1}}\cdots\frac{\partial\bar{x}^{\alpha_p}}{\partial x^{i_p}}\frac{\partial x^{l_1}}{\partial\bar{x}^{\beta_1}}\cdots\frac{\partial x^{l_q}}{\partial\bar{x}^{\beta_q}}C^{i_1\ldots i_p}_{l_1\ldots l_q}. \tag{6}$$

This shows that $C^{i_1\ldots i_p}_{l_1\ldots l_q}$ is also a tensor of contravariant rank p and covariant rank q.

In a similar way, we can define the subtraction of two tensors. Provided two tensors are of the same type, their difference is defined as a tensor each component of which is the difference of the corresponding components of the two tensors, i.e.,

$$A^{i_1\ldots i_p}_{j_1\ldots j_q}-B^{i_1\ldots i_p}_{j_1\ldots j_q}=D^{i_1\ldots i_p}_{j_1\ldots j_q}, \tag{7}$$

which can be easily shown to be a tensor.

Outer product of tensors

Consider two tensors, say A^{ij}_k and B^p_q. A^{ij}_k has total rank three and hence has N^3 components, while B^p_q has total rank two and hence has N^2 components. If each component of one tensor is multiplied by every component of the other, the resulting set of quantities again gives a tensor whose rank is the sum of the ranks of the two original tensors. We shall prove this for the above two tensors.

The tensors A^{ij}_k and B^p_q transform according to the following equations:

$$\bar{A}^{\alpha\beta}_{\gamma}=\frac{\partial\bar{x}^{\alpha}}{\partial x^{i}}\frac{\partial\bar{x}^{\beta}}{\partial x^{j}}\frac{\partial x^{k}}{\partial\bar{x}^{\gamma}}A^{ij}_k\,, \tag{8a}$$

$$\bar{B}^{\rho}_{\sigma}=\frac{\partial\bar{x}^{\rho}}{\partial x^{p}}\frac{\partial x^{q}}{\partial\bar{x}^{\sigma}}B^p_q\,. \tag{8b}$$

Multiplying the respective components on both sides of the above equation and rearranging terms, we find

$$\bar{A}^{\alpha\beta}_{\gamma}\bar{B}^{\rho}_{\sigma}=\frac{\partial\bar{x}^{\alpha}}{\partial x^{i}}\frac{\partial\bar{x}^{\beta}}{\partial x^{j}}\frac{\partial x^{k}}{\partial\bar{x}^{\gamma}}\frac{\partial\bar{x}^{\rho}}{\partial x^{p}}\frac{\partial x^{q}}{\partial\bar{x}^{\sigma}}A^{ij}_k B^p_q. \tag{9}$$

The expressions which appear in Eq. (9) are the products of the components of the tensors A^{ij}_k and B^p_q in the barred and the unbarred coordinate systems. Defining

$$C^{ijp}_{kq}=A^{ij}_k B^p_q,\quad \bar{C}^{\alpha\beta\rho}_{\gamma\sigma}=\bar{A}^{\alpha\beta}_{\gamma}\bar{B}^{\rho}_{\sigma}. \tag{10}$$

Eq. (9) reduces to

$$\overline{C}^{\alpha\beta\rho}_{\gamma\sigma} = \frac{\partial \overline{x}^\alpha}{\partial x^i}\frac{\partial \overline{x}^\beta}{\partial x^j}\frac{\partial \overline{x}^\rho}{\partial x^p}\frac{\partial x^k}{\partial \overline{x}^\gamma}\frac{\partial x^q}{\partial \overline{x}^\sigma}\, C^{ijp}_{kq}, \tag{11}$$

which shows that C^{ijp}_{kq} is a tensor of contravariant rank 3, covariant rank 2, and total rank 5. It has therefore N^5 components each of which is the product of one component of A^{ij}_k with one of B^p_q.

This is known as the *outer product* or the *Kronecker product* of two tensors. The tensor C^{ijp}_{kq} in the above case is said to be the outer product of the tensors A^{ij}_k and B^p_q. In general, *the contravariant rank of the outer-product tensor is the sum of the contravariant ranks and the covariant rank is the sum of the covariant ranks of the tensors whose outer product it is.*

The concept of the outer product of tensors can be easily extended to more than two tensors.

EXAMPLE 1: Find the outer product of the three tensors A^i_j, B_k and C^{lm}_{np}.

Solution: The outer product is given by

$$D^{ilm}_{jknp} = A^i_j\, B_k C^{lm}_{np}, \tag{12}$$

which is a tensor of total rank 7 with contravariant rank 3 and covariant rank 4.

In taking the outer product of any number of tensors, care should be taken to use distinct indices. For example, it would be *wrong* to write the outer product of A^i_j, B_k and C^{lm}_{np} as $A^i_j B_k C^{lm}_{kp}$, because the index k is repeated.

Inner product of tensors

Consider the two tensors A^{ij}_k and B^p_q introduced in connection with the outer product of tensors. Consider the set of functions $A^{ij}_k B^k_q$ where i, j and q are free indices but a summation over k from 1 to N is implied according to the convention. Since there are only three free indices, there will be N^3 such functions. We shall show that these N^3 functions are the components of a tensor of total rank 3.

For this purpose, put $\rho=\gamma$ in Eq. (9) and sum over γ, which gives

$$\overline{A}^{\alpha\beta}_\gamma \overline{B}^\gamma_\sigma = \frac{\partial \overline{x}^\alpha}{\partial x^i}\frac{\partial \overline{x}^\beta}{\partial x^j}\frac{\partial x^k}{\partial \overline{x}^\gamma}\frac{\partial \overline{x}^\gamma}{\partial x^p}\frac{\partial x^q}{\partial \overline{x}^\sigma}\, A^{ij}_k\, B^p_q. \tag{13}$$

The summation over γ on the right hand side can be easily evaluated by using Eq. (15.23b), which in this case gives

$$\frac{\partial x^k}{\partial \bar{x}^\gamma}\frac{\partial \bar{x}^\gamma}{\partial x^p}=\delta^k_p. \tag{14}$$

This reduces Eq. (13) to

$$\bar{A}^{\alpha\beta}_{\gamma}\bar{B}^{\gamma}_{\sigma}=\frac{\partial \bar{x}^\alpha}{\partial x^i}\frac{\partial \bar{x}^\beta}{\partial x^j}\frac{\partial x^q}{\partial \bar{x}^\sigma}\delta^k_p A^{ij}_k B^p_q. \tag{15}$$

Now p is a dummy index and summation over it on the right hand side can be evaluated. Owing to the appearance of the delta symbol, it is clear that in the summation over p from 1 to N, only that term will survive for which $p=k$. This gives

$$\bar{A}^{\alpha\beta}_{\gamma}\bar{B}^{\gamma}_{\sigma}=\frac{\partial \bar{x}^\alpha}{\partial x^i}\frac{\partial \bar{x}^\beta}{\partial x^j}\frac{\partial x^q}{\partial \bar{x}^\sigma}A^{ij}_k B^k_q. \tag{16}$$

This shows that the N^3 entities $A^{ij}_k B^k_q$ transform like the components of a tensor of contravariant rank 2 and covariant rank 1. We may denote the components of the new tensor in the barred and the unbarred coordinate systems by

$$\bar{C}^{\alpha\beta}_{\sigma}=\bar{A}^{\alpha\beta}_{\gamma}\bar{B}^{\gamma}_{\sigma},\ C^{ij}_q=A^{ij}_k B^k_q. \tag{17}$$

The tensor C^{ij}_q is said to be the *inner product* of the two tensors A^{ij}_k and B^p_q.

The inner product of two tensors could be taken in a variety of ways. Thus, $A^{ij}_k B^p_i$ and $A^{ij}_k B^p_j$ are examples of the inner product of A^{ij}_k and B^p_q. We could also equate two sets of indices. Thus, the sets of functions $A^{ij}_k B^k_i$ and $A^{ij}_k B^k_j$ are also tensors, each of contravariant rank 1, because there is only one free contravariant index in either case.

In taking the inner product of two tensors, it is important that *one contravariant index of one tensor should be equated to one covariant index of the other*. No index should occur more than twice. For example, it would be *wrong* to write the inner product as $A^{ij}_k B^i_q$ because two contravariant indices have been equated (see the following example), or as $A^{ik}_k B^k_k$ because k is repeated four times (see Exercise 3).

EXAMPLE 2: If A^{ij}_k and B^p_q are tensors, show that $A^{ij}_k B^i_q$ is not a tensor.

Solution: It can be shown that if one index of a tensor is equated to a similar index of the other while taking the inner product of the two tensors, the resulting set of functions is *not* a tensor. Thus, consider the set of entities $A^{ij}_k B^i_q$. Here again we have N^3 functions because there are three free indices j, k and q. However, these do not transform like the components of a tensor of rank 3.

To see this, we equate ρ with α in Eq. (9) to find

$$\bar{A}_{\gamma}^{\alpha\beta}\bar{B}_{\sigma}^{\alpha}=\frac{\partial\bar{x}^{\alpha}}{\partial x^{i}}\frac{\partial\bar{x}^{\beta}}{\partial x^{j}}\frac{\partial x^{k}}{\partial\bar{x}^{\gamma}}\frac{\partial\bar{x}^{\alpha}}{\partial x^{p}}\frac{\partial x^{q}}{\partial\bar{x}^{\sigma}}A_{k}^{ij}B_{q}^{p}. \tag{18}$$

The term containing α on the right hand side is

$$\frac{\partial\bar{x}^{\alpha}}{\partial x^{i}}\frac{\partial\bar{x}^{\alpha}}{\partial x^{p}}.$$

There is no general method of evaluating this summation and Eq. (18) shows that the transformation is not tensorial in character.

Contraction of a tensor

Consider a tensor A_{lm}^{ijk} of contravariant rank 3 and covariant rank 2, which has N^5 components. Let any one of the contravariant indices be equated to any one of the covariant indices (and, of course, summed over from 1 to N). In other words, consider the set of functions, say, A_{im}^{ijk}. Here i is a dummy index while j, k and m are free indices. By convention, we have

$$A_{im}^{ijk}=A_{1m}^{1jk}+A_{2m}^{2jk}+\ldots+A_{Nm}^{Njk}. \tag{19}$$

The entity A_{im}^{ijk} evidently has N^3 components. We shall show that it is a tensor of total rank 3.

For this purpose, let us write the transformation of the tensor A_{lm}^{ijk}:

$$\bar{A}_{\rho\sigma}^{\alpha\beta\gamma}=\frac{\partial\bar{x}^{\alpha}}{\partial x^{i}}\frac{\partial\bar{x}^{\beta}}{\partial x^{j}}\frac{\partial\bar{x}^{\gamma}}{\partial x^{k}}\frac{\partial x^{l}}{\partial\bar{x}^{\rho}}\frac{\partial x^{m}}{\partial\bar{x}^{\sigma}}A_{lm}^{ijk}. \tag{20}$$

Putting $\rho=\alpha$ in the above equation and summing over α from 1 to N, we have

$$\begin{aligned}\bar{A}_{\alpha\sigma}^{\alpha\beta\gamma}&=\frac{\partial\bar{x}^{\alpha}}{\partial x^{i}}\frac{\partial\bar{x}^{\beta}}{\partial x^{j}}\frac{\partial\bar{x}^{\gamma}}{\partial x^{k}}\frac{\partial x^{l}}{\partial\bar{x}^{\alpha}}\frac{\partial x^{m}}{\partial\bar{x}^{\sigma}}A_{lm}^{ijk}\\&=\frac{\partial\bar{x}^{\beta}}{\partial x^{j}}\frac{\partial\bar{x}^{\gamma}}{\partial x^{k}}\frac{\partial x^{m}}{\partial\bar{x}^{\sigma}}\delta_{i}^{l}A_{lm}^{ijk}=\frac{\partial\bar{x}^{\beta}}{\partial x^{j}}\frac{\partial\bar{x}^{\gamma}}{\partial x^{k}}\frac{\partial x^{m}}{\partial\bar{x}^{\sigma}}A_{im}^{ijk}.\end{aligned} \tag{21}$$

This shows that A_{im}^{ijk} is a tensor of contravariant rank 2 and covariant rank 1.

This process is known as the *contraction of a tensor*. In general, when a tensor is contracted by equating one of its contravariant indices with one of its covariant indices, the resulting entity is a tensor whose contravariant and covariant ranks are reduced by one each, thus reducing the total rank by two.

It should be clear that a tensor can be contracted in various ways. Thus, A_{li}^{ijk}, A_{jm}^{ijk}, A_{lk}^{ijk}, etc., are all contracted forms of the tensor A_{lm}^{ijk}.

A tensor can be repeatedly contracted. Thus, the tensor A_{lm}^{ijk} of total rank 5, on contraction, gives the tensor A_{im}^{ijk} of total rank 3, which can be further contracted to give the tensor A_{ij}^{ijk} or A_{ik}^{ijk} of contravariant rank 1.

It should be evident that the inner product of tensors can be thought of as their outer product followed by contraction. Thus, the inner product $A_k^{ij} B_q^k = C_q^{ij}$ [see Eq. (17)] can be obtained by first taking the outer product $A_k^{ij} B_q^p = D_{kq}^{ijp}$, then contracting this tensor by equating the indices p and k, and finally identifying C_q^{ij} with D_{pq}^{ijp}.

Finally, it should be understood that if two *similar* indices of a tensor are equated, the resulting entity is not a tensor. Thus, if D_{kq}^{ijp} is a tensor, D_{kq}^{iip} and D_{kk}^{ijp} are not tensors [see Exercise 4].

Symmetric and antisymmetric tensors

If the elements of a contravariant tensor A^{ij} of rank 2 satisfy the equations

$$A^{ij} = A^{ji}, \tag{22}$$

(which should be understood as being true for all values of i and j between 1 and N as per convention), A^{ij} is said to be a *symmetric tensor* of contravariant rank two. Similarly, a covariant tensor A_{ij} of rank two is said to be symmetric if

$$A_{ij} = A_{ji}. \tag{23}$$

A tensor A^{ij} or A_{ij} is said to be antisymmetric if its elements satisfy

$$A^{ij} = -A^{ji}, \quad A_{ij} = -A_{ji}. \tag{24}$$

EXAMPLE 3: If a contravariant tensor of rank two is symmetric in one coordinate system, show that it is symmetric in any coordinate system.

Solution: Let A^{ij} be the components of a tensor of rank two in the x^i-coordinate system, in which it is symmetric, so that $A^{ij} = A^{ji}$. Consider the transformation of the components of A^{ij} to some other coordinate system, say $\bar{x}^\alpha$. We have

$$\bar{A}^{\alpha\beta} = \frac{\partial \bar{x}^\alpha}{\partial x^i} \frac{\partial \bar{x}^\beta}{\partial x^j} A^{ij}. \tag{25}$$

The component $\bar{A}^{\beta\alpha}$ is given by interchanging α and β in the above equation, which yields

$$\bar{A}^{\beta\alpha}=\frac{\partial\bar{x}^{\beta}}{\partial x^{i}}\frac{\partial\bar{x}^{\alpha}}{\partial x^{j}}A^{ij}. \tag{26}$$

Interchanging the dummy indices i and j and remembering that $A^{ij}=A^{ji}$, we have

$$\bar{A}^{\beta\alpha}=\frac{\partial\bar{x}^{\beta}}{\partial x^{j}}\frac{\partial\bar{x}^{\alpha}}{\partial x^{i}}A^{ji}=\frac{\partial\bar{x}^{\beta}}{\partial x^{j}}\frac{\partial\bar{x}^{\alpha}}{\partial x^{i}}A^{ij}. \tag{27}$$

The right hand sides of Eqs. (25) and (27) are identical and hence we get

$$\bar{A}^{\alpha\beta}=\bar{A}^{\beta\alpha}, \tag{28}$$

showing that $\bar{A}^{\alpha\beta}$ is a symmetric tensor in the $\bar{x}^{\alpha}$ coordinate system. *Thus the property of symmetry is an intrinsic property of a tensor and is independent of the choice of the coordinate system.* The same result holds good for the property of antisymmetry.

For a general tensor of arbitrary rank, *symmetry and antisymmetry can only be defined for a pair of similar indices.* Thus, for example, a tensor A^{ijk}_{lmp} is said to be *symmetric in the first two contravariant indices* if

$$A^{ijk}_{lmp}=A^{jik}_{lmp}, \tag{29}$$

or is said to be *symmetric in the first and the third covariant indices* if

$$A^{ijk}_{lmp}=A^{ijk}_{pml}, \tag{30}$$

etc. It is important to specify the positions of the indices rather than the indices themselves. Thus, it would be meaningless to say that A^{ijk}_{lmp} is symmetric in i and j because we could as well denote the tensor in the form A^{lmi}_{kjp}.

In a similar fashion, antisymmetry of an arbitrary tensor can be defined for any pair of similar indices.

We have mentioned that symmetry and antisymmetry can be defined only for *similar* indices, *not* when one index is covariant and the other contravariant. This is proved in the following example.

EXAMPLE 4: If a mixed tensor A^{i}_{j} is symmetric in the x^{i}-coordinate system, find out whether it will be symmetric in any other coordinate system or not.

Solution: It is given that $A^{i}_{j}=A^{j}_{i}$ in the x^{i}-coordinate system. Let $\bar{x}^{\alpha}$ be some other coordinate system. The components A^{i}_{j} transform according to

$$\bar{A}^{\alpha}_{\beta}=\frac{\partial\bar{x}^{\alpha}}{\partial x^{i}}\frac{\partial x^{j}}{\partial\bar{x}^{\beta}}A^{i}_{j}. \tag{31}$$

Interchanging α and β and also i and j in the above equation, we have

$$\bar{A}^{\beta}_{\alpha}=\frac{\partial\bar{x}^{\beta}}{\partial x^{j}}\frac{\partial x^{i}}{\partial\bar{x}^{\alpha}}A^{j}_{i}=\frac{\partial\bar{x}^{\beta}}{\partial x^{j}}\frac{\partial x^{i}}{\partial\bar{x}^{\alpha}}A^{i}_{j}. \tag{32}$$

It is evident that the right hand sides of Eqs. (31) and (32) are not equal, in general, and hence $\bar{A}^{\alpha}_{\beta}\neq\bar{A}^{\beta}_{\alpha}$.

Thus, we can state, in general, that *the property of symmetry or antisymmetry of a tensor between a pair of similar indices is invariant under coordinate transformations, but the property of symmetry or antisymmetry of a tensor between a pair of dissimilar indices is not invariant under coordinate transformations.*

The Kronecker delta

In this and the previous sections, we have used the Kronecker delta symbol δ^{i}_{j} [defined in Eq. (15.19)] as a mixed tensor of rank two. That it is indeed a tensor of the said nature can be seen as follows.

Let the Kronecker delta symbol be denoted by $\bar{\delta}^{\alpha}_{\beta}$ in the barred coordinate system; we have

$$\bar{\delta}^{\alpha}_{\beta}=\frac{\partial\bar{x}^{\alpha}}{\partial\bar{x}^{\beta}}=\frac{\partial\bar{x}^{\alpha}}{\partial x^{i}}\frac{\partial x^{i}}{\partial\bar{x}^{\beta}}=\frac{\partial\bar{x}^{\alpha}}{\partial x^{i}}\frac{\partial x^{j}}{\partial\bar{x}^{\beta}}\delta^{i}_{j}, \tag{33}$$

showing that the Kronecker delta transforms like a mixed tensor of rank two. Since the only values the delta symbol takes are 1 and 0, it is seen that it is a constant tensor, that is, has the same components in any coordinate system. If, for a moment, we use Latin indices for the barred coordinates also, we have

$$\bar{\delta}^{i}_{j}=\delta^{i}_{j}. \tag{34}$$

The fully antisymmetric tensor

It is often convenient to define a tensor $\epsilon_{i_1 i_2\cdots i_N}$ of rank N in an N-dimensional space in the following way. Since each of the indices i_k can take N values $(1\leqslant i_k\leqslant N)$, the tensor has N^N components. The tensor is antisymmetric in any two of its indices. Moreover, $\epsilon_{i_1 i_2\cdots i_N}=+1$ or -1 according as whether $(i_1\, i_2\ldots i_N)$ is an even or odd permutation of $(1\,2\ldots N)$. From the complete antisymmetry of the tensor, it follows that if any two indices are equal to each other, the corresponding component of the tensor vanishes. The tensor thus has only $N!$ nonvanishing components. This tensor is called the *fully antisymmetric tensor of rank N*.

In particular, if we choose cartesian coordinates x, y, z in a three-dimensional space, the only nonvanishing components of the fully antisymmetric tensor of rank three are

$$\epsilon_{xyz}=\epsilon_{yzx}=\epsilon_{zxy}=1, \quad \epsilon_{yxz}=\epsilon_{xzy}=\epsilon_{zyx}=-1, \tag{35}$$

while all other components such as ϵ_{xxz}, ϵ_{xxx}, etc., are zero. In other words, ϵ_{ijk} equals $+1$ if $(i\ j\ k)$ is an even permutation of $(x\ y\ z)$, -1 if it is an odd permutation of $(x\ y\ z)$, and zero in all other cases.

The fully antisymmetric tensor of rank three defined above can be conveniently used to abbreviate many well-known mathematical equations. We shall discuss two typical examples below.

EXAMPLE 5: Express (a) the vector product of two vectors, and (b) the commutation relations between the components of angular momentum in quantum mechanics by using the fully antisymmetric tensor of rank three.

Solution:

(a) *Vector product of two vectors*: Let $\mathbf{u}=(u_x, u_y, u_z)$ and $\mathbf{v}=(v_x, v_y, v_z)$ be two three-dimensional vectors. Their vector product is a vector having components

$$\begin{aligned}(\mathbf{u}\times\mathbf{v})_x&=u_y v_z-u_z v_y,\\ (\mathbf{u}\times\mathbf{v})_y&=u_z v_x-u_x v_z,\\ (\mathbf{u}\times\mathbf{v})_z&=u_x v_y-u_y v_x.\end{aligned} \tag{36}$$

These can be written as a single equation in the form[1]

$$(\mathbf{u}\times\mathbf{v})_i=\epsilon_{ijk}u_j v_k, \tag{37}$$

where i stands for any of the three indices x, y, z and the summation over j and k is over the three indices x, y, z.

(b) *Commutation relations between angular momentum components*: Unlike in classical mechanics, angular momentum in quantum mechanics is an operator whose components do not commute with each other. In fact, if $\mathbf{L}=(L_x, L_y, L_z)$ is a quantum mechanical

[1]It would seem that the indices in Eq. (37) and the following Eq. (39) are not properly placed, as all the indices have been shown to be covariant. However, it will be shown later in Example 18.7 and Exercise 18.7 that the distinction between covariant and contravarient indices vanishes when a cartesian coordinate system is used.

angular momentum operator, its components satisfy the commutation relations[2]

$$[L_x, L_y]=i\hbar L_z, [L_y, L_z]=i\hbar L_x, [L_z, L_x]=i\hbar L_y. \tag{38}$$

where $\hbar=h/2\pi$, h being the Planck's constant. These equations can be abbreviated to the single equation

$$(L_i, L_j)=i\hbar\varepsilon_{ijk}L_k, \tag{39}$$

where, again, i, j, k stand for any of x, y, z.

There are some simple rules for checking the correctness of indices in a tensor equation. These are explained below.

1. *A free index should match in all terms throughout the equation.* This means that if a free index occurs as a contravariant (covariant) index in one term, it should occur as a contravariant (covariant) index in each term of the equation.

2. *A dummy index should match in each term of the equation separately.* A dummy index may occur only in some of the terms of an equation. When it occurs in a term, it should occur twice, once in a contravariant position and once in a covariant position.

3. *No index should occur more than twice in any term.*

4. When a coordinate differential such as ∂x^i occurs in a term, i is to be regarded as a contravariant index if ∂x^i occurs in the numerator and as a covariant index if it occurs in the denominator. Thus, in an expression such as $\partial x^i/\partial \bar{x}^\alpha$, i is a contravariant index while α is a covariant index.

Exercise 16

1. If A^i_j is a mixed tensor of rank two, show that A^i_i is also a tensor.

*2. In an N-dimensional space, how many different expressions are represented by $A^{ip}_j B^k_{ir} C^{rt}_{sk}$? When each expression is written out explicitly, how many terms does it contain ?

3. If A^{ij}_k and B^{pq}_r are tensors, show that $A^{ij}_i B^{pi}_i$ is not a tensor.

4. If A^{ij}_{kl} is a tensor, show that A^{ii}_{kl} and A^{ij}_{kk} are not tensors.

5. If A^{ijk}_{lm} is a tensor, show that A^{ijk}_{jk}, A^{ijk}_{kj} , A^{ljm}_{lm} and A^{lmk}_{lm} are all contravariant vectors.

*6. Show that any covariant or contravariant tensor of the second rank can be expressed as the sum of a symmetric tensor and an antisymmetric tensor of the same rank and type.

[2]See, for example, E. Merzbacher, *Quantum Mechanics* (Wiley, New York, 1961), Eqs. (9.5).

7. If A^{ij} is an antisymmetric tensor and B_i a vector, show that $A^{ij}B_iB_j=0$.

8. If **a**, **b**, **c** are three-dimensional vectors, show that their scalar triple product can be written as $(\mathbf{a}\times\mathbf{b}).\mathbf{c}=\varepsilon_{ijk}a_ib_jc_k$, where a_i, b_i, c_i are the cartesian components of **a**, **b**, **c**, respectively, and summation convention is used.

17. Quotient law

It is often required to find out whether the functions of a given set are the components of a tensor or not. The direct method, of course, requires us to find out whether the functions transform like the components of a tensor under coordinate transformations. However, this can be laborious and troublesome in practice. The *quotient law* provides a simpler alternative.

The quotient law states that *if the inner product of an entity with an arbitrary tensor*[1] *is a tensor, the entity is a tensor.*

It will be sufficient to prove the quotient law for a particular case. Suppose we wish to find out whether the N^3 functions $A(i, j, k)$ of a given set are the components of a tensor. Suppose it is known that the inner product of $A(i, j, k)$ with an arbitrary tensor B^{pq} is a contravariant tensor of the first rank, i.e.,

$$A(i, j, k)B^{jk} = C^i \tag{1}$$

is a tensor, where summation over j and k on the left hand side is implied. Let $\bar{A}(\alpha, \beta, \gamma)$ be the N^3 functions in the barred coordinate system. They satisfy a relation similar to Eq. (1), that is,

$$\bar{A}(\alpha, \beta, \gamma)\bar{B}^{\beta\gamma} = \bar{C}^{\alpha}. \tag{2}$$

Writing for $\bar{B}^{\beta\gamma}$ and $\bar{C}^{\alpha}$ in terms of the unbarred components, the above equation becomes

$$\begin{aligned} \bar{A}(\alpha, \beta, \gamma)\frac{\partial \bar{x}^{\beta}}{\partial x^j}\frac{\partial \bar{x}^{\gamma}}{\partial x^k}B^{jk} &= \frac{\partial \bar{x}^{\alpha}}{\partial x^i}C^i \\ &= \frac{\partial \bar{x}^{\alpha}}{\partial x^i}A(i, j, k)B^{jk}, \end{aligned} \tag{3}$$

where we have used Eq. (1) for C^i on the right hand side. Eq. (3) can be further written as

$$\left[\bar{A}(\alpha, \beta, \gamma)\frac{\partial \bar{x}^{\beta}}{\partial x^j}\frac{\partial \bar{x}^{\gamma}}{\partial x^k} - \frac{\partial \bar{x}^{\alpha}}{\partial x^i}A(i, j, k)\right]B^{jk} = 0. \tag{4}$$

Since this equation must be true for an arbitrary tensor B^{jk}, it follows that the expression in the square brackets must identically vanish, so that

$$\bar{A}(\alpha, \beta, \gamma)\frac{\partial \bar{x}^{\beta}}{\partial x^j}\frac{\partial \bar{x}^{\gamma}}{\partial x^k} = \frac{\partial \bar{x}^{\alpha}}{\partial x^i}A(i, j, k). \tag{5}$$

Inner multiplication by $(\partial x^j/\partial \bar{x}^{\rho})(\partial x^k/\partial \bar{x}^{\sigma})$ on both sides of the above equation yields

[1]An arbitrary tensor is one whose elements are not subject to any conditions and are independent of each other.

$$\bar{A}(\alpha, \rho, \sigma) = \frac{\partial \bar{x}^\alpha}{\partial x^i} \frac{\partial x^j}{\partial \bar{x}^\rho} \frac{\partial x^k}{\partial \bar{x}^\sigma} A(i, j, k), \tag{6}$$

which shows that $A(i, j, k)$ is a tensor of contravariant rank one and covariant rank two, which can be written as A^i_{jk}, so that Eq. (6) can be written in the tensor notation as

$$\bar{A}^\alpha_{\rho\sigma} = \frac{\partial \bar{x}^\alpha}{\partial x^i} \frac{\partial x^j}{\partial \bar{x}^\rho} \frac{\partial x^k}{\partial \bar{x}^\sigma} A^i_{jk}. \tag{7}$$

It is important in the use of the quotient law that the tensor with which inner product is taken should be an *arbitrary* tensor. To appreciate this, let us consider the following example, which is a slight variation of the proof of the quotient law discussed above.

EXAMPLE 1: Let $A(i, j, k)$ be a set of N^3 functions whose inner product with a *symmetric* tensor B^{jk} yields a tensor C^i. What can you conclude about $A(i, j, k)$?

Solution: As before, we can still arrive at Eq. (4). However, we can no longer deduce Eq. (5) from Eq. (4) because B^{jk} is *not* an arbitrary tensor. To understand this, let us write Eq. (4) as

$$D(\alpha, j, k)B^{jk} = 0, \tag{8}$$

where $D(\alpha, j, k)$ stands for the expression in square brackets in Eq. (4). Since summation over j and k is implicit in Eq. (8), for every term of the type $D(\alpha, j, k)B^{jk}$ (no summation over j and k), there will be a term $D(\alpha, k, j)B^{kj} = D(\alpha, k, j)B^{jk}$ (no summation over j and k). Therefore the conclusion that can be drawn is that $D(\alpha, j, k) + D(\alpha, k, j) = 0$. In the present case, this gives

$$\bar{A}(\alpha, \beta, \gamma) \frac{\partial \bar{x}^\beta}{\partial x^j} \frac{\partial \bar{x}^\gamma}{\partial x^k} + \bar{A}(\alpha, \beta, \gamma) \frac{\partial \bar{x}^\beta}{\partial x^k} \frac{\partial \bar{x}^\gamma}{\partial x^j}$$
$$= \frac{\partial \bar{x}^\alpha}{\partial x^i} A(i, j, k) + \frac{\partial \bar{x}^\alpha}{\partial x^i} A(i, k, j), \tag{9}$$

or, an interchanging β and γ in the second term on the left hand side of the above equation,

$$[\bar{A}(\alpha, \beta, \gamma) + \bar{A}(\alpha, \gamma, \beta)] \frac{\partial \bar{x}^\beta}{\partial x^j} \frac{\partial \bar{x}^\gamma}{\partial x^k} = \frac{\partial \bar{x}^\alpha}{\partial x^i} [A(i, j, k) + A(i, k, j)]. \tag{10}$$

On taking inner product with $(\partial x^j/\partial \bar{x}^\rho)(\partial x^k/\partial \bar{x}^\sigma)$, as before, we arrive at the result that $A^i_{jk} \equiv A(i, j, k) + A(i, k, j)$, but not necessarily $A(i, j, k)$, is a tensor. Moreover, it is evident that the tensor A^i_{jk} is symmetric in its covariant indices in this case.

Conjugate symmetric tensors of second rank

Let A_{ij} be a symmetric covariant tensor of the second rank such that, when A_{ij} is expressed as a matrix, $\det(A_{ij})\neq 0$. Let B^{ij} denote the cofactor of the element A_{ij} in the matrix (A_{ij}) divided by the determinant of (A_{ij}), that is,

$$B^{ij}=(\text{cofactor of } A_{ij})/\det(A_{ij}). \tag{11}$$

We shall now show that B^{ij} are the components of a contravariant symmetric tensor.

The result of Eq. (4.5), in the present tensorial notation, can be written as

$$A_{ij}B^{ik}=\delta_j^k. \tag{12}$$

We cannot directly apply the quotient law to this equation because A_{ij} is a symmetric tensor, not an arbitrary tensor. So let C^i be an arbitrary contravariant vector and let

$$D_i=A_{ij}C^j. \tag{13}$$

Treated as a system of N linear equations in N unknowns (the N components C^j), this shows that Eq. (13) has unique solutions for C^j [because $\det(A_{ij})\neq 0$] in terms of D_i. Hence D_i is an arbitrary covariant vector. Taking the inner product of D_i with B^{ik}, we have

$$D_iB^{ik}=A_{ij}B^{ik}C^j=\delta_j^kC^j=C^k. \tag{14}$$

We can now apply the quotient law to this equation. The inner product of B^{ik} with an arbitrary vector D_i gives a vector C^k. Hence B^{ik} is a tensor. Moreover, since in a symmetric matrix A_{ij}, the cofactor of A_{ij} is the same as that of A_{ji}, the defining Eq. (11) shows that B^{ij} is also symmetric.

If we apply this process to B^{ij}, that is, define $E_{ij}=(\text{cofactor of } B^{ij})/\det(B^{ij})$, it can be shown [Exercise 2] that E_{ij} is the same as A_{ij}.

Two tensors A_{ij} and B^{ij} related to each other by Eq. (12) are known as *conjugate tensors*. It is evident that when A_{ij} is treated as a matrix, the elements B^{ij} of the conjugate tensor are the elements of the inverse matrix of A_{ij}. This also makes it clear that a symmetric tensor of the second rank has a conjugate tensor if and only if its determinant is nonzero.

Exercise 17

1. The distance between two points with coordinates x^i and x^i+dx^i in an N-dimensional space can be expressed as $ds^2=g_{ij}\,dx^idx^j$. Remembering that ds^2 is an invariant and dx^i are the components of the contravariant displacement vector, show that $g_{ij}+g_{ji}$ must be a symmetric covariant tensor of rank two.

*2. Let A_{ij} be a symmetric covariant tensor and let B^{ij}=(cofactor of A_{ij})/det (A_{ij}), as in the text. Further, define E_{ij} =(cofactor of B^{ij})/det (B^{ij}). Show that $E_{ij}=A_{ij}$.

3. If B^k is an arbitrary tensor and $A_{ij}B^k$ is a tensor, show that A_{ij} is a tensor.

4. Assuming that all the entities denoted by capital letters in the following equations are tensors, rewrite these equations in a proper tensorial form:

*(a) $A(i, j, k)B^j{}_k+C(r, i)D_r=E(r, s, t)F^{ri}{}_{st}$;

(b) $A(i, j, k)B^l{}_{ij}-C(r, s, k, t)D_{rs}E^{lt}=F(l, i, j)G^j{}_{ki}$;

(c) $A(i, l, k, p)B^p{}_{lj}+C(j, k, l, m)D^{im}{}_l+E(i, j)F_k=0$.

18. The fundamental tensor

In this section, we now introduce the concept of a *metric* (or *distance*) in the N-dimensional space. Let x^i and x^i+dx^i be the coordinates of two neighbouring points in an N-dimensional space with respect to some coordinate system. Let ds denote the distance between the two points. If the squared distance ds^2 can be expressed as a quadratic form[1] in the infinitesimals dx^i, that is,

$$ds^2 = g_{ij}\, dx^i\, dx^j, \tag{1}$$

where the coefficients g_{ij} may be functions of x^i, with the only restriction that the determinant of the coefficient matrix be nonsingular [that is, $g \equiv \det(g_{ij}) \neq 0$], the space is said to be a *Riemannian space.* This is illustrated below by a few examples. In particular, if the coefficients g_{ij} are all independent of x^i, the space becomes a *Euclidean space.*

EXAMPLE 1: Obtain the metric for a two-dimensional plane in terms of cartesian coordinates.

Solution: In a two-dimensional plane with cartesian coordinates x, y, the distance between the two points (x, y) and $(x+dx, y+dy)$ can be written as

$$ds^2 = dx^2 + dy^2. \tag{2}$$

This is of the form of Eq. (1) with

$$g_{11} = g_{22} = 1,\ g_{12} = g_{21} = 0, \tag{3}$$

giving

$$g \equiv \det(g_{ij}) = \begin{vmatrix} 1 & 0 \\ 0 & 1 \end{vmatrix} = 1. \tag{4}$$

EXAMPLE 2: Obtain the metric for a two-dimensional plane in terms of polar coordinates.

Solution: In the same two-dimensional plane, if polar coordinates r, θ are chosen, the distance between the two points (r, θ) and $(r+dr, \theta+d\theta)$ can be written as (see Fig. 9)

$$ds^2 = dr^2 + r^2 d\theta^2. \tag{5}$$

This is again of the form of Eq. (1) with

$$g_{11} = 1,\ g_{22} = r^2,\ g_{12} = g_{21} = 0, \tag{6}$$

[1]See Section 11 for the definition of a quadratic form.

giving

$$g=\begin{vmatrix} 1 & 0 \\ 0 & r^2 \end{vmatrix}=r^2. \tag{7}$$

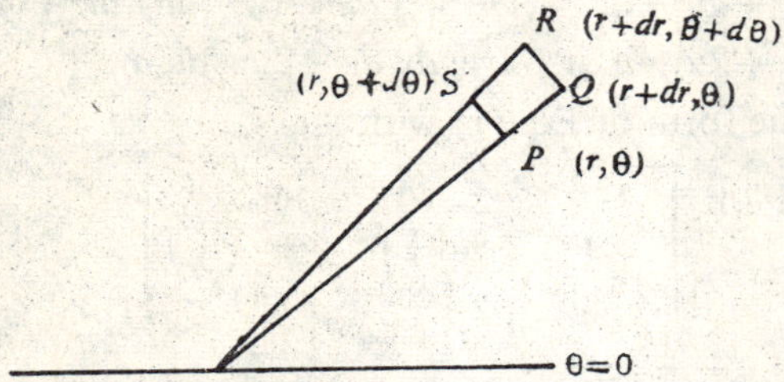

Figure 9. The distance between the two points $P=(r, \theta)$ and $R=(r+dr, \theta+d\theta)$ is given by $PR^2=PQ^2+QR^2$. But $PQ=dr$, while $QR=(r+dr)\, d\theta \simeq r\, d\theta$, neglecting second order quantities.

EXAMPLE 3: Find the metric for (a) the three-dimensional Euclidean space, and (b) the surface of a sphere of constant radius a in terms of spherical polar coordinates:

Solution: (a) Choosing spherical polar coordinates r, θ, ϕ in a three-dimensional space, the distance ds between the two points (r, θ, ϕ) and $(r+dr, \theta+d\theta, \phi+d\phi)$ can be written as

$$ds^2=dr^2+r^2\, d\theta^2+r^2 \sin^2\theta\, d\phi^2. \tag{8}$$

Comparing this with Eq. (1), we have, in this case,

$$g_{11}=1, g_{22}=r^2, g_{33}=r^2 \sin^2\theta, g_{ij}=0 \text{ for } i\neq j. \tag{9}$$

The determinant of the coefficient matrix is

$$g=\begin{vmatrix} 1 & 0 & 0 \\ 0 & r^2 & 0 \\ 0 & 0 & r^2 \sin^2\theta \end{vmatrix}=r^4 \sin^2\theta. \tag{10}$$

(b) In particular, if r is a constant, we get the distance between two points (θ, ϕ) and $(\theta+d\theta, \phi+d\phi)$ on the surface of a sphere of constant radius $r=a$ as

$$ds^2=a^2\,(d\theta^2+\sin^2\theta\, d\phi^2), \tag{11}$$

with

$$g_{11}=a^2, g_{22}=a^2 \sin^2\theta, g_{12}=g_{21}=0, \tag{12}$$

and

$$g=\begin{vmatrix} a^2 & 0 \\ 0 & a^2 \sin^2\theta \end{vmatrix}=a^4 \sin^2\theta. \tag{13}$$

EXAMPLE 4: Consider a coordinate system (u, v, w) which is related to the cartesian coordinates by

$$x = vw,\ y = uw,\ z = uv. \tag{14}$$

Obtain the metric in terms of u, v, w.

Solution: The element of distance is found to be

$$ds^2 = dx^2 + dy^2 + dz^2 = (v^2 + w^2)\, du^2 + (w^2 + u^2)\, dv^2 + (u^2 + v^2)\, dw^2$$
$$+ 2uv\, du\, dv + 2vw\, dv\, dw + 2uw\, du\, dw, \tag{15}$$

which is of the form of Eq. (1) with

$$[g_{ij}] = \begin{bmatrix} v^2 + w^2 & uv & uw \\ uv & u^2 + w^2 & vw \\ uw & vw & u^2 + v^2 \end{bmatrix}, \tag{16a}$$

with

$$g = 4u^2v^2w^2. \tag{16b}$$

From Exercise 17.1, it is seen that $g_{ij} + g_{ji}$ is a symmetric covariant tensor of rank two. Now, it is evident from the quadratic form in Eq. (1) that there is no loss of generality in choosing

$$g_{ij} = g_{ji}, \tag{17}$$

that is, in choosing $[g_{ij}]$ to be a symmetric matrix. This makes g_{ij} a symmetric covariant tensor of rank two. It is called the *fundamental tensor* or the *metric tensor*.

In an N-dimensional space, a coordinate system in terms of which $g_{ij} = 0$ for $i \neq j$ is called an *orthogonal coordinate system*. Further, a system in which $g_{ii} = 1$ for $1 \leqslant i \leqslant N$ (no summation over i) and $g_{ij} = 0$ for $i \neq j$ is called a *cartesian coordinate system*.

Contravariant metric tensor

In the previous section, we have defined conjugate symmetric tensors and have noted that given a covariant symmetric tensor A_{ij} with nonzero determinant, it is possible to find a contravariant tensor B^{ij}, also symmetric, such that the inner product of the two tensors satisfies Eq. (17.12). We define the contravariant tensor conjugate to g_{ij} and denote it by g^{ij}, so that according to the definition of Eq. (17.11), we have, in this case,

$$g^{ij} = g^{ji} = (\text{cofactor of } g_{ij})/g, \tag{18}$$

and Eq. (17.12) reduces to

$$g_{ij}\, g^{jk} = \delta_i^k\, . \tag{19}$$

It is convenient to regard g_{ij} and g^{ij} not as two distinct tensors, but as the covariant and the contravariant components, respectively, of the *same* tensor, that is the metric tensor. Notice that the matrix $[g^{ij}]$ is just the inverse of the matrix $[g_{ij}]$.

EXAMPLE 5: Obtain the contravariant components of the metric tensor in a three-dimensional space in terms of spherical polar coordinates.

Solution: From Example 3, the desired components can be easily found to be

$$g^{11}=1,\ g^{22}=1/r^2,\ g^{33}=1/(r^2 \sin^2 \theta),\ g^{ij}=0 \text{ for } i\neq j. \tag{20}$$

Associate tensors

Let A^i be an arbitrary contravariant vector. The inner product of A^i with the covariant metric tensor g_{ij} will be a covariant vector. This inner product is denoted by

$$A^i g_{ij}=A_j, \tag{21}$$

and A^i and A_j are thought of as the contravariant and the covariant components, respectively, of the *same* vector. This is consistent with the interpretation of the contravariant and the covariant components given in Section 15 (page 164). Once again, taking the inner product of both sides of the above equation with the contravariant metric tensor g^{ij} and using Eq. (19), we find

$$A_j g^{jk}=A^i g_{ij} g^{jk}=A^i\ \delta_i^k=A^k, \tag{22}$$

showing that the relation between A_j and A^i is reciprocal. The tensors A^i and A_i are called *associate tensors.*

Raising and lowering of indices

The above procedure provides us with a method of changing the convariant indices of a tensor to contravariant indices and vice versa.

Thus let A^{ij} be a contravariant tensor of rank two. On taking the scalar product of A^{ij} with the metric tensor g_{ij} successively, we can define the following tensors:

$$A^{ij}\, g_{jk}=A^{i}_{.k}\,, \tag{23a}$$

$$A^{ij}\, g_{ik}=A_{k.}^{\ j}, \tag{23b}$$

$$A^{i}_{.k}\, g_{il}=A_{l.}^{\ j}\, g_{jk}=A^{ij}\, g_{il}\, g_{jk}=A_{lk}, \tag{23c}$$

where we have used a dot in the vacant place below a contravariant index of a mixed tensor to avoid confusion. Thus we have four tensors A^{ij}, $A^{i}_{.j}$, $A_{j.}^{\ i}$ and A_{ij}, all associate of each other. A^{ij} are the contravariant components, A_{ij} the covariant components, and $A^{i}_{.j}$ and $A_{j.}^{\ i}$ are the *two types* of mixed components of the same tensor. That $A^{i}_{.j}$ and $A_{j.}^{\ i}$ are not the same is easily shown. For

$$A^{i}_{\cdot j}=A^{ik}g_{kj},\ A_{j\cdot}^{\ i}=A^{ki}g_{kj}, \tag{24}$$

which clearly shows that

$$A^{i}_{\cdot j}\neq A_{j\cdot}^{\ i} \tag{25}$$

unless $A^{ik}=A^{ki}$, i.e., unless A^{ik} is a symmetric tensor.

This process is called the *lowering of an index.* When the inner product of a tensor (which has at least one contravariant index) is taken with g_{ij}, one contravariant index of the tensor is lowered to a covariant position.

The reverse process in which the inner product of a tensor is taken with the contravariant metric tensor g^{ij} raises a covariant index to a contravariant position and hence is known as the *raising of an index.*

It is worth noting that in Eq. (23a), the *second* index of the tensor A^{ij} is lowered, giving the tensor $A^{i}_{\cdot k}$, whereas in Eq. (23b), the *first* index of A^{ij} is lowered, giving $A_{k\cdot}^{\ i}$.

The inverse relationship can be obtained from Eqs. (24) by taking the inner product with g^{ij}. It is left as an exercise to show that

$$A^{i}_{\cdot k}g^{kj}=A^{ij},\ A_{k\cdot}^{\ i}g^{kj}=A^{ji},\ A_{ij}g^{ik}g^{jl}=A^{kl}. \tag{26}$$

In particular, if we treat Eq. (19) as raising an index of g_{ij}, or as lowering an index of g^{ij}, and correspondingly define the mixed metric tensor, it becomes

$$g_{i\cdot}^{\ k}=g^{k}_{\cdot i}=\delta^{k}_{\cdot i}. \tag{27}$$

Order of indices

It is fairly elementary to see that the order of the contravariant indices of an arbitrary tensor among themselves is important, and similarly, the order of its covariant indices among themselves is important. However, the above discussion shows that *the order of the contravariant and the covariant indices in relation to each other is also important*. So far, we have been denoting arbitrary mixed tensors in the form, say, A^{ij}_{kl} without regard to the order of indices. However, a simple analysis shows that the order of the four indices i, j, k, l, irrespective of whether they appear as superscripts or subscripts, has a significance.

Thus, let A^{ijkl} be a contravariant tensor of rank four. Let us successively lower the first, the second, the third and the fourth index of the tensor to obtain its mixed components in the form

$$A_{i\cdot\cdot\cdot}^{\ jkl}=A^{pjkl}g_{pi}, \tag{28a}$$

$$A^{i\ \ kl}_{\cdot j\cdot\cdot}=A^{ipkl}g_{pj}, \tag{28b}$$

$$A^{ij\ \ l}_{\cdot\cdot k\cdot}=A^{ijpl}g_{pk}, \tag{28c}$$

$$A^{ijk}_{\cdot\cdot\cdot l}=A^{ijkp}g_{pl}. \tag{28d}$$

Interchanging i and j in Eq. (28b), it becomes

$$A^{j\ \ kl}_{\cdot i\cdot\cdot}=A^{jpkl}g_{pi}. \tag{29}$$

Comparison of Eqs. (29) and (28a) shows that

$$A_{i\cdot\cdot\cdot}^{\ jkl}\neq A^{j\ \ kl}_{\cdot i\cdot\cdot}, \tag{30}$$

unless A^{jpkl} is symmetric in its first two indices. The inequality (30) shows clearly that it makes a difference whether the subscript occurs first or the superscript occurs first.

Lowering one more index, we obtain

$$A_{ij}^{\ \ kl}{}_{\cdot\cdot}=A^{pqkl}\,g_{pi}\,g_{qj}, \tag{31a}$$

$$A_{i\cdot k\cdot}^{\ j\ l}=A^{pjql}\,g_{pi}\,g_{qk}, \tag{31b}$$

$$A_{i\cdot\cdot l}^{\ jk}=A^{pjkq}\,g_{pi}\,g_{ql}, \tag{31c}$$

$$A^{i\ k}_{\cdot j\cdot l}=A^{ipkq}\,g_{pj}\,g_{ql}, \tag{31d}$$

etc., as the different types of components each of contravariant rank two and covariant rank two. If all the four indices are lowered, we obtain the covariant tensor of rank four given by

$$A_{ijkl}=A^{pqrs}\,g_{pi}\,g_{qj}\,g_{rk}\,g_{sl}, \tag{32}$$

which is, of course, unique.

EXAMPLE 6: If A^{ij} and B^{ij} are two tensors, show that

$$A^{ij}\,B_{ij}=A_{ij}\,B^{ij}. \tag{33}$$

Solution: We can write

$$A^{ij}=A_{kl}\,g^{ki}\,g^{lj},$$

$$B_{ij}=B^{mn}\,g_{mi}\,g_{nj}. \tag{34}$$

Taking the inner product of the respective sides of Eqs. (34), we have

$$A^{ij}\,B_{ij}=A_{kl}\,B^{mn}\,g^{ki}\,g_{mi}\,g^{lj}\,g_{nj}$$

$$=A_{kl}\,B^{\cdot\cdot n}\,\delta^{k}_{\cdot m}\,\delta^{l}_{\cdot n} \qquad \text{[by Eq. (19)]}$$

$$=A_{kl}\,B^{kl}=A_{ij}\,B^{ij},$$

which is the desired result.

EXAMPLE 7: Show that in a cartesian coordinate system, the contravariant and the covariant components of a vector are identical.

Solution: Let A^i and A_i be respectively the contravariant and the covariant components of a vector with respect to a cartesian coordinate system. They are then related by

$$A_i = A^j g_{ji}. \tag{35}$$

For a cartesian coordinate system, $g_{ii}=1$ (no summation over i) and $g_{ij}=0$ if $i \neq j$. This reduces Eq. (35) to

$$\begin{aligned} A_i &= A^i g_{ii} \text{ (no summation over } i) \\ &= A^i, \end{aligned} \tag{36}$$

showing that there is no distinction between the covariant and the contravariant components.

A generalization of this result to arbitrary tensors is left to Exercise 7.

Before closing this section, we note that the magnitude A of a vector A_i is defined by

$$A^2 = A_i A^i = g_{ij} A^i A^j = g^{ij} A_i A_j. \tag{37}$$

Similarly, the angle θ between two vectors A_i and B_i is defined by

$$AB \cos\theta = A_i B^i = g_{ij} A^i B^j = g^{ij} A_i B_j, \tag{38}$$

where A and B are the magnitudes of A_i and B_i, respectively, or

$$\cos\theta = A_i B^i / [A_j A^j B_k B^k]^{1/2}. \tag{39}$$

Exercise 18

1. Obtain Eqs. (26) from Eqs. (24).

2. Referring to Exercise 15.4, show that the covariant components of velocity in spherical polar coordinates are

$$\frac{dr}{dt}, \; r^2 \frac{d\theta}{dt}, \; r^2 \sin^2\theta \frac{d\phi}{dt}.$$

*3. Using the covariant components of the gradient of a scalar field in sperical polar coordinates obtained in Exercise 15.6, find the contravariant components in the same coordinate system.

*4. Let z^1 and z^2 be plane oblique coordinates with an angle α between the two axes. Determine g^{ij} and g_{ij} for these coordinates.

5. Show that in the case of an orthogonal coordinate system, a covariant component of a tensor is related only to the *corresponding* contravariant component of the tensor.

6. The contravariant components of the acceleration vector in spherical polar coordinates are given in Eqs. (15.50). Obtain the covariant components of the acceleration vector in the same coordinate system.

7. Show that with respect to a cartesian coordinate system, the distinction among the contravariant, the covariant and the mixed components of a tensor vanishes.

*8. Define the covariant tensor δ_{jk} by lowering the contravariant index of the mixed Kronecker delta $\delta^{l}_{\cdot j}$. In a three-dimensional space, obtain the components δ_{jk} in (a) the cartesian coordinates, (b) the spherical polar coordinates, and (c) the coordinates defined in Example 4. [Hence note that though δ_{jk} is a Kronecker delta in the cartesian coordinate system, it is not a Kronecker delta in any other coordinate system.]

9. Show that

$$A_{ijk}B^{jlk}=A_{i\cdot k}^{\;j}B_{j\cdot\cdot}^{\;lk}=A_{i\cdot\cdot}^{\;jk}B_{j\cdot k}^{\;l}=A_{ij\cdot}^{\;\;k}B_{\cdot\cdot k}^{jl}.$$

19. Tensors in nonrelativistic physics

In Section 15, we have already pointed out a few occasions where tensors of the second rank occur in physics. In this section, we shall discuss a few more cases which also require the use of tensors of higher than the second rank.

Stress, strain and Hook's law

Stress is defined as the force per unit area of a solid. It is evident that the stress at a given point in the solid will depend on the orientation of the element of area at that point. In order to specify the stress at a point in the solid completely, therefore, one must specify the stresses on three mutually perpendicular elements of area at the point.

Let P be a point in the solid at which the stress is to be specified. Choose cartesian coordinate axes x, y, z in the body with origin at the point P and consider three elements of area $d\sigma_x$, $d\sigma_y$, $d\sigma_z$, respectively in the yz, zx and xy planes, and each containing the point P (see Fig. 10). Now the net stress at the point P will be specified by a force $\mathbf{F}_1$ acting on the area $d\sigma_x$, a force $\mathbf{F}_2$ acting on $d\sigma_y$, and a force

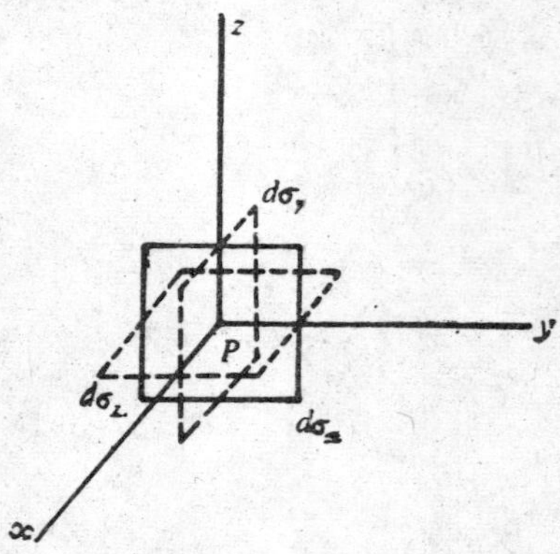

Figure 10. Three mutually perpendicular elements of area at the point P.

$\mathbf{F}_3$ acting on $d\sigma_z$. Each of these forces can be further resolved into its cartesian components. For example, F_{1x} would be the force acting on $d\sigma_x$ along the x direction, that is, perpendicular to the area $d\sigma_x$, while F_{1y} and F_{1z} would be the forces acting on $d\sigma_x$ respectively along the y and z directions. Since stress is force per unit area, let us define the corresponding stress components by

$$X_x = F_{1x}/d\sigma_x,\ Y_x = F_{1y}/d\sigma_x,\ Z_x = F_{1z}/d\sigma_x. \tag{1}$$

These three stress components acting on the element of area $d\sigma_x$ are illustrated in Fig. 11.

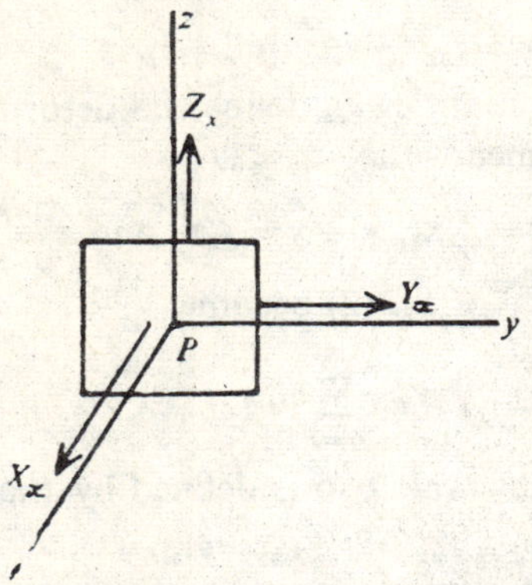

Figure 11. The three stress components X_x, Y_x, Z_x acting on the element of area $d\sigma_x$.

In a similar manner, we can introduce the stress components X_y, Y_y, Z_y, which define the force per unit area acting on $d\sigma_y$, and three more stress components X_z, Y_z, Z_z, which give the force per unit area acting on $d\sigma_z$. These nine components are the cartesian components of the *stress tensor* which is a second rank tensor.

In fact, we could have arrived at this conclusion directly from the definition of stress which is force per unit area. Force is a vector and area is a vector and the force is not necessarily normal to the area; this makes stress a tensor of the second rank.

We shall denote the cartesian components of the stress tensor by X_{ij}, where $i, j = 1, 2, 3$. Identifying x, y, z with 1, 2, 3, we see that, for example, X_{11} stands for X_x, X_{12} for X_y, X_{31} for Z_x, etc.

The stress in a body gives rise to strain or deformation of the body. The strain at any point in the body can be specified as follows. Imagine an orthonormal coordinate system $\hat{\mathbf{x}}$, $\hat{\mathbf{y}}$, $\hat{\mathbf{z}}$ in the unstrained body at the point P under consideration. After a deformation of the body, the axes are distorted in orientation and length. Let the new axes be denoted by $\mathbf{x}'$, $\mathbf{y}'$, $\mathbf{z}'$. If the stress and the resulting strain are small, it may be assumed that the deviations of $\mathbf{x}'$ from $\hat{\mathbf{x}}$, of $\mathbf{y}'$ from $\hat{\mathbf{y}}$ and of $\mathbf{z}'$ from $\hat{\mathbf{z}}$ are small. The new vectors can then be written as linear combinations of the original basic vectors in the form[1]

$$\begin{aligned} \mathbf{x}' &= (1+\varepsilon_{xx})\,\hat{\mathbf{x}}+\varepsilon_{xy}\hat{\mathbf{y}}+\varepsilon_{xz}\hat{\mathbf{z}}, \\ \mathbf{y}' &= \varepsilon_{yx}\hat{\mathbf{x}}+(1+\varepsilon_{yy})\,\hat{\mathbf{y}}+\varepsilon_{yz}\hat{\mathbf{z}}, \\ \mathbf{z}' &= \varepsilon_{zx}\hat{\mathbf{x}}+\varepsilon_{zy}\hat{\mathbf{y}}+(1+\varepsilon_{zz})\hat{\mathbf{z}}. \end{aligned} \tag{2}$$

The dimensionless quantities ε_{ij} define the deformation. The assumption of small strain means that $\varepsilon_{ij} \ll 1$.

If we define $\mathbf{x}'-\hat{\mathbf{x}}=\Delta\mathbf{x}_1$, $\mathbf{y}'-\hat{\mathbf{y}}=\Delta\mathbf{x}_2$ and $\mathbf{z}'=\hat{\mathbf{z}}=\Delta\mathbf{x}_3$, and denote $(\hat{\mathbf{x}}, \hat{\mathbf{y}}, \hat{\mathbf{z}})$ by $(\mathbf{x}_1, \mathbf{x}_2, \mathbf{x}_3)$, Eqs. (2) become

$$\Delta\mathbf{x}_i=\sum_{j=1}^{3}\varepsilon_{ij}\,\mathbf{x}_j. \tag{3}$$

The *strain tensor* e_{ij} of rank two is defined by the relations

$$\begin{gathered} e_{xx}=\varepsilon_{xx},\ e_{yy}=\varepsilon_{yy},\ e_{zz}=\varepsilon_{zz}, \\ e_{xy}=e_{yx}=\varepsilon_{xy}+\varepsilon_{yx},\ e_{yz}=e_{zy}=\varepsilon_{yz}+\varepsilon_{zy},\ e_{zx}=e_{xz}=\varepsilon_{zx}+\varepsilon_{xz}. \end{gathered} \tag{4}$$

For small stresses and strains, the Hooke's law states that the stress is 'proportional' to the strain; strictly speaking, *the stress components are linearly related to the strain components.* In tensorial notation, this relation could be expressed as[2]

$$X_{ij}=C_{ijkl}e_{kl}, \tag{5}$$

summation over repeated indices being implied. The $3^4=81$ coefficients C_{ijkl} are called *elastic stiffness constants* or *moduli of elasticity*, and are evidently the components of a fourth rank tensor. The relation given in Eq. (5) can be inverted to express the strain components as linear functions of the stress components in the form

$$e_{ij}=S_{ijkl}X_{kl}, \tag{6}$$

where the components S_{ijkl} are called the *elastic compliance constants* or simply *elastic constants.* Substitution of Eq. (5) into Eq. (6) shows that the tensors C_{ijkl} and S_{ijkl} are related by

$$C_{ijkl}S_{klmn}=\delta_{im}\delta_{jn}. \tag{7}$$

In a sence, the two tensors are reciprocals of each other.[3]

[1]See, for example, C. Kittel, *Introduction to Solid State Physics*, 4th ed. (Wiley, New York, 1971), Chap. 4.

[2]In accordance with the results of Example 18.7 and Exercise 18.7, all the indices in the following equations have been written as covariant indices. The summation convention, nevertheless, remains in operation.

[3]Compare this with the conjugate tensors defined in Section 17.

It turns out, as can be readily verified, that the strain tensor and the stress tensor are symmetric, so that

$$X_{ij}=X_{ji},\ e_{ij}=e_{ji}. \tag{8}$$

This reduces the number of independent elements of either tensor to six. At the same time, the number of independent elements of the elastic stiffness tensor C_{ijkl} or the elastic compliance tensor S_{ijkl} becomes $6^2=36$. It can be further shown that these tensors are symmetric under the interchange of the first two and the last two indices, that is,

$$C_{ijkl}=C_{klij},\ S_{ijkl}=S_{klij}. \tag{9}$$

This brings down the number of independent elastic constants *for any solid* to $6(6+1)/2=21$. If the solid has any internal symmetries, the number of independent elastic constants is further reduced. A cubic crystal, for example, has only three independent elastic constants which are conventionally chosen to be

$$C_{11}\equiv C_{xxxx},\ C_{12}\equiv C_{xxyy},\ C_{44}\equiv C_{yzyz}. \tag{10}$$

Each of the remaining components either equals one of the above three or identically vanishes.

Piezoelectricity and dielectric susceptibility

When an electric field **E** is applied to a crystal, an electric polarization[4] **P** develops across the crystal. If the polarization **P** is always in the same direction as the applied electric field **E**, which would be the case in an isotropic medium, the ratio $P/E=\chi$ is called the *dielectric susceptibility* of the medium. However, in anisotropic solids, the electric polarization that develops is not always parallel to the electric field. For small fields, the components of polarization are linearly related to those of the electric field, so that the modified relationship becomes

$$P_i=\chi_{ij}E_j. \tag{11}$$

The coefficients χ_{ij} are the components of a second rank tensor known as the *dielectric susceptibility tensor* of the medium.[5]

There are certain crystals which develop an electric polarization under an external mechanical stress (even in the absence of an external electric field). Such crystals are called *piezoelectric crystals* and

[4]Electric polarization is defined as the electric dipole moment per unit volume.

[5]See, for example, E. Fatuzzo and W.J. Merz, *Ferroelectricity* (North-Holland, Amsterdam, 1967), p. 44.

quartz is an excellent example of these. For small stresses, the components of electric polarization are linear functions of the stress tensor, so that the relation can be written in the form

$$P_i = d_{ijk} X_{jk}. \tag{12}$$

The coefficients d_{ijk} are the components of a tensor of rank three known as the *piezoelectric strain coefficient tensor*. Since the tensor X_{ij} has only six independent elements, the tensor d_{ijk} has only $3 \times 6 = 18$, rather than $3^3 = 27$, independent components for an arbitrary medium. For a medium with internal symmetries, the number of independent components is much less.

If a piezoelectric crystal is simultaneously subjected to an external electric field and an external mechanical stress, the resulting electric polarization will have contributions from both the sources. For small fields and stresses, it can be assumed that there is little interaction between the two agents so that the contributions to the electric polarization can be taken to be additive. The total electric polarization is then given by

$$P_i = d_{ijk} X_{jk} + \chi_{ij} E_j. \tag{13}$$

Moment of inertia tensor

When a system consisting of a number of point masses is in a rotatory motion, its total angular momentum is given by[6]

$$\mathbf{L} = \sum_i m_i (\mathbf{r}_i \times \mathbf{v}_i), \tag{14}$$

where m_i is the mass of the i-th point mass, $\mathbf{r}_i$ its position vector with respect to some origin and $\mathbf{v}_i$ its linear velocity. If $\boldsymbol{\omega}$ is the angular velocity vector of the body, the linear velocity of the i-th point mass is given by $\mathbf{v}_i = \boldsymbol{\omega} \times \mathbf{r}_i$. Using this in Eq. (14), the angular momentum becomes

$$\mathbf{L} = \sum_i m_i [\mathbf{r}_i \times (\boldsymbol{\omega} \times \mathbf{r}_i)] = \sum_i m_i [\boldsymbol{\omega} r_i^2 - \mathbf{r}_i (\mathbf{r}_i . \boldsymbol{\omega})], \tag{15}$$

where we have used the vector identity $\mathbf{a} \times (\mathbf{b} \times \mathbf{c}) = (\mathbf{a} . \mathbf{c})\mathbf{b} - (\mathbf{a} . \mathbf{b})\mathbf{c}$. Let the cartesian components of the various vectors occurring in Eq. (15) be written as $\mathbf{L} = (L_x, L_y, L_z)$, $\mathbf{r}_i = (x_i, y_i, z_i)$ and $\boldsymbol{\omega} = (\omega_x, \omega_y, \omega_z)$. It could then be written in the form

[6]See H. Goldstein, *Classical Mechanics* (Addison-Wesley, Reading, Mass., 1968), Chap. 5.

$$\begin{aligned} L_x &= I_{xx}\omega_x + I_{xy}\omega_y + I_{xz}\omega_z, \\ L_y &= I_{yx}\omega_x + I_{yy}\omega_y + I_{yz}\omega_z, \\ L_z &= I_{zx}\omega_x + I_{zy}\omega_y + I_{zz}\omega_z, \end{aligned} \tag{16}$$

with

$$\begin{gathered} I_{xx} = \sum_i m_i(r_i^2 - x_i^2),\ I_{yy} = \sum_i m_i(r_i^2 - y_i^2),\ I_{zz} = \sum_i m_i(r_i^2 - z_i^2), \\ I_{xy} = I_{yx} = -\sum_i m_i x_i y_i,\ I_{yz} = I_{zy} = -\sum_i m_i y_i z_i, \\ I_{zx} = I_{xz} = -\sum_i m_i z_i x_i. \end{gathered} \tag{17}$$

Eqs. (16) show that the angular momentum $\mathbf{L}$ is not necessarily parallel to the angular velocity $\boldsymbol{\omega}$. Eqs. (16) could be written briefly in the tensor notation as

$$L_i = I_{ij}\omega_j. \tag{18}$$

Since L_i and ω_i are vectors, it follows from the quotient law that I_{ij} is a tensor of the second rank. This symmetric tensor with components given by Eqs. (17) is called the *moment of inertia tensor* of the body.

20. Tensor calculus

It was shown in Example 15.3(b) that the partial derivatives of a scalar field with respect to the coordinates are the components of a covariant vector. In general, however, differentiation of a tensor (except that of rank zero) with respect to the coordinates does not give a tensor. It is nevertheless possible to construct expressions containing the partial derivatives of a tensor and those of the metric tensor transforming like a tensor. Such expressions will be defined and discussed in this section.

It will be convenient to introduce the following notation for the partial derivatives of the unbarred coordinates with respect to the barred coordinates and vice versa:[1]

$$x^i_{,\alpha} \equiv \partial x^i / \partial \bar{x}^\alpha, \; \bar{x}^\alpha_{,i} \equiv \partial \bar{x}^\alpha / \partial x^i. \tag{1}$$

In terms of this notation, Eqs. (15.23) become

$$x^i_{,\alpha} \bar{x}^\alpha_{,j} = \delta^i_{.j}; \; \bar{x}^\alpha_{,i} x^i_{,\beta} = \delta^\alpha_{.\beta}. \tag{2}$$

Further, the second partial derivatives will be denoted in a similar way by

$$x^i_{,\alpha\beta} = \partial^2 x^i / \partial \bar{x}^\alpha \partial \bar{x}^\beta, \tag{3}$$

etc. It is evident that the order of the indices to the right of the comma is immaterial. Finally, the coordinate derivatives of a scalar ϕ and a tensor, say $A^{ij}_{..k}$, will be denoted by

$$\phi_{,i} = \partial \phi / \partial x^i, \; \phi_{,ij} = \partial^2 \phi / \partial x^i \partial x^j, \; A^{ij}_{..k,l} = \partial A^{ij}_{..k} / \partial x^l, \text{ etc.} \tag{4}$$

This notation will be used in the rest of this chapter.

Differentiation of a tensor

If $\phi(x^i)$ is a scalar field, its derivatives with respect to the coordinates, $A_i = \phi_{,i}$, transform like the components of a covariant vector, i.e.,

$$\bar{A}_\alpha = x^i_{,\alpha} A_i, \tag{5}$$

which is the inverse transformation of Eq. (15.36). Differentiating both sides of the above equation with respect to $\bar{x}^\beta$, we find

$$\bar{A}_{\alpha,\beta} = x^i_{,\alpha\beta} A_i + x^i_{,\alpha} A_{i,\beta}$$

[1]Note that $x^i_{,\alpha}$ and $\bar{x}^\alpha_{,i}$ are *not* tensors. It is to be remembered in using this notation that Latin letters denote unbarred coordinates while Greek letters denote barred coordinates.

$$=x^{i}_{,\alpha\beta}A_i+x^{i}_{,\alpha}x^{j}_{,\beta}A_{i,j}. \tag{6}$$

The second term on the right hand side of the above equation has a tensorial character, but the appearance of the first term shows that the functions $A_{i,j}$ do not transform like the components of a second rank tensor.

In a similar way, it can be seen that the coordinate derivatives of any tensor (excluding a scalar) do not transform like the components of a tensor.

To find a 'derivative' which is a tensor, we shall first have to define new symbols involving the derivatives of the metric tensor.

Christoffel symbols

The *Christoffel three-index symbols,* or simply the *Christoffel symbols,* of the first and the second kinds are respectively defined by[2]

$$[ij,k]=\tfrac{1}{2}(g_{ik,j}+g_{jk,i}-g_{ij,k}); \tag{7a}$$

$$\Gamma^{k}{}_{ij}=g^{km}\,[ij,m]=\tfrac{1}{2}\,g^{km}(g_{im,j}+g_{jm,i}-g_{ij,m}). \tag{7b}$$

Since the metric tensor is symmetric, it is evident from the above equations that the Christoffel symbols have the following symmetry:

$$[ji,k]=[ij,k],\ \Gamma^{k}{}_{ji}=\Gamma^{k}{}_{ij}. \tag{8}$$

EXAMPLE 1: Prove that

$$g_{ij,k}=[ik,j]+[jk,i]. \tag{9}$$

Solution: From the definition of the Christoffel symbols of the first kind, we have

$$[ik,j]=\tfrac{1}{2}(g_{ij,k}+g_{kj,i}-g_{ik,j}); \tag{10a}$$

$$[jk,i]=\tfrac{1}{2}(g_{ji,k}+g_{ki,j}-g_{jk,i}). \tag{10b}$$

Adding the respective sides of the above two equations and noting that g_{ij} is a symmetric tensor, the desired result immediately follows.

The Christoffel symbols are not tensors. This is shown in the following example.

EXAMPLE 2: How do the Christoffel symbols of the first and the second kinds transform under coordinate transformations?

Solution: To obtain the transformation of the Christoffel symbols, we use the defining Eqs. (7) and the transformation of the metric tensor. The metric tensor transforms according to

[2]The Christoffel symbol of the second kind is also denoted by $\left\{ {k \atop ij} \right\}$.

$$\bar{g}_{\alpha\beta}=x^i{}_{,\alpha}x^j{}_{,\beta}g_{ij}. \tag{11}$$

Differentiating this with respect to $\bar{x}^\gamma$, we get

$$\begin{aligned}\bar{g}_{\alpha\beta,\gamma}&=[x^i{}_{,\alpha\gamma}x^j{}_{,\beta}+x^i{}_{,\alpha}x^j{}_{,\beta\gamma}]\,g_{ij}+x^i{}_{,\alpha}x^j{}_{,\beta}g_{ij,\gamma}\\&=[x^i{}_{,\alpha\gamma}x^j{}_{,\beta}+x^i{}_{,\alpha}x^j{}_{,\beta\gamma}]\,g_{ij}+x^i{}_{,\alpha}x^j{}_{,\beta}x^k{}_{,\gamma}g_{ij,k}.\end{aligned} \tag{12}$$

Cyclic permutation of the three indices α, β and γ in Eq. (12) gives two more similar equations:

$$\bar{g}_{\beta\gamma,\alpha}=[x^i{}_{,\beta\alpha}x^j{}_{,\gamma}+x^i{}_{,\beta}x^j{}_{,\gamma\alpha}]\,g_{ij}+x^i{}_{,\beta}x^j{}_{,\gamma}x^k{}_{,\alpha}g_{ij,k}; \tag{13}$$

$$\bar{g}_{\gamma\alpha,\beta}=[x^i{}_{,\gamma\beta}x^j{}_{,\alpha}+x^i{}_{,\gamma}x^j{}_{,\alpha\beta}]\,g_{ij}+x^i{}_{,\gamma}x^j{}_{,\alpha}x^k{}_{,\beta}g_{ij,k}. \tag{14}$$

Now, add the two Eqs. (13) and (14), subtract Eq. (12) and divide by 2. The left hand side gives the Christoffel symbol of the first kind in the barred coordinate system:

$$\tfrac{1}{2}(\bar{g}_{\beta\gamma,\alpha}+\bar{g}_{\gamma\alpha,\beta}-\bar{g}_{\alpha\beta,\gamma})=\overline{[\alpha\beta,\gamma]}. \tag{15}$$

Since g_{ij} is a symmetric tensor, the indices i and j could be interchanged in the coefficient of g_{ij} on the right hand side of Eq. (12). This term then becomes

$$[x^j{}_{,\alpha\gamma}x^i{}_{,\beta}+x^j{}_{,\alpha}x^i{}_{,\beta\gamma}]\,g_{ij}.$$

In the last terms on the right hand sides of Eqs. (13) and (14), we perform respectively the index changes $i\to j\to k\to i$ and $i\to k\to j\to i$. Finally, performing the operation mentioned earlier, we arrive at the result

$$\overline{[\alpha\beta,\gamma]}=x^i{}_{,\alpha\beta}x^j{}_{,\gamma}g_{ij}+x^i{}_{,\alpha}x^j{}_{,\beta}x^k{}_{,\gamma}\,[ij,k]. \tag{16}$$

This is the law of transformation of the Christoffel symbols of the first kind from one coordinate system to another. The presence of the first term on the right hand side of the above equation shows that the Christoffel symbol of the first kind is not a tensor.

The transformation of the Christoffel symbol of the second kind can be obtained by taking the inner product of Eq. (16) with $\bar{g}^{\gamma\rho}$, which gives

$$\begin{aligned}\bar{\Gamma}^\rho{}_{\alpha\beta}\equiv\bar{g}^{\gamma\rho}\,\overline{[\alpha\beta,\gamma]}&=x^i{}_{,\alpha\beta}x^j{}_{,\gamma}\bar{g}^{\gamma\rho}g_{ij}+x^i{}_{,\alpha}x^j{}_{,\beta}x^k{}_{,\gamma}\bar{g}^{\gamma\rho}\,[ij,k]\\&=x^i{}_{,\alpha\beta}x^j{}_{,\gamma}\bar{x}^\gamma{}_{,k}\bar{x}^\rho{}_{,l}g^{kl}g_{ij}+x^i{}_{,\alpha}x^j{}_{,\beta}x^k{}_{,\gamma}\bar{x}^\gamma{}_{,m}\bar{x}^\rho{}_{,n}g^{mn}\,[ij,k]\\&=x^i{}_{,\alpha\beta}\bar{x}^\rho{}_{,l}\,\delta^j{}_{.k}g^{kl}g_{ij}+x^i{}_{,\alpha}x^j{}_{,\beta}\bar{x}^\rho{}_{,n}\delta^k{}_{.m}g^{mn}\,[ij,k];\end{aligned} \tag{17}$$

or

$$\bar{\Gamma}^\rho{}_{\alpha\beta}=x^i{}_{,\alpha\beta}\bar{x}^\rho{}_{,i}+x^i{}_{,\alpha}x^j{}_{,\beta}\bar{x}^\rho{}_{,n}\Gamma^n{}_{ij}. \tag{18}$$

This gives the law of transformation of the Christoffel symbols of the second kind. Owing to the presence of the first term on the right hand side, this is also not tensorial in character.

None the less, the nature of the second term on the right hand sides of Eqs. (16) and (18) suggests that in the two kinds of Christoffel symbols $[ij, k]$ and $\Gamma^r{}_{mn}$, the indices i, j, k, m and n occur as covariant indices while only r appears as a contravariant index. This rule is very helpful in checking the correctness of indices in an equation containing Christoffel symbols of the two kinds.

Covariant derivative

Taking the inner product of Eq. (18) with $x^k{}_{,\rho}$ and rearranging terms, we obtain

$$x^k{}_{,\alpha\beta}=x^k{}_{,\rho}\bar{\Gamma}^\rho{}_{\alpha\beta}-x^i{}_{,\alpha}x^j{}_{,\beta}\Gamma^k{}_{ij}. \tag{19}$$

This gives us the second partial derivatives of the x^k with respect to the $\bar{x}^\alpha$ in terms of their first partial derivatives and the Christoffel symbols of the second kind.

Referring back to Eq. (6), we substitute for the first term on its right hand side from Eq. (19) (with suitable change of indices, of course). This gives

$$\begin{aligned}\bar{A}_{\alpha,\beta}&=x^i{}_{,\rho}A_i\bar{\Gamma}^\rho{}_{\alpha\beta}-x^i{}_{,\alpha}x^i{}_{,\beta}A_k\Gamma^k{}_{ij}+x^i{}_{,\alpha}x^j{}_{,\beta}A_{i,j}\\&=\bar{A}_\rho\bar{\Gamma}^\rho{}_{\alpha\beta}+x^i{}_{,\alpha}x^j{}_{,\beta}(A_{i,j}-A_k\Gamma^k{}_{ij}),\end{aligned} \tag{20}$$

or

$$(\bar{A}_{\alpha,\beta}-\bar{A}_\rho\bar{\Gamma}^\rho{}_{\alpha\beta})=x^i{}_{,\alpha}x^j{}_{,\beta}(A_{i,j}-A_k\,\Gamma^k{}_{ij}). \tag{21}$$

If we define

$$A_{i;j}\equiv A_{i,j}-A_k\Gamma^k{}_{ij}, \tag{22a}$$

so that

$$\bar{A}_{\alpha;\beta}=\bar{A}_{\alpha,\beta}-A_\rho\bar{\Gamma}^\rho{}_{\alpha\beta}, \tag{22b}$$

Eq. (21) becomes

$$\bar{A}_{\alpha;\beta}=x^i{}_{,\alpha}x^j{}_{,\beta}A_{i;j}. \tag{23}$$

This shows that we have finally obtained N^2 functions defined by Eqs. (22) containing the partial derivatives of a covariant vector and the Christoffel symbols of the second kind transforming like the components of a covariant second rank tensor. The function[3] $A_{i;j}$ is said to be the j-th *covariant derivative* of the vector A_i.

[3]It is important to distinguish between the symbols $A_{i,j}$ and $A_{i;j}$. The former denotes the ordinary derivative of A_i with respect to x^j while the latter stands for the j-th covariant derivative of A_i. The latter is a tensor, while the former is not. Different notations in this regard are in vogue in different books.

One could start with a contravariant vector A^i, follow the same steps as above and arrive at the result that

$$(\bar{A}^{\alpha}{}_{,\beta}+\bar{A}^{\rho}\bar{\Gamma}^{\alpha}{}_{\beta\rho})=\bar{x}^{\alpha}{}_{,i}x^{j}{}_{,\beta}(A^{i}{}_{,j}+A^{k}\Gamma^{i}{}_{jk}), \tag{24}$$

showing that the mixed tensor

$$A^{i}_{.;j}\equiv A^{i}{}_{,j}+A^{k}\Gamma^{i}{}_{jk} \tag{25}$$

of rank two can be treated as the covariant derivative of the contravariant vector A^i.

The covariant derivative of a tensor of rank zero (scalar) is identical to its ordinary derivative. Thus, if ϕ is a scalar function,

$$\phi_{;j}=\phi_{,j}\equiv\partial\phi/\partial x^{j}. \tag{26}$$

Finally, the covariant derivative of a tensor of arbitrary rank can be defined by

$$A^{i_1 i_2 \ldots i_p}_{j_1 j_2 \ldots j_q;k}\equiv A^{i_1 i_2 \ldots i_p}_{j_1 j_2 \ldots j_q,k}+\sum_{r=1}^{p} A^{i_1 \ldots i_{r-1}\mu i_{r+1}\ldots i_p}_{j_1 j_2 \ldots j_q}\Gamma^{i_r}_{\mu k}$$

$$-\sum_{s=1}^{q} A^{i_1 i_2 \ldots i_p}_{j_1 \ldots j_{s-1}\mu j_{s+1}\ldots j_q}\Gamma^{\mu}_{j_s k}, \tag{27}$$

which can be shown to be a tensor whose contravariant rank is the same as that of the original tensor and whose covariant rank is increased by unity.

EXAMPLE 3: Write down the covariant derivatives of the tensors A_{ij}, A^{ij} and $A^{i}_{.j}$.

Solution: As special cases of Eq. (27), we have

$$A_{ij;k}=A_{ij,k}-A_{pj}\Gamma^{p}{}_{ik}-A_{ip}\Gamma^{p}{}_{jk}; \tag{28a}$$

$$A^{ij}_{..;k}=A^{ij}_{..,k}+A^{pj}\Gamma^{i}{}_{pk}+A^{ip}\Gamma^{j}{}_{pk}; \tag{28b}$$

$$A^{i}_{.j;k}=A^{i}_{.j,k}+A^{p}_{.j}\Gamma^{i}{}_{pk}-A^{i}_{.p}\Gamma^{p}{}_{jk}. \tag{28c}$$

EXAMPLE 4: Show that the covariant derivative of (a) the metric tensor, and (b) the Kronecker delta are identically zero.

Solution: (a) The covariant derivative of a covariant tensor of rank two is given in Eq. (28a). Replacing A_{ij} by g_{ij} in that equation, we find that

$$g_{ij;k}=g_{ij,k}-g_{pj}\Gamma^{p}{}_{ik}-g_{ip}\Gamma^{p}{}_{jk}. \tag{29}$$

It can be shown by using the first part of Eq. (7b) that

$$g_{pj}\Gamma^{p}{}_{ik}=[ik,j],\ g_{ip}\Gamma^{p}{}_{jk}=[jk,i]. \tag{30}$$

This reduces Eq. (29) to

$$g_{ij;k}=g_{ij,k}-[ik,j]-[jk,i]=0, \tag{31}$$

where we have used the result of Eq. (9).

For the contravariant metric tensor, we replace A^{ij} by g^{ij} in Eq. (28b) to find that the covariant derivative of the contravariant metric tensor is given by

$$\begin{aligned} g^{ij}_{..;k} &= g^{ij}_{..,k} + g^{pj}\Gamma^{i}_{pk} + g^{ip}\Gamma^{j}_{pk} \\ &= g^{ij}_{..,k} + g^{pj}g^{qi}[pk, q] + g^{ip}g^{jq}[pk, q] \\ &= g^{ij}_{..,k} + g^{pj}g^{qi}\{[pk, q] + [qk, p]\} \\ &= g^{ij}_{..,k} + g^{pj}g^{qi}g_{pq,k} \text{ [by Eq. (9)]}. \end{aligned} \tag{32}$$

Now, consider the expression

$$\begin{aligned} (g^{pj}g^{qi}g_{pq})_{,k} &= g^{qi}g_{pq}g^{pj}{}_{,k} + g^{pj}g_{pq}g^{qi}{}_{,k} + g^{pj}g^{qi}g_{pq,k} \\ &= \delta^{i}_{.p}g^{pj}_{..,k} + \delta^{j}_{.q}g^{qi}_{..,k} + g^{pj}g^{qi}g_{pq,k} \\ &= 2g^{ij}_{..,k} + g^{pj}g^{qi}g_{pq,k}. \end{aligned} \tag{33}$$

On the other hand,

$$g^{pj}g^{qi}g_{pq} = g^{pj}\delta^{i}_{.p} = g^{ij}, \tag{34}$$

so that

$$(g^{pj}g^{qi}g_{pq})_{,k} = g^{ij}_{..,k}. \tag{35}$$

Comparing Eqs. (33) and (35), we see that the right hand side of Eq. (32) is identically zero, proving that

$$g^{ij}_{..;k} = 0. \tag{36}$$

The results of Eqs. (31) and (36) are known as *Ricci's lemma.*

(b) On replacing $A^{i}_{.j}$ by $\delta^{i}_{.j}$ in Eq. (28c), we see that

$$\begin{aligned} \delta^{i}_{.j;k} &= \delta^{i}_{.j,k} + \delta^{p}_{.j}\Gamma^{i}_{pk} - \delta^{i}_{.p}\Gamma^{p}_{jk} \\ &= \delta^{i}_{.j,k} + \Gamma^{i}_{jk} - \Gamma^{i}_{jk}. \end{aligned} \tag{37}$$

Noting that the components of the Kronecker delta tensor are independent of coordinates, so that $\delta^{i}_{.j,k} = 0$, we have the desired result that

$$\delta^{i}_{.j;k} = 0. \tag{38}$$

EXAMPLE 5: Find the Christoffel symbols of the first and the second kinds in the two-dimensional space of the surface of a sphere in terms of polar coordinates.

Solution: The summation convention will be dropped in the solution of this example. The Christoffel symbol of the first kind is given by

$$[ij, k] = \tfrac{1}{2}(g_{ik,j} + g_{jk,i} - g_{ij,k}). \tag{39}$$

We choose the polar coordinates to be $x^1 \equiv \theta$, $x^2 \equiv \phi$. The components of the covariant metric tensor g_{ij} for the space under consideration are given in Eqs. (18.12).

The indices i, j, k can take only two values, 1 and 2. We have therefore the following three categories of Christoffel symbols.

CATEGORY 1: $i=j=k$; Eq. (39) reduces to

$$[ii, i]=\tfrac{1}{2}g_{ii,i}=\tfrac{1}{2}\,\partial g_{ii}/\partial x^i. \tag{40}$$

From Eqs. (18.12), we see that

$$\partial g_{11}/\partial\theta=0,\ \partial g_{22}/\partial\phi=0; \tag{41}$$

herefore

$$[ii, i]=0,\ i=1, 2. \tag{42}$$

CATEGORY 2: $i=k\neq j$; Eq. (39) reduces to

$$[ij, i]=\tfrac{1}{2}g_{ii,j}=\tfrac{1}{2}\partial g_{ii}/\partial x^j, \tag{43}$$

since $g_{ij}=0$ for $i\neq j$. From Eqs. (18.12), we have

$$\partial g_{11}/\partial\phi=0,\ \partial g_{22}/\partial\theta=2a^2\sin\theta\cos\theta, \tag{44}$$

where a is the radius of the sphere, giving

$$[12, 1]=[21, 1]=0,\ [12, 2]=[21, 2]=a^2\sin\theta\cos\theta. \tag{45}$$

CATEGORY 3: $i=j\neq k$; Eq. (39) in this case reduces to

$$[ii, k]=-\tfrac{1}{2}g_{ii,k}=-\tfrac{1}{2}\partial g_{ii}/\partial x^k, \tag{46}$$

so that

$$[11, 2]=0,\ [22, 1]=-a^2\sin\theta\cos\theta. \tag{47}$$

Combining the above results, we find that the eight Christoffel symbols of the first kind are given by

$$[ii, i]=0,\ i=1, 2; \tag{48a}$$

$$[12, 1]=[21, 1]=0; \tag{48b}$$

$$[12, 2]=[21, 2]=a^2\sin\theta\cos\theta; \tag{48c}$$

$$[11, 2]=0; \tag{48d}$$

$$[22, 1]=-a^2\sin\theta\cos\theta. \tag{48e}$$

The Christoffel symbols of the second kind are defined in Eq. (7b). If the metric tensor is diagonal, as in the present case, this equation reduces to

$$\Gamma^k_{ij}=g^{kk}\,[ij, k], \tag{49}$$

where k is a free index. The contravariant components of the metric tensor are obtained from Eq. (18·20) and are found to be

$$g^{11}=1/a^2,\ g^{22}=1/(a^2\sin^2\theta),\ g^{12}=g^{21}=0. \tag{50}$$

Using these and Eqs. (48) in Eq. (49), we find that the Christoffel symbols of the second kind are given by

$$\Gamma^i_{ii}=0,\ i=1, 2; \tag{51a}$$

$$\Gamma^1{}_{12}=\Gamma^1{}_{21}=0; \tag{51b}$$

$$\Gamma^2{}_{12}=\Gamma^2{}_{21}=\cot\theta; \tag{51c}$$

$$\Gamma^2{}_{11}=0; \tag{51d}$$

$$\Gamma^1{}_{22}=-\sin\theta\cos\theta. \tag{51e}$$

It is seen that the Christoffel symbols of the second kind are independent of the radius of the sphere. Also, only three of the eight symbols are different from zero.

EXAMPLE 6: Prove that

$$\Gamma^i{}_{ij}=(\ln\sqrt{g})_{,j}=\partial(\ln\sqrt{g})/\partial x^j. \tag{52}$$

Solution: From the relation

$$\Gamma^i{}_{kj}=g^{ih}\,[kj,\,h], \tag{53}$$

we have

$$\Gamma^i{}_{ij}=g^{ih}\,[ij,\,h]=\tfrac{1}{2}g^{ih}\,(g_{ih,j}+g_{jh,i}-g_{ij,h}). \tag{54}$$

Since i and h are both dummy indices in the above equation and $g^{ih}=g^{hi}$, we can interchange i and h in the last factor inside the parentheses on the right hand side of Eq. (54). It then reduces to

$$\begin{aligned}\Gamma^i{}_{ij}&=\tfrac{1}{2}g^{ih}g_{ih,j}\\&=(1/2g)\sum_{i,h}(\text{cofactor of } g_{ih})\,(\partial g_{ih}/\partial x^j),\end{aligned} \tag{55}$$

where use has been made of the fact that g^{ih} is the cofactor of g_{ih} divided by g, and the summation over i and h has been explicitly written. Comparing this with Eq. (4.6), we get

$$\Gamma^i{}_{ij}=(1/2g)\partial g/\partial x^j=\tfrac{1}{2}\partial(\ln g)/\partial x^j=\partial(\ln\sqrt{g})/\partial x^j. \tag{56}$$

Intrinsic derivative

Let C be a certain curve described by the parametric equations $x^k\equiv x^k(\tau)$ in a Riemannian space. Let $A^{i_1\ldots i_p}_{j_1\ldots j_q}$ be a tensor field defined on the points of C, that is, whose components are continuous functions of the parameter τ along C:

$$A^{i_1\ldots i_p}_{j_1\ldots j_q}\equiv A^{i_1\ldots i_p}_{j_1\ldots j_q}(x^k)\equiv A^{i_1\ldots i_p}_{j_1\ldots j_q}(\tau). \tag{57}$$

Then the *intrinsic derivative*[4] of the tensor along C is defined by

$$\frac{DA^{i_1\ldots i_p}_{j_1\ldots j_q}}{D\tau}=A^{i_1\ldots i_p}_{j_1\ldots j_q;k}\,\frac{dx^k}{d\tau}. \tag{58}$$

Since $A^{i_1\ldots i_p}_{j_1\ldots j_q;k}$ is a tensor and dx^k is a tensor (and hence $dx^k/d\tau$ is a

[4]Also known as the *absolute derivative*.

tensor), it follows from the quotient law that the intrinsic derivative defined by Eq. (58) is a tensor of the same rank and type as the original tensor.

The intrinsic derivative assumes particularly simple expressions for a scalar and a vector. If $\phi(x^i)$ is a scalar field, so that $\phi \equiv \phi(\tau)$ along the curve C, then the intrinsic derivative of ϕ along C is

$$\frac{D\phi}{D\tau} = \phi_{;k} \frac{dx^k}{d\tau} = \frac{\partial \phi}{\partial x^k} \frac{dx^k}{d\tau} = \frac{d\phi}{d\tau}, \tag{59}$$

where use has been made of Eq. (26). This shows that the intrinsic derivative of a scalar field is the same as its ordinary derivative along a curve. Similarly, if A^i is a contravariant vector, we find, on using Eqs. (58) and (25), that the intrinsic derivative of A^i along C is given by

$$\frac{DA^i}{D\tau} = A^i_{;k} \frac{dx^k}{d\tau} = \frac{\partial A^i}{\partial x^k} \frac{dx^k}{d\tau} + A^j \Gamma^i_{kj} \frac{dx^k}{d\tau}$$

$$= \frac{dA^i}{d\tau} + \Gamma^i_{kj} A^j \frac{dx^k}{d\tau}. \tag{60}$$

Finally, the intrinsic derivative of a covariant vector is given by

$$\frac{DA_i}{D\tau} = \frac{dA_i}{d\tau} - \Gamma^j_{ik} A_j \frac{dx^k}{d\tau}. \tag{61}$$

The above equations are also sometimes written in the differential form as

$$DA^i = dA^i + \Gamma^i_{kj} A^j dx^k; \tag{62a}$$

$$DA_i = dA_i - \Gamma^j_{ik} A_j dx^k. \tag{62b}$$

For a tensor of higher rank, say, for example $A^{ij}_{..kl}$, the intrinsic derivative can be written as

$$\frac{DA^{ij}_{..kl}}{D\tau} = A^{ij}_{..kl;m} \frac{dx^m}{d\tau}$$

$$= \frac{\partial A^{ij}_{..kl}}{\partial \tau} + [\Gamma^i_{pm} A^{pj}_{..kl} + \Gamma^j_{pm} A^{ip}_{..kl} - \Gamma^p_{km} A^{ij}_{..pl} - \Gamma^p_{lm} A^{ij}_{..kp}] \frac{dx^m}{d\tau}. \tag{63}$$

From Eqs. (31), (36) and (38), we have

$$Dg_{ij}/D\tau = g_{ij;k}\, dx^k/d\tau \equiv 0; \tag{64a}$$

$$Dg^{ij}/D\tau = g^{ij}_{..;k}\, dx^k/d\tau \equiv 0; \tag{64b}$$

$$D\delta^i_{.j}/D\tau = \delta^i_{.j;k}\, dx^k/d\tau \equiv 0; \tag{64c}$$

that is, the intrinsic derivatives of the metric tensor and the Kronecker delta along any curve are identically zero.

EXAMPLE 7: A fluid in motion in a plane has the velocity vector field given by

$$v^i=(x^2, y^2) \tag{65}$$

in cartesian coordinates. Find the covariant derivative of the vector field in polar coordinates.

Solution: Let us choose the usual polar cordinates r, θ, and let u^i denote the velocity vector in polar coordinates. A simple transformation from the cartesian components v^i to the polar components u^i according to Eq. (15.24) gives

$$u^1=r^2 (\cos^3 \theta+\sin^3 \theta),\ u^2=r \sin \theta \cos \theta (\sin \theta-\cos \theta). \tag{66}$$

The covariant derivative is given by

$$u^i_{.;j}=\partial u^i/\partial x^j+u^k\Gamma^i{}_{jk}, \tag{67}$$

where $x^1=r$, $x^2=\theta$. The Christoffel symbols of the second kind in polar coordinates are

$$\Gamma^1{}_{11}=\Gamma^2{}_{22}=\Gamma^2{}_{11}=\Gamma^1{}_{12}=\Gamma^1{}_{21}=0,$$
$$\Gamma^1{}_{22}=-r,\ \Gamma^2{}_{12}=\Gamma^2{}_{21}=1/r. \tag{68}$$

Using these in Eq. (67), the components of the covariant derivative in polar coordinates are found to be

$$u^r_{.;r}=\partial u^1/\partial r+u^k\,\Gamma^1{}_{1k}=\partial u^1/\partial r$$
$$=2r(\cos^3 \theta+\sin^3 \theta); \tag{69}$$
$$u^r_{.;\theta}=\partial u^1/\partial\theta+u^k\,\Gamma^1{}_{2k}=\partial u^1/\partial\theta+u^2\Gamma^1{}_{22}$$
$$=2r^2\sin \theta \cos \theta (\sin \theta-\cos \theta); \tag{70}$$
$$u^\theta_{.;r}=\partial u^2/\partial r+u^k\,\Gamma^2{}_{1k}=\partial u^2/\partial r+u^2\Gamma^2{}_{12}$$
$$=2 \sin \theta \cos \theta (\sin \theta-\cos \theta); \tag{71}$$
$$u^\theta_{.;\theta}=\partial u^2/\partial\theta+u^k\,\Gamma^2{}_{2k}=\partial u^2/\partial\theta+u^1\Gamma^2{}_{21}$$
$$=2r \sin \theta \cos \theta (\sin \theta+\cos \theta). \tag{72}$$

Ordinary differentiation of the product of two functions satisfies the distributive law. For example, $(uv)'=uv'+u'v$, where u and v are functions of some variable and the primes denote differentiation with respect to that variable. It can be shown that covariant differentiation of the product of two or more tensors also satisfies the same distributive law. This is done in the following example.

EXAMPLE 8: Show that

$$(A^i_{.j}\, B^{kl})_{;q}=A^i_{.j;q}\, B^{kl}+A^i_{.j}\, B^{kl}_{..;q}\,. \tag{73}$$

Solution: Let us denote the outer-product tensor by $C^{ikl}_{.j..}=A^i_{.j}\, B^{kl}$. Then

$$C^{i\,kl}_{.j..;q}=C^{i\,kl}_{.j..,q}+C^{m\,kl}_{.j..}\,\Gamma^i{}_{mq}+C^{i\,ml}_{.j..}\,\Gamma^k{}_{mq}+C^{i\,\,km}_{.j..}\,\Gamma^l{}_{mq}-C^{i\,\,kl}_{.m..}\,\Gamma^m{}_{jq}$$

$$=(A^i_{.j}\,B^{kl})_{,q}+A^m_{.j}B^{kl}\Gamma^i_{mq}+A^i_{.j}\,B^{ml}\,\Gamma^k_{mq}+A^i_{.j}\,B^{km}\,\Gamma^l_{mq}-A^i_{.m}\,B^{kl}\,\Gamma^m_{jq}$$

$$=A^i_{.j,q}\,B^{kl}+A^i_{.j}\,B^{kl}_{..,q}+A^m_{.j}\,B^{kl}\,\Gamma^i_{mq}+A^i_{.j}\,B^{ml}\,\Gamma^k_{mq}$$

$$+A^i_{.j}\,B^{km}\,\Gamma^l_{mq}-A^i_{.m}\,B^{kl}\,\Gamma^m_{jq}$$

$$=[A^i_{.j,q}+A^m_{.j}\,\Gamma^i_{mq}-A^i_{.m}\,\Gamma^m_{jq}]\,B^{kl}$$

$$+A^i_{.j}\,[B^{kl}_{..,q}+B^{ml}\,\Gamma^k_{mq}+B^{km}\,\Gamma^l_{mq}]$$

$$=A^i_{.j;q}\,B^{kl}+A^i_{.j}\,B^{kl}_{..;q}, \tag{74}$$

which proves the desired result.

Exercise 20

1. Write down the covariant derivatives of the tensors (a) $A^i_{.jk}$, (b) $A^{ij}_{..kl}$, *(c) $A^{ijk}_{...l}$, and (d) $A^{ij}_{..klm}$ with respect to x^q.

2. Write down the intrinsic derivatives of the tensors mentioned in Exercise 1 with respect to τ along a curve $x^i \equiv x^i(\tau)$.

*3. If A_i is a covariant vector, show that $A_{i,j}-A_{j,i}$ is a tensor.

*4. If A_{ij} is an antisymmetric tensor, show that $A_{ij,k}+A_{jk,i}+A_{ki,j}$ is a tensor, which is antisymmetric in any pair of indices.

5. Compute the Christoffel symbols of the first and the second kinds in a plane in terms of polar coordinates r, θ.

6. Compute the nonvanishing Christoffel symbols in terms of cylindrical and spherical polar coordinates in a three-dimensional Euclidean space.

7. Compute all the Christoffel symbols for a space whose line element is given by

$$ds^2=-e^{\lambda}\,dr^2-r^2d\theta^2-r^2\sin^2\theta\,d\phi^2+e^{\mu}\,dt^2.$$

8. The velocity vector field of a fluid in motion in a plane is $v^i=(x, 2y)$ in cartesian coordinates. Find its covariant derivative in polar coordinates.

9. Find the intrinsic derivative of the vector field of Exercise 8 along the spiral $r=a\,\theta+b$, $a>0$, $b>0$.

*10. Obtain the Christoffel symbols of both kinds for a space with the metric $ds^2=f(u, v)du^2+h(u, v)dv^2$.

*11. Obtain the Christoffel symbols of both kinds for a space with the metric $ds^2=R\,du^2+2S\,du\,dv+T\,dv^2$, where R, S, T are functions of u and v and $RT-S^2$ does not identically vanish.

12. Prove the following identities:

(a) $(A_{ij}\,B^k_{.lm})_{;q}=A_{ij;q}\,B^k_{.lm}+A_{ij}\,B^k_{.lm;q}$;

(b) $(A_{i.}^{\;j}\,B^{ik})_{;q}=A_{i.;q}^{\;j}\,B^{ik}+A_{i.}^{\;j}\,B^{ik}_{..;q}$;

(c) $(A_iB^jC_k)_{;q}=A_{i;q}\,B^jC_k+A_iB^j_{.;q}\,C_k+A_iB^jC_{k;q}$.

*13. Prove in general that intrinsic differentiation of the product of two or more tensors satisfies the distributive law.

21. Kinematics in a Riemannian space

Geodesics

In a Euclidean space, the shortest path between any two points is a straight line joining the two points. In a Riemannian space, however, the shortest path between two points may be a curved path. On the surface of a sphere, for example, the shortest path connecting any two points is the smaller arc of the great circle passing through the two points. As long as we are concerned with small regions of space, the approximations of Euclidean geometry hold very well. But suppose we wish to find the shortest path from the earth to a star. Which methods are we to apply? We could send a ray of light to the star, or receive on earth a ray of light from the star, and say that the path of the light ray is the shortest path. But it is known that even a ray of light follows a curved path in a gravitational field—in fact, this was one of the major effects which provided a verification of Einstein's general theory of relativity. Hence we are forced to drop the concept of a straight line and adopt that of a *geodesic.*

Consider two points P and Q in an N-dimensional Riemannian space. There is an infinite number of paths connecting P and Q. Let A be one such path, as shown in Fig. 12. Since A is a one-dimensional curve, the coordinates of any point on it can be specified in terms of a single continuous parameter, say τ. In other words, the equation of curve A can be symbolically written as

$$x^i \equiv x^i(\tau). \tag{1}$$

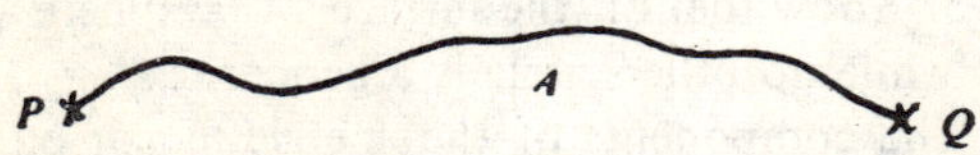

Figure 12. Interval between P and Q along a curve A.

Let the end points P and Q of the path correspond respectively to the values τ_1 and τ_2 of the parameter.

Consider two points on path A close to each other with coordinates x^i and x^i+dx^i. Let the parameter take values τ and $\tau+d\tau$ corresponding to these two points, respectively. The interval ds between the two points is given by

$$ds^2=g_{ij}dx^i dx^j=g_{ij}\frac{dx^i}{d\tau}\frac{dx^j}{d\tau}\,d\tau^2. \tag{2}$$

The total interval between P and Q along the path A is therefore

$$s_{PQ}=\int_P^Q \left[g_{ij}\frac{dx^i}{d\tau}\frac{dx^j}{d\tau}\right]^{1/2} d\tau. \tag{3}$$

By applying the Euler-Lagrange variational principle, it is possible to obtain the differential equations of a path along which the interval between P and Q is an extremum. *A path along which the interval between two points in a Riemannian space is an extremum is called a geodesic* connecting the two points. Without giving the mathematical detail, we shall mention that the variation of Eq. (3) leads to the equations

$$\frac{d^2x^i}{ds^2}+\Gamma^i_{\ jk}\frac{dx^j}{ds}\frac{dx^k}{ds}=0, \tag{4}$$

where s itself is chosen as the parameter. This, of course, represents a set of N coupled differential equations. Eq. (4) is known as the *geodesic equation.*

EXAMPLE 1: Show that in a Euclidean space, the geodesics are straight lines.

Solution: In a Euclidean space, g_{ij} are independent of the coordinates x^i, so that the Christoffel symbols identically vanish, and Eq. (4) reduces to

$$d^2x^i/ds^2=0, \tag{5}$$

with the solution

$$x^i=a_is+b_i, \tag{6}$$

where a_i and b_i are constants independent of s, and hence of x^i. Eq. (6) clearly represents a straight line.

EXAMPLE 2: Show that on the surface of a sphere all great circles are geodesics while no other circle is a geodesic.

Solution: The components of the metric tensor on the surface of a sphere of radius a are given in Eqs. (18.12), while the Christoffel symbols of the second kind are given in Eqs. (20.51). Since the only nonvanishing symbols are

$$\Gamma^2_{\ 12}=\Gamma^2_{\ 21}=\cot\theta,\ \Gamma^1_{\ 22}=-\sin\theta\cos\theta, \tag{7}$$

the geodesic equation reduces to (with $\theta=x^1$, $\phi=x^2$)

$$\frac{d^2\theta}{ds^2}-\sin\theta\cos\theta\left(\frac{d\phi}{ds}\right)^2=0,$$

$$\frac{d^2\phi}{ds^2}+2\cot\theta\frac{d\theta}{ds}\frac{d\phi}{ds}=0. \tag{8}$$

(a) Consider a great circle on the surface of the sphere and choose the z-axis ($\theta=0$) to be normal to the plane of the circle, so that the great circle is the equator. The parametric equation of the great circle is

$$\theta=\pi/2,\ \phi=cs+d,\ c\neq 0. \tag{9}$$

where c and d are independent of s, θ and ϕ. It is evident that Eqs. (9) satisfy Eqs. (8), so that the great circle is a geodesic. Since the choice of the polar axis $\theta=0$ is arbitrary, it follows that any great circle is a geodesic.

(b) Consider a circle on the sphere such that the plane of the circle does not pass through the centre of the sphere. Again, choose the direction $\theta=0$ to be normal to the plane of the circle so that the parametric equation of the circle is

$$\theta=\theta_0,\ \phi=ps+q,\ 0<\theta_0<\pi,\ \theta_0\neq\pi/2,\ p\neq 0, \tag{10}$$

where p and q are independent of s, θ and ϕ. On substitution, it can be seen that the second of Eqs. (8) is satisfied but the first one reduces to

$$p^2 \sin\theta_0 \cos\theta_0=0. \tag{11}$$

The conditions on θ_0 and p [see Eqs. (10)] show that this equation cannot be satisfied, proving that the circle is not a geodesic.

There is one and only one geodesic passing through two given points in a Riemannian space, except when it is possible to find a coordinate system which makes $g=0$ at both the points. For example, if we choose two diametrically opposite points on the surface of a sphere then choosing the polar axis to pass through these two points makes $g=0$ at both of them [see Eq. (18.13)]. In this case, there is an infinite number of geodesics passing through two such points.

Einstein's law

The Newton's first law of motion states that a body continues to remain in a state of rest or of uniform rectilinear motion as long as it is not acted upon by an external force. Although this law held its sway for two centuries, a little reflection shows that it contains a circular argument. For, if we ask the question as to what force is, we are compelled to say, within the domain of Newtonian mechanics, that it is some entity which changes or tries to change the state of rest or of uniform motion of a body. When we throw a stone it follows a curved path; so we invoke the force of gravitation. If we roll an object on a plane, its velocity decreases and it comes to a stop after travelling a certain distance, we explain this by invoking a force of friction between the object and the plane.

The second law of Newton also has a drawback. While studying the motion of a planet around a sun, shall we consider the rest mass or the relativistic mass of the planet in calculating the gravitational attraction on it? Should the distance between them be taken to be that measured by an observer stationary relative to the sun or the planet or relative to some other body? These simple problems show that there is vagueness and inconsistency in the classical description of Kepler's and Newton's laws.

Einstein proposed to deal with these difficulties by doing away with the concept of the gravitational field through which one body acts on another at a distance, and, instead, introducing the concept of *curved* space-time. It was assumed that the presence of matter changes the geometry of the space-time continuum by causing it to bend and that a body moves in such a curved Riemannian space in a curved path which is the geodesic for that space. Newton's laws of motion thus gave way to the Einstein's law which states that *every body moves along a geodesic in the four-dimensional space-time continuum provided it does not undergo collision with another body and is not acted upon by electromagnetic forces.* Thus the presence of matter with its accompanying gravitational field could be described simply by a new geometry. The underlying hypothesis is known as the *principle of equivalence* according to which *the presence of gravitational fields can be completely described by the metric tensor* $g_{ij}(x^k)$.

Note that electromagnetic forces have to be excluded from the Einstein's law. This is because it has not yet been possible to describe the effect of electromagnetic fields on charged particles in an equivalent geometrical way. Einstein himself spent many years of his later life trying to construct a *unified field theory* which could account for both gravitational and electromagnetic fields by a new geometry.

The importance of geodesics in the general theory of relativity then becomes clear. A planet moves around the sun in a particular orbit not because it is attracted towards the sun but because this path is the geodesic in the space-time continuum in the presence of the sun, i.e., a path of extremum interval between any two points along it. Whether it is a planet moving around a sun, a rocket going from the earth to the Jupitor or a ray of light, every object is simply moving along a geodesic whose nature is governed by the distribution of mass in the space-time.

One paradox is likely to arise. Consider the path of the earth around the sun and let $P=(x, y, z)$ and $Q=(x', y', z')$ be two points on this path. When the earth is at the point P, let us suppose that a ray of light and a rocket start in such a way that both pass through the point Q at suitable later instants. Since the velocity of the earth is the lowest of the three, its path will be the most curved while that of the ray of light will be the least curved, as shown in Fig. 13. Since each of the three objects must move along a geodesic, it seems apparently that there are three (or, in fact, an infinite number of) geodesics

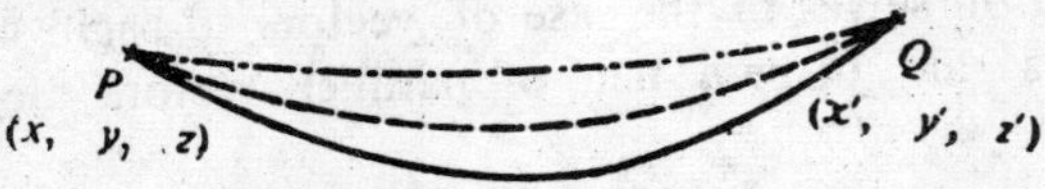

Figure 13. The paths of earth (a), a racket (b), and a ray of light (c) between two points (x, y, z) and (x', y', z') near a massive sun (not shown).

connecting P and Q, which is contrary to the fact that there is one and only one geodesic between two arbitrary points in a Riemannian space. However, this paradox arises from the fact that we are actually concentrating only on the three-dimensional spatial projection of the four-dimensional motion of the objects. Thus, although the earth, the rocket and the ray of light all start from $P=(x, y, z)$ at the same time t, they will reach $Q=(x', y', z')$ at different times, say t_1, t_2 and t_3, respectively. There is a unique geodesic between the two space-time points (x, y, z, t) and (x', y', z', t_1) which will be the path of the earth in the space-time continuum, a unique geodesic for the rocket between (x, y, z, t) and (x', y', z', t_2), and finally, a unique geodesic for the ray of light going from the point (x, y, z, t) to the point (x', y', z', t_3), etc.

Parallel displacement

In a Euclidean space, a vector field $A_i(x^j)$ is said to be a field of parallel vectors if the components A_i are constants, that is, $A_{i,j}=0$. Or, suppose that a vector A_i defined at a point x^j undergoes a displacement dx^j to the point x^j+dx^j, where the vector becomes $A_i+\partial A_i$. We shall say that A_i undergoes a *parallel displacement* if $\partial A_i/\partial x^j=0$. For example, a uniform electric field is a field of parallel vectors; the gravitational field of the earth in a sufficiently small region of space is a field of parallel vectors.

Our object is to generalize the concept of parallel displacement of vectors and tensors in Riemannian spaces. We have shown in the previous section that if A_i is a vector, its coordinate derivative $\partial A_i/\partial x^j$ is not a tensor, but the intrinsic derivative DA_i/Ds along a curve $x^j \equiv x^j(s)$ is a tensor. (It should be remembered, of course, that the intrinsic derivative is defined only along a curve.) To retain the tensor character, therefore, we must define the parallel displacement of a tensor in a Riemannian space differently. Thus, we shall say that *a tensor is said to be parallely displaced along a curve* $x^i \equiv x^i(s)$ *provided the intrinsic derivative of the tensor along the curve vanishes* [see Fig. 14].

Restricting ourselves to the case of vectors to begin with, a vector field $A_i(x^j)$ is said to be a field of parallel vectors along a curve $x^i \equiv x^i(s)$ if

$$DA_i/Ds = 0 \tag{12}$$

along the curve. Referring to Eq. (20.61), this means that we have a field of parallel vectors if

$$\frac{dA_i}{ds} = \Gamma^j_{ik} A_j \frac{dx^k}{ds} \tag{13a}$$

along the curve, or

$$dA_i = \Gamma^j_{ik} A_j dx^k. \tag{13b}$$

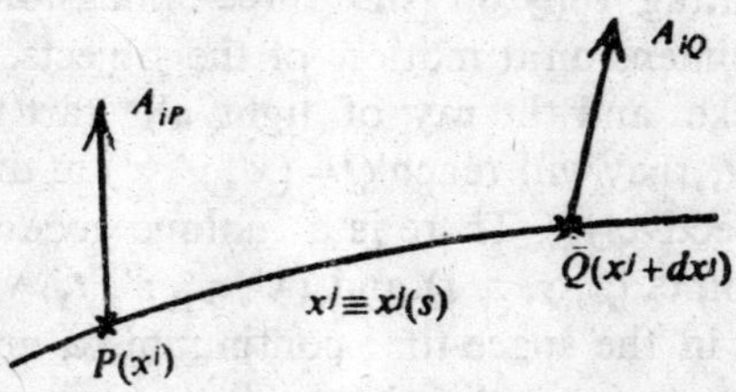

Figure 14. The vector A_{iP} at $P \equiv (x^j)$ will be parallel to the vector $A_{iQ} = A_{iP} + dA_{iP}$ at $Q \equiv (x^j + dx^j)$ if $DA_{iP}/Ds = 0$ along a curve $x^j \equiv x^j(s)$.

Similarly, we have a field of parallel contravariant vectors A^i along a curve $x^j \equiv x^j(s)$ if

$$DA^i/Ds = 0, \tag{14}$$

or

$$\frac{dA^i}{ds} = -\Gamma^i_{jk} A^j \frac{dx^k}{ds} \tag{15a}$$

along the curve, or

$$dA^i = -\Gamma^i_{jk} A^j dx^k. \tag{15b}$$

EXAMPLE 3: Show that the magnitude of a vector and the angle between two vectors remaininvariant under parallel displacement.

Solution: (a) According to Eqs. (18.37), the magnitude of a vector is given by

$$A^2 = g_{ij} A^i A^j. \tag{16}$$

Differentiating both sides with respect to a parameter τ and noting that the intrinsic derivative of a scalar is identical to its ordinary derivative [Eq. (20.59)], we have

$$2A\,\frac{dA}{d\tau}=\frac{d}{d\tau}\,(g_{ij}A^iA^j)=\frac{D}{D\tau}\,(g_{ij}A^iA^j)$$
$$=A^iA^j\,Dg_{ij}/D\tau+g_{ij}(A^jDA^i/D\tau+A^iDA^j/D\tau)$$
$$=A^iA^j\,Dg_{ij}/D\tau+2g_{ij}A^iDA^j/D\tau. \tag{17}$$

For parallel displacement, the intrinsic derivatives of A^j as well as g_{ij} vanish [see Eq. (20.64a)] giving $dA/d\tau=0$, or A=constant, proving the result.

(b) The angle θ between two vectors A_i and B_i is given by [see Eqs. (18.38)]

$$AB\cos\theta=g_{ij}A^iB^j. \tag{18}$$

Under parallel displacement, the magnitudes A and B are constant, as shown in part (a) above. Differentiating both sides of the above equation with respect to τ, a similar treatment shows that $d(\cos\theta)/d\tau=0$, giving θ=constant, the desired result.

Now consider a point $P\equiv(x^i)$ and two displacements dx^i and δx^i from P giving rise to the two points $Q\equiv(x^i+dx^i)$ and $R\equiv(x^i+\delta x^i)$, respectively, as shown in Fig. 15. Imagine the segment $PQ=\delta x^i$ being parallely displaced such that as its one end P goes to Q, the other end R goes to S. The displacement QS will be given by

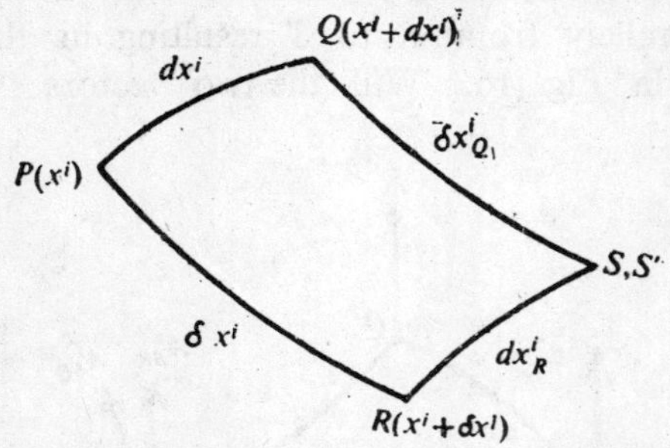

Figure 15. The parallel displacements of the segment δx^i along PQ and that of dx^i along PR give the same end point $S'=S$.

$$QS\equiv\delta x^i{}_Q=\delta x^i+d(\delta x^i), \tag{19}$$

which makes the coordinates of the points S

$$S\equiv[x^i+dx^i+\delta x^i{}_Q]\equiv[x^i+dx^i+\delta x^i+d(\delta x^i)], \tag{20}$$

where, by Eq. (15b), we have

$$d(\delta x^i)=-\Gamma^i{}_{jkP}\,\delta x^j\,dx^k. \tag{21}$$

The subscript P on the Christoffel symbol means that the symbol is to be evaluated at P. Similarly, imagine the segment $PQ=dx^i$ being parallely displaced such that as its one end P goes to R, the other end Q goes to S'. The displacement RS' will be given by

$$RS' \equiv dx^i{}_R = dx^i + \delta(dx^i), \tag{22}$$

which makes the coordinates of the point S'

$$S' \equiv [x^i + \delta x^i + dx^i{}_R] \equiv [x^i + \delta x^i + dx^i + \delta(dx^i)], \tag{23}$$

where

$$\delta(dx^i) = -\Gamma^i{}_{jkP}\, dx^j\, \delta x^k = -\Gamma^i{}_{jkP}\, dx^k\, \delta x^j, \tag{24}$$

which follows by interchanging j and k and using the symmetry property of the Christoffel symbol. Eqs. (21) and (24) show that $d(\delta x^i) = \delta(dx^i)$, indicating that the point S' coincides with S. The closed figure $PQRS$ can be regarded as a generalized infinitesimal parallelogram with opposite sides parallel to each other in a Riemannian space.

Curvature of Riemannian space

Consider an infinitesimal parallelogram $PQRS$ as described above with adjacent sides $PQ=dx^i$ and $PR=\delta x^i$. Imagine a contravariant vector $A^i{}_P$ defined at the point P being parallely displaced in the following two ways: (a) Displace $A^i{}_P$ parallely from P to Q resulting in $A^i{}_Q$ and then displace $A^i{}_Q$ parallely from Q to S giving the vector $A^i{}_{SQ}$. (b) Displace $A^i{}_P$ parallely from P to R giving $A^i{}_R$ and then displace $A^i{}_R$ parallely from R to S resulting in the vector $A^i{}_{SR}$. This is shown in Fig. 16. Will the two vectors $A^i{}_{SQ}$ and $A^i{}_{SR}$ at

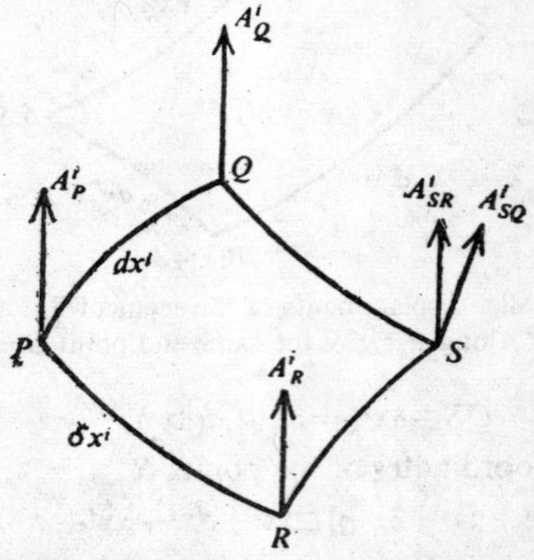

Figure 16. The parallel displacement of a vector $A^i{}_P$ at P from P to S along two different paths gives rise to two different vectors.

S be equal? If not, what does the difference between them depend on?

A simple treatment shows that the two vectors $A^i{}_{SR}$ and $A^i{}_{SQ}$ differ from each other. The vector $A^i{}_Q$ obtained by a parallel displacement of $A^i{}_P$ is given by

$$A^i{}_Q = A^i{}_P + dA^i{}_P, \tag{25}$$

where

$$dA^i{}_P = -\Gamma^i{}_{jkP}\, A^j{}_P\, dx^k, \tag{26}$$

so that

$$A^i{}_Q = A^i{}_P - \Gamma^i{}_{jkP}\, A^j{}_P\, dx^k. \tag{27}$$

This vector is parallely displaced from Q to S resulting in the vector $A^i{}_{SQ}$ which will be given by

$$A^i{}_{SQ} = A^i{}_Q + \delta A^i{}_Q = A^i{}_Q - \Gamma^i{}_{lhQ} A^l{}_Q \delta x^h. \tag{28}$$

The Christoffel symbol depends on the metric tensor which in turn is a function of coordinates. For a small displacement $PQ = dx^i$, we can therefore write

$$\Gamma^i{}_{lhQ} = \Gamma^i{}_{lhP} + \Gamma^i{}_{lhP,m}\, dx^m, \tag{29}$$

where $\Gamma^i{}_{lhP,m} = \partial \Gamma^i{}_{lh}/\partial x^m$ at the point P. Substituting Eqs. (29) and (27) in Eq. (28), we have

$$\begin{aligned} A^i{}_{SQ} &= A^i{}_P - \Gamma^i{}_{jkP}\, A^j{}_P\, dx^k \\ &\qquad - [\Gamma^i{}_{lhP} + \Gamma^i{}_{lhP,m}\, dx^m]\,[A^l{}_P - \Gamma^l{}_{jkP}\, A^j{}_P\, dx^k]\,\delta x^h \\ &= A^i - \Gamma^i{}_{jk}\, A^j\, dx^k - \Gamma^i{}_{lh}\, A^l\, \delta x^h - \Gamma^i{}_{lh,m}\, A^l\, dx^m\, \delta x^h \\ &\qquad + \Gamma^i{}_{lh}\, \Gamma^l{}_{jk}\, A^j\, dx^k\, \delta x^h + 0[(dx^k)^3], \end{aligned} \tag{30}$$

where we have dropped the subscript P from the right hand side of the final expression because all the quantities are to be evaluated at P.

The vector $A^i{}_{SR}$ obtained by a parallel displacement along the path PRS will be simply given by an interchange of dx^i and δx^i in the expression for $A^i{}_{SQ}$:

$$\begin{aligned} A^i{}_{SR} &= A^i - \Gamma^i{}_{jk}\, A^j\, \delta x^k - \Gamma^i{}_{lh}\, A^l\, dx^h \\ &\qquad - \Gamma^i{}_{lh,m}\, A^l\, \delta x^m\, dx^h + \Gamma^i{}_{lh}\, \Gamma^l{}_{jk}\, A^j\, \delta x^k\, dx^h + 0[(dx^k)^3]. \end{aligned} \tag{31}$$

Subtracting one from the other, we see that the first three terms on the right hand sides of Eqs. (30) and (31) cancel, giving, to second order in dx^k,

$$\begin{aligned} A^i{}_{SR} - A^i{}_{SQ} &= \Gamma^i{}_{lh,m}\, A^l\, dx^m\, \delta x^h - \Gamma^i{}_{lh,m}\, A^l\, \delta x^m\, dx^h \\ &\qquad + \Gamma^i{}_{lh}\, \Gamma^l{}_{jk}\, A^j\, \delta x^k\, dx^h - \Gamma^i{}_{lh}\, \Gamma^l{}_{jk}\, A^j\, dx^k\, \delta x^h. \end{aligned} \tag{32}$$

Replacing l by j and m by k in the first two terms on the right hand side and then interchanging k and h in the second and the third terms, the above equation can be written as

$$A^i{}_{SR}-A^i{}_{SQ}=R_{j\cdot hk}^{\;i}\, A^j\, dx^k\, \delta x^h \tag{33}$$

where[1]

$$R_{j\cdot hk}^{\;i}=\Gamma^i{}_{jh,k}-\Gamma^i{}_{jk,h}+\Gamma^i{}_{lk}\,\Gamma^l{}_{jh}-\Gamma^i{}_{lh}\Gamma^l{}_{jk}. \tag{34}$$

Since A^j, dx^k and δx^h are arbitrary vectors, it follows from the quotient law that $R_{j\cdot kh}^{\;i}$ is a tensor of rank four. It is known as the *Riemann-Christoffel curvature tensor.* It is independent of the vector A^i and depends only on the metric tensor and its first and second derivatives.

In a Euclidean space, it is possible to choose a cartesian coordinate system in which all the Christoffel symbols vanish identically. Hence it can be seen that *the Riemann-Christoffel curvature tensor identically vanishes for a Euclidean space.*

If we start with a covariant vector A_i instead of a contravariant vector, it can be shown by a similar treatment that its parallel displacement along two different paths gives the result

$$A_{iSR}-A_{iSQ}=-R_{i.hk}^{\;j}\, A_j\, dx^k\, \delta x^h. \tag{35}$$

This shows that the result of parallel displacement of a vector (and, in general, a tensor) from one point to another in a Riemannian space depends on the path chosen. It follows that if a tensor is parallely displaced along a closed curve until we come back to the starting point, the resulting vector will not necessarily be the same as the original vector in a non-Euclidean space.

EXAMPLE 4: Show that the field of tangent vectors to a geodesic is a field of parallel vectors.

Solution: Let $x^i\equiv x^i(s)$ be a geodesic in a Riemannian space. Then the coordinates of the points on the geodesic satisfy the geodesic equation

$$\frac{d^2x^i}{ds^2}+\Gamma^i{}_{jk}\frac{dx^j}{ds}\frac{dx^k}{ds}=0. \tag{36}$$

If A^i is a tangent vector to the geodesic, the components A^i are given by

$$A^i=dx^i/ds. \tag{37}$$

This reduces Eq. (36) to

[1]Different authors denote the expression on the right hand side of Eq. (34) in different ways such as $R^l{}_{.jhk}$, $R_{hkj}{}^{.i}$, etc. This is of little consequence; what is important, however, is that a definition once adopted should be followed consistently.

$$\frac{dA^i}{ds}+\Gamma^i{}_{jk}\,A^j\,\frac{dx^k}{ds}=0, \tag{38a}$$

or

$$DA^i/Ds=0 \tag{38b}$$

along the geodesic. Hence it follows from the definition of parallel displacement that the field of tangent vectors to a geodesic is a field of parallel vectors.

EXAMPLE 5: Show that the parallel displacement of a vector taken all around a circle on the surface of a sphere does not lead back to the same vector except when the circle is a great circle, that is, a geodesic.

Solution: Choose spherical polar coordinates $x^1=\theta$ and $x^2=\phi$ on the surface of a sphere so that the only nonvanishing Christoffel symbols are given by Eqs. (20.51c) and (20.51e). Choose a small circle normal to the polar axis so that its parametric equation is $\theta=\alpha$, $\phi=ps+q$, $p\neq0$, with $d\theta/ds=0$. Consider a vector $A^i=(A^1, A^2)$ at a point on the circle which may be taken to be $\phi=0$. When this vector is displaced parallely along the circle, the change in its components is given, from Eq. (15a), by

$$\frac{dA^1}{ds}=-\Gamma^1{}_{jk}A^j\frac{dx^k}{ds}=\sin\alpha\cos\alpha\,A^2\,\frac{d\phi}{ds},$$

$$\frac{dA^2}{ds}=-\Gamma^2{}_{jk}A^j\frac{dx^k}{ds}=-\cot\alpha\,A^1\,\frac{d\phi}{ds}. \tag{39}$$

Now, for any function f of θ and ϕ through s, we have

$$\frac{df}{ds}=\frac{\partial f}{\partial\theta}\frac{d\theta}{ds}+\frac{\partial f}{\partial\phi}\frac{d\phi}{ds}=\frac{\partial f}{\partial\phi}\frac{d\phi}{ds}, \tag{40}$$

which reduces Eqs. (39) to

$$\partial A^1/\partial\phi=\sin\alpha\cos\alpha\,A^2,\ \partial A^2/\partial\phi=-\cot\alpha\,A^1. \tag{41}$$

Differentiating once again with respect to ϕ and substituting one in the other, Eqs. (41) give rise to the decoupled equations

$$\partial^2A^1/\partial\phi^2=\cos^2\alpha\,A^1,\ \partial^2A^2/\partial\phi^2=-\cos^2\alpha\,A^2, \tag{42}$$

whose solution, consistent with Eqs. (41), is

$$A^1=\sin\alpha\,[c\cos(\phi\cos\alpha)+d\sin(\phi\cos\alpha)],$$

$$A^2=d\cos(\phi\cos\alpha)-c\sin(\phi\cos\alpha). \tag{43}$$

At $\phi=0$, the components of the vector are given by

$$A^1(\phi=0)=c\sin\alpha,\ A^2(\phi=0)=d, \tag{44}$$

whereas, after parallel displacements once around the circle, the components become

$$A^1(\phi=2\pi)=\sin\alpha\,[c\cos(2\pi\cos\alpha)+d\sin(2\pi\cos\alpha)],$$

$$A^2(\phi=2\pi)=d\cos(2\pi\cos\alpha)-c\sin(2\pi\cos\alpha), \tag{45}$$

which are not the same as those given by Eqs. (44) except when $\alpha=\pi/2$, that is, when the displacement is along a geodesic.

Exercise 21

1. Consider a great circle on the surface of a sphere and choose the polar axis to lie in the plane of the great circle so that the circle is a meridian. With this choice of coordinates, show that any meridian on a sphere is a geodesic.
2. Show that the parallel displacement of a vector along any closed curve in a Euclidean space leads back to the original vector.
3. Obtain the geodesic equations in cylindrical coordinates and spherical polar coordinates for a three-dimensional Euclidean space.

*4. Obtain the geodesic equations for a two-dimensional space with metric $ds^2=f\,du\,dv$, where f is a function of u and v.

22. Riemann-Christoffel curvature tensor

A space is Euclidean (or flat) if it is possible to find a cartesian coordinate system everywhere in it; if this is not possible, the space is non-Euclidean (or curved). Given the metric $ds^2 = g_{ij}\, dx^i\, dx^j$ of a space in terms of a certain coordinate system x^i, how does one know whether the space is flat or not ? It is impracticable to try all possible coordinate transformations in order to determine whether one can bring the metric in the cartesian form $ds^2 = \Sigma(dx^i)^2$. For example, consider the metric $ds^2 = (dx^1)^2 + (x^1)^2(dx^2)^2$. It would be tedious to try various transformations with the aim of bringing this metric into the cartesian form. However, a simple look in this particular case tells that this is just the metric in terms of polar coordinate $x^1 = r$, $x^2 = \theta$ of a two-dimensional flat space. In the general case, what is desired is a simple criterion for determining whether a space is flat or curved. The Riemann-Christoffel curvature tensor provides such a criterion. Without proof,[1] we shall state that *a space is flat if and only if the Riemann-Christoffel curvature tensor identically vanishes at every point of the space* irrespective of the coordinate system chosen. It is this fact which makes the curvature tensor[2] very important in the general theory of relativity.

Commutation law of covariant differentiation

In the previous section, we have introduced the curvature tensor by considering the parallel displacement of a vector along two different paths. There is one more way in which the curvature tensor arises.

Ordinary differentiation with respect to coordinates is commutative. For example, if an entity A is a function of the coordinates x^i, then $\partial^2 A/\partial x^i \partial x^j = \partial^2 A/\partial x^j \partial x^i$, since x^i and x^j $(j \neq i)$ are independent of each other. However, we shall see that the covariant differentiation of tensors is not generally commutative.

[1]For proof, the reader may refer to any suitable book on the general theory of relativity, e.g., C. Mollar, *The Theory of Relativity* (Oxford University Press, London, 1966); J. L. Synge, *Relativity: The General Theory* (North-Holland, Amsterdam, 1960).

[2]Hereafter, we shall refer to the Riemann-Christoffel curvature tensor simply as the *curvature tensor.*

Let A_i be a covariant vector. Its covariant derivative is given by [see Eq. (20.22a)]

$$A_{i;j}=A_{i,j}-\Gamma^l_{ij}A_l, \tag{1}$$

which is a covariant tensor of rank two. One more covariant differentiation with respect to x^k and the use of Eq. (20.28a) gives

$$A_{i;jk}\equiv(A_{i;j})_{;k}=(A_{i;j})_{,k}-\Gamma^h_{ik}A_{h;j}-\Gamma^h_{jk}A_{i;h}. \tag{2}$$

Substituting Eq. (1) into Eq. (2) we have

$$\begin{aligned}A_{i;jk}&=A_{i,kj}-\Gamma^l_{ij,k}A_l-\Gamma^l_{ij}A_{l,k}\\&\quad-\Gamma^h_{ik}(A_{h,j}-\Gamma^l_{hj}A_l)-\Gamma^h_{jk}(A_{i,h}-\Gamma^l_{ih}A_l)\\&=A_{i,kj}-\Gamma^l_{ij}A_{l,k}-\Gamma^h_{ik}A_{h,j}-\Gamma^h_{jk}A_{i,h}\\&\quad-\Gamma^l_{ij,k}A_l+\Gamma^h_{ik}\Gamma^l_{hj}A_l+\Gamma^h_{jk}\Gamma^l_{ih}A_l.\end{aligned} \tag{3}$$

Writing the above expression on interchanging j and k, we find

$$\begin{aligned}A_{i;kj}&=A_{i,jk}-\Gamma^l_{ik}A_{l,j}-\Gamma^h_{ij}A_{h,k}-\Gamma^h_{kj}A_{i,h}\\&\quad-\Gamma^l_{ik,j}A_l+\Gamma^h_{ij}\Gamma^l_{hk}A_l+\Gamma^h_{kj}\Gamma^l_{ih}A_l.\end{aligned} \tag{4}$$

Subtracting one from the other, we have

$$A_{i;jk}-A_{i;kj}=-R^{\ l}_{i.jk}A_l, \tag{5}$$

where

$$R^{\ l}_{i.jk}=\Gamma^l_{ij,k}-\Gamma^l_{ik,j}+\Gamma^h_{ij}\Gamma^l_{hk}-\Gamma^h_{ik}\Gamma^l_{hj} \tag{6}$$

is the curvature tensor which is identical to that arrived at in Eq. (21.34).

Starting from a contravariant vector A^i rather than a covariant one, we arrive at

$$A^i_{.;jk}-A^i_{.;kj}=R^{\ i}_{l.jk}A^l. \tag{7}$$

For a tensor of arbitrary rank, the result comes out to be

$$\begin{aligned}A^{i_1\ldots i_r}_{j_1\ldots j_s;kh}-A^{i_1\ldots i_r}_{j_1\ldots j_s;hk}&=\sum_{\mu=1}^{r}R_{l\,.\,kh}^{\ i_\mu}A^{i_1..i_{\mu-1}li_{\mu+1}..i_r}_{j_1\ldots j_s}\\&\quad-\sum_{\mu=1}^{s}R_{j_\mu.kh}^{\ \ \ l}A^{i_1\ldots i_r}_{j_1..j_{\mu-1}lj_{\mu+1}..j_s}.\end{aligned} \tag{8}$$

This shows that covariant differentiation is not commutative in a non-Euclidean space.

Covariant curvature tensor

The completely *covariant curvature tensor* R_{injk} is defined by lowering the contravariant index in Eq. (6), i.e.,

$$R_{injk}=g_{ln}R^{\ l}_{i.jk}. \tag{9}$$

To work out the expression for the covariant curvature tensor, we note that

$$[ij, n]_{,k}=(g_{nl}\,\Gamma^l{}_{ij})_{,k}=g_{nl,k}\,\Gamma^l{}_{ij}+g_{nl}\,\Gamma^l{}_{ij,k}. \qquad (10)$$

Using Eqs. (6) and (10) in Eq. (9), we have

$$\begin{aligned}R_{injk}&=[ij, n]_{,k}-g_{nl,k}\,\Gamma^l{}_{ij}-[ik, n]_{,j}+g_{nl,j}\,\Gamma^l{}_{ik}\\&\quad+g_{ln}(\Gamma^h{}_{ij}\,\Gamma^l{}_{hk}-\Gamma^h{}_{ik}\,\Gamma^l{}_{hj})\\&=[ij, n]_{,k}-[ik, n]_{,j}+\Gamma^h{}_{ik}(g_{nh,j}-[hj, n])\\&\quad-\Gamma^h{}_{ij}(g_{nh,k}-[hk, n])\\&=[ij, n]_{,k}-[ik, n]_{,j}+\Gamma^h{}_{ik}[nj, h]-\Gamma^h{}_{ij}\,[nk, h], \qquad (11a)\end{aligned}$$

or

$$R_{injk}=[ij, n]_{,k}-[ik, n]_{,j}+g^{lh}\{[ik, l]\,[nj, h]-[ij, l]\,[nk, h]\}. \qquad (11b)$$

Symmetries of the curvature tensor

From both the definitions of $R_{i\cdot jk}^{\;l}$, Eq. (6), and of R_{injk}, Eq. (11), it is evident that *the curvature tensor is antisymmetric in the interchange of the third and the fourth indices,* i.e.,

(a) $$R_{i\cdot jk}^{\;l}=-R_{i\cdot kj}^{\;l};\quad R_{injk}=-R_{inkj}. \qquad (12)$$

It can be shown that the covariant curvature tensor is also *antisymmetric in the interchange of the first two indices.* To prove this, it is convenient to write explicitly the first two terms in Eq. (11b). This gives

$$\begin{aligned}&[ij, n]_{,k}-[ik, n]_{,j}\\&\quad=\tfrac{1}{2}(g_{in,kj}+g_{jn,ki}-g_{ij,kn})-\tfrac{1}{2}(g_{in,jk}+g_{kn,ji}-g_{ik,jn})\\&\quad=\tfrac{1}{2}(g_{jn,ki}+g_{ik,jn}-g_{ij,kn}-g_{kn,ij}). \qquad (13)\end{aligned}$$

Interchanging i and n, it is evident that

$$[nj, i]_{,k}-[nk, i]_{,j}=-\{[ij, n]_{,k}-[ik, n]_{,j}\}, \qquad (14)$$

showing that the first two terms on the right hand side of Eq. (11) are separately antisymmetric in the interchange of i and n. Interchanging i and n in the remaining part of Eq. (11b) gives

$$\begin{aligned}&g^{lh}\{[nk, l]\,[ij, h]-[nj, l]\,[ik, h]\}\\&\quad=g^{hl}\,\{[nk, h]\,[ij, l]-[nj, h]\,[ik, l]\}\\&\quad=-g^{lh}\,\{[ik, l]\,[nj, h]-[ij, l]\,[nk, h]\}. \qquad (15)\end{aligned}$$

All this finally leads to the desired result

(b) $$R_{injk}=-R_{nijk}. \qquad (16)$$

From these antisymmetry relations, it follows that components of the form R_{iijk}, R_{injj}, R_{iijj} and R_{iiii} (no summation) of the curvature tensor vanish identically.

The above result (b) can be proved more easily if we assume the results of Eq. (20.31) and Eq. (8). Writing Eq. (8) for a covariant tensor of rank two, we have

$$A_{ij;kh}-A_{ij;hk}=-R_{i.kh}^{\,l}A_{lj}-R_{j.kh}^{\,l}A_{il}. \tag{17}$$

Replacing A_{ij} by the metric tensor g_{ij} and remembering that its covariant derivative vanishes, Eq. (17) reduces to

$$0=R_{i\cdot kh}^{\,l}g_{lj}+R_{j\cdot kh}^{\,l}g_{il}=R_{ijkh}+R_{jikh}, \tag{18}$$

where Eq. (9) has been used. Eq. (18) gives the desired result.

Next, it can be shown that the covariant curvature tensor is *symmetric in the interchange of the two index pairs*, i. e., under the interchange $(in)\leftrightarrow(jk)$. Carrying out the interchanges $i\leftrightarrow j$ and $n\leftrightarrow k$ in Eq. (13), it is easily seen that the expression remains unchanged, i. e.,

$$[ji, k]_{,n}-[jn, k]_{,i}=[ij, n]_{,k}-[ik, n]_{,j}. \tag{19}$$

Similarly, carrying out these interchanges in the second half of the expression on the right hand side of Eq. (11b), we see that

$$g^{lh}\{[jn, l][ki, h]-[ji, l][kn, h]\}$$
$$=g^{lh}\{[ik, l][nj, h]-[ij, l][nk, h]\}, \tag{20}$$

remembering that $g^{hl}=g^{lh}$ and $[ki, l]=[ik, l]$. Combining Eqs. (19) and (20), we have

(c) $$R_{injk}=R_{jkin}. \tag{21}$$

Lastly, we have the cyclic property of the curvature tensor: *If any one index of the tensor is held fixed while the remaining three indices are cyclically permuted and the components added, the result is zero*, i.e.,

(d) $$R_{injk}+R_{ijkn}+R_{iknj}=0, \tag{22}$$

where the first index is held fixed. To prove this, we again write the explicit expressions for the three components appearing in Eq. (22) and add. Using Eq. (13) in Eq. (11), we have

$$R_{injk}=\tfrac{1}{2}(g_{jn,ki}+g_{ik,jn}-g_{ij,kn}-g_{kn,ij})$$
$$+g^{lh}\{[ik, l][nj, h]-[ij, l][nk, h]\}. \tag{23}$$

On performing the cyclic permutation $n\to j\to k\to n$ twice, we have

$$R_{ijkn}=\tfrac{1}{2}(g_{kj,ni}+g_{in,kj}-g_{ik,nj}-g_{nj,ik})$$
$$+g^{lh}\{[in, l][jk, h]-[ik, l][jn, h]\}, \tag{24}$$

$$R_{iknj}=\tfrac{1}{2}(g_{nk,ji}+g_{ij,nk}-g_{in,jk}-g_{jk,in})$$
$$+g^{lh}\{[ij, l][kn, h]-[in, l][kj, h]\}. \tag{25}$$

Eq. (22) follows immediately on adding the above three equations.

It will be convenient to write these relations among the components of the curvature tensor together:

$$R_{injk}=-R_{nijk}=-R_{inkj}=R_{jkin}, \tag{26a}$$

$$R_{injk}+R_{ijkn}+R_{iknj}=0. \tag{26b}$$

These are the four *independent* relations among the components of the curvature tensor. Combining these with each other, various other relations can be obtained which are given below:

$$\begin{aligned} &\text{(i)} && R_{injk}=-R_{kjin}=-R_{jkni}=R_{kjni}; \\ &\text{(ii)} && R_{injk}+R_{jnki}+R_{knij}=0 \quad \text{(second index fixed)}; \\ &\text{(iii)} && R_{injk}+R_{nkji}+R_{kijn}=0 \quad \text{(third index fixed)}; \\ &\text{(iv)} && R_{injk}+R_{njik}+R_{jink}=0 \quad \text{(fourth index fixed)}. \end{aligned} \tag{27}$$

Number of independent components

The curvature tensor, being of rank four, has a total of N^4 components in an N-dimensional space. However, we have seen that many of these identically vanish and the others are related to each other in various ways. This considerably reduces the number of independent components of the curvature tensor.

To calculate this number, firstly we note that the nonvanishing components of the curvature tensor can be divided into three categories for the present purpose: (i) components with two *distinct* indices, of the form R_{inin} $(i\neq n)$ and others related to it, such as R_{inni}, R_{nini}, etc.; (ii) components with three *distinct* indices, of the form R_{inji}; and (iii) components with four *distinct* indices, such as R_{injk}. Secondly, we note that unless all the four indices are distinct, the cyclic symmetry relation, Eq. (22), reduces to one of the earlier three relations. This can be seen very easily by equating any two indices in Eq. (22). For example, putting $n=i$ in it, we have

$$R_{iijk}+R_{ijki}+R_{ikij}=0. \tag{28}$$

By relation (b), Eq. (16), we have $R_{iijk}=0$, and by relation (c), Eq. (19), we have $R_{ikij}=R_{ijik}$. This reduces Eq. (28) to

$$R_{ijki}=-R_{ijik}, \tag{29}$$

which is of the form of relation (a), Eq. (12). Thus when two of the four indices are equal, relation (d) has been reduced to relation (a) by the use of relations (b) and (c). In other words, unless all the four indices are distinct, relation (d) is not independent of relations (a), (b) and (c).

Now we are ready to calculate the number of independent components of the three categories listed above.

(i) *Two distinct indices*, R_{inin}: The first index can be chosen in

N ways and the second can be chosen in $N-1$ ways. However, since $R_{inin}=R_{nini}$, the order of the two indices is immaterial, and we must divide the total number of ways by a factor of 2. This gives the number of independent components of the first category [which is just the number of distinct index pairs (i, n) with $i \neq n$] as

$$N_I=\tfrac{1}{2}N(N-1). \tag{30}$$

There are no other relations which further reduce this number.

(ii) *Three distinct indices*, R_{inji}: The first index pair can be chosen in $\frac{1}{2}N(N-1)$ ways as calculated in (i) above. Having chosen two indices, the third distinct index can be chosen in $N-2$ ways. Therefore, the number of distinct components of the second category is

$$N_{II}=\tfrac{1}{2}N(N-1)(N-2). \tag{31}$$

There are no other relations which further reduce this number.

(iii) *Four distinct indices*, R_{injk}: The first index pair can be chosen in $\frac{1}{2}N(N-1)$ ways. Having chosen the first index pair, the second pair can be chosen in $\frac{1}{2}(N-2)(N-3)$ ways. But, since the index pairs can be interchanged owing to relation (c), Eq. (19), we must divide this number by a factor of 2. This makes the number of ways of choosing two distinct index pairs $N(N-1)(N-2)(N-3)/8$. However, now that all the four indices are distinct, the cyclic relation (d) becomes effective and shows that only two-thirds of these components are independent. Hence, finally, the number of independent components of this category is

$$N_{III}=N(N-1)(N-2)(N-3)/12. \tag{32}$$

Adding the three contributions, we get the number of algebraically independent components of the curvature tensor to be

$$N_I+N_{II}+N_{III}=N^2(N^2-1)/12. \tag{33}$$

The number of independent components of the curvature tensor for a few small values of N is listed below:

Dimensionality of space:	1	2	3	4	5
Number of independent components:	0	1	6	20	50

The fact that the number of independent components of R_{injk} in a one-dimensional space is zero means that a one-dimensional space cannot be 'curved' in the sense of the general theory of relativity, i.e., in the sense of being a non-Euclidean space. In other words, every one-dimensional space is a Euclidean space. This follows directly from the fact that the most general metric in a one-dimensional space (with coordinate x) is of the form $ds^2=f(x)\,dx^2$, which can be reduced to cartesian form $ds^2=du^2$ by the coordinate transformation $f^{1/2}(x)\,dx=du$.

In a two-dimensional space, the indices i, n, j, k can take only the two values 1 and 2. Out of the total of 16 components, the only nonvanishing components of the curvature tensor are R_{1212}, R_{1221}, R_{2112} and R_{2121}, all of which are related to each other through the relations (a), (b) and (c). There is only one algebraically independent component which is chosen to be R_{1212} by convention.

EXAMPLE 1: Find the components of the curvature tensor in the two-dimensional space of the surface of a sphere.

Solution: On the surface of a sphere of radius a, there are only four nonvanishing components of the curvature tensor, as discussed above. They are related to each other by

$$R_{1212}=R_{2121}=-R_{1221}=-R_{2112}. \tag{34}$$

Using Eqs. (20.48) and Eq. (11), we have

$$\begin{aligned}R_{1212}&=\partial[11,2]/\partial\phi-\partial[12,2]/\partial\theta+g^{lh}\{[12,l]\,[12,h]\\&\qquad\qquad-[11,l]\,[22,h]\}\\&=-\partial(a^2\sin\theta\cos\theta)/\partial\theta+a^4\sin^2\theta\cos^2\theta/(a^2\sin^2\theta)\\&=a^2\sin^2\theta.\end{aligned} \tag{35}$$

Similarly, from Eqs. (20.51) and Eq. (6), we have

$$\begin{aligned}R_{1\cdot12}^{\,2}&=\partial\Gamma^2{}_{11}/\partial\phi-\partial\Gamma^2{}_{12}/\partial\theta+\Gamma^h{}_{11}\,\Gamma^2{}_{h2}-\Gamma^h{}_{12}\,\Gamma^2{}_{h1}\\&=\operatorname{cosec}^2\theta-\cot^2\theta=1.\end{aligned} \tag{36}$$

Exercise 22

1. Obtain the relations given in Eqs. (27) among the components of the curvature tensor by using those of Eqs. (26).

2. Find the number of nonvanishing components of the curvature tensor for $N=3$ and $N=4$. Write down a suitable set of $N^2(N^2-1)/12$ algebraically independent components of the tensor for the above cases.

*3. A two-dimensional space has the metric $ds^2=f(u,v)du^2+h(u,v)dv^2$. Find the component R_{1212} of the curvature tensor for this space [see Exercise 20.10]. Consider the following special cases and verify that these results agree with those of direct calculations: (a) $f=1$, $h=u^2$ [polar coordinates in a two-dimensional flat space]; (b) $f=a^2$, $h=a^2\sin^2 u$ [polar coordinates on the surface of a sphere; see Example 1 and Exercise 4 below].

4. Show that the component R_{1212} of the curvature tensor for a two-dimensional space with the metric $ds^2=dx^2+f(x, y)dy^2$ equals $-\frac{1}{2}\partial^2 f/\partial x^2 +(1/4f)(\partial f/\partial x)^2$.

*5. Consider a space with the metric $ds^2=2S\,du\,dv$. Find R_{1212} for this space. Obtain the nonvanishing component of the mixed curvature tensor $R^{j}_{i\cdot kl}$ for the space. Show that if S is independent of at least one of the two coordinates u and v, then the space is flat.

Bibliography for chapter 2

BOAS, M. L., *Mathematical Methods in the Physical Sciences*, New York: Wiley (1966), Chap. 10.

BORISENKO, A. I., and I. E. TARAPOV, *Vector and Tensor Analysis with Applications*, Englewood Cliffs: Prentice-Hall (1968).

EISELE, J. A., and R. M. MASON, *Applied Matrix and Tensor Analysis*, New York: Wiley (1970).

FLÜGGE, W., *Tensor Analysis and Continuum Mechanics*, Berlin: Springer-Verlag (1972).

HAY, G. E., *Vector and Tensor Analysis*, New York: Dover (1953).

KYRALA, A., *Theoretical Physics: applications of vectors, matrices, tensors and quaternions*, Philadelphia: Saunders (1967).

LASS, H., *Vector and Tensor Analysis*, New York: McGraw-Hill (1950).

LICHNEROWICZ, A., *Elements of Tensor Calculus, 4th ed.*, London: Methuen (1962).

MARGENAU, H., and G. M. Murphy, *Mathematics for Physics and Chemistry, 2nd ed.*, New York: Van Nostrand (1962), pp. 161–171.

MERCIER, J. L., *An Introduction to Tensor Calculus*, Groningen: Wolters Noordhoff (1971).

SOKOLNIKOFF, I. S., *Tensor Analysis*, New York: Wiley (1967).

SPAIN, B., *Tensor Calculus, 3rd ed.*, Edinburgh: Oliver & Boyd (1965).

SPIEGEL, M. R., *Vector Analysis*, New York: McGraw-Hill (1959), Chap. 8.

SYNGE, J. L., and A. SCHILD, *Tensor Calculus*, Toronto: University of Toronto Press (1969).

WEATHERBURN, C. E., *An Introduction to Riemannian Geometry and the Tensor Calculus*, London: Cambridge University Press (1966).

Appendix I: The logic of necessity and sufficiency

One often comes across two statements related to each other by phrases such as 'necessary for', 'sufficient for', 'if', 'only if', 'if and only if', etc. It is important to understand their meaning clearly and to avoid their wrong use.

Consider the following simple statements:

(i) If $a>b$, then $b<a$;

(ii) If $b<a$, then $a>b$.

It is evident that both the statements are true. Therefore, we can combine the two and express it in the form "$a>b$ if, and only if, $b<a$".

As another example, consider the following two statements:

(i) A: **U** is a unitary matrix;

(ii) B: det **U** has unit magnitude.

When we combine these two sentences in the statement

'*B* if *A*' ("det **U** has unit magnitude if **U** is a unitary matrix"),

we get a *true* statement. However, the statement '*A* if *B*' ("**U** is a unitary matrix if det **U** has unit magnitude") is obviously a *false* statement, because the determinant of a nonunitary matrix may also have unit magnitude.

The above statements (i) and (ii) can also be combined by using the phrase 'only if' by saying that '*A* only if *B*' ("**U** is a unitary matrix *only if* det **U** has unit magnitude"), which is also *true*. Thus we see that if the statement '*B* if *A*' is true, then '*A* only if B' is also true. On the other hand, '*A* if *B*' does not logically follow from '*B* if *A*'.

The above relations can also be written in a different language. Thus '*B* if *A*' is the same as saying that '*B* is *necessary* (though *not sufficient*) for *A*' ("for **U** to be a unitary matrix, it is *necessary*, though *not sufficient*, that det **U** has unit magnitude"). Conversely, we can translate the statement '*A* only if *B*' as '*A* is *sufficient*, (though *not necessary*) for *B*' ("that **U** is a unitary matrix is *sufficient*, though *not necessary*, for det **U** to have unit magnitude").

Symbolically, the above statements can be written as follows:

$A \Rightarrow B$: *A implies B* ("**U** is a unitary matrix *implies* that det **U** has unit magnitude");

$B \nRightarrow A$: *B does not imply A* ("that det **U** has unit magnitude *does not imply* that **U** is unitary");

$B \Leftarrow A$: *B is implied by A* ("that det **U** has unit magnitude *is implied by* (the fact that) **U** is a unitary matrix");

$A \nLeftarrow B$: *A is not implied by B* (that **U** is a unitary matrix *is not implied by* (the fact that) det **U** has unit magnitude").

The three methods of stating the logical relationships discussed above can be expressed in a tabular form as follows:

if, only if	*necessary, sufficient*	$\Rightarrow$, $\Leftarrow$
(a) *A* if *B*	*A* is necessary for *B* or *B* is sufficient for *A*	$A \Leftarrow B$ (*A* is implied by B) or $B \Rightarrow A$ (*B* implies *A*).
(b) *A* only if *B*	*A* is sufficient for *B* or *B* is necessary for *A*	$A \Rightarrow B$ (*A* implies *B*) or $B \Leftarrow A$ (*B* is implied by *A*).
(c) *A* if, and only if, *B*	*A* is necessary and sufficient for *B*	$A \Leftrightarrow B$ (*A* implies, and is implied by, *B*).

Appendix II : On the order of a finite nonsingular matrix

Let **A** and **B** be two finite matrices which satisfy the equations

$$\mathbf{AB}=\mathbf{I}_1,\ \mathbf{BA}=\mathbf{I}_2, \tag{1}$$

simultaneously, where $\mathbf{I}_1$ and $\mathbf{I}_2$ are unit matrices. Our object is to show that **A**, and hence **B**, are square matrices.

The proof will follow the argument that if **A** is a rectangular matrix, there exists no matrix **B** such that both the Eqs. (1) are satisfied simultaneously.

Let $\mathbf{A}\equiv[a_{ij}]$ be of order $m\times n$ and $\mathbf{B}\equiv[b_{ij}]$ of order $n\times m$, so that $\mathbf{I}_1$ and $\mathbf{I}_2$ in Eqs. (1) are unit matrices of order m and n, respectively. Consider first the case $m>n$. Let $\mathbf{b}_i=\{b_{1i}\ b_{2i}\ldots b_{ni}\}$ be the i-th column vector of **B** and let $\mathbf{e}_i=\{0\ \ 0\,.\,.\,1_i\,.\,.\,0\}$ be a unit vector with unity in the i-th position. The first of Eqs. (1) is then equivalent to the m equations

$$\mathbf{A}\mathbf{b}_i=\mathbf{e}_i,\ 1\leqslant i\leqslant m. \tag{2}$$

For each value of i, this represents a set of m equations in the n unknowns b_{ki}, $1\leqslant k\leqslant n$.

Let r be the rank of **A** and s the rank of the augmented matrix $\mathbf{D}_i=[\mathbf{A}\,\vdots\,\mathbf{e}_i]$. It is evident, in the present case, that $r\leqslant n$ and $s=r$ or $r+1$. This implies that **A** has at least one square submatrix of order r whose determinant is different from zero; let this be denoted by **P**. Then it is evident that for some value of i between 1 and m, there exists an augmented matrix $\mathbf{D}_i$ containing a square submatrix of order $r+1$ of the form

$$\mathbf{Q}=\left[\begin{array}{c|c}\mathbf{P} & \mathbf{0}\\ \hline \mathbf{X} & 1\end{array}\right], \tag{3}$$

where **0** is an $r\times 1$ null vector and **X** is a $1\times r$ row vector. It follows from Eq. (3) that $\det\mathbf{Q}=\det\mathbf{P}$, so that $\mathbf{D}_i$ contains a square submatrix of order $r+1$ whose determinant is nonzero, so that $s=r+1$. The discussion of Section 8 shows that, in this case, Eq. (2) has no solution for $\mathbf{b}_i$. Hence if **A** is of order $m\times n$ with $m>n$, there exists no matrix **B** which satisfies the equation $\mathbf{AB}=\mathbf{I}$.

In the second case when $m<n$, consider the transpose of the second of Eqs. (1):

$$\tilde{\mathbf{A}}\tilde{\mathbf{B}}=\mathbf{I}_2. \tag{4}$$

$\tilde{\mathbf{A}}$ is of order $n\times m$ with $n>m$, and the above result shows that there exists no matrix $\tilde{\mathbf{B}}$, and hence **B**, which satisfies Eq. (4).

Combining the two results, it follows that if **A** is a rectangular matrix, there exists no matrix **B** which satisfies both the Eqs. (1) simultaneously. For a square matrix **A**, we have already shown in Section 5 that a matrix **B** satisfying Eqs. (1) exists provided det $\mathbf{A} \neq 0$, and have shown how to obtain it. All this goes to show that *a finite nonsingular matrix must be square.*

Using this result, it can also be shown that finite orthogonal and unitary matrices must be square.

Answers to selected exercises

Exercise 1

1. (a) $\sqrt{15}, \dfrac{1}{\sqrt{15}}(2, 1, 3, -1)$; (c) $\sqrt{2}, (1/\sqrt{2}, i/\sqrt{2})$.

2. (a) Linearly independent; (c) linearly dependent.

5. (a) 7; (c) for no value of x (the three vectors are linearly independent for any value of x).

7. (a) True; (b) false: it is possible to find a set of n vectors any two of which are linearly independent of each other but, as a set, they are linearly dependent.

9. $x+w=y+z$.

10. On using the vector identity $\mathbf{b}\times(\mathbf{c}\times\mathbf{d})=(\mathbf{b}.\mathbf{d})\mathbf{c}-(\mathbf{b}.\mathbf{c})\mathbf{d}$, we have

$$\mathbf{A}^2\mathbf{r}=\mathbf{A}(\mathbf{A}\mathbf{r})=\mathbf{A}(\mathbf{a}\times\mathbf{r})=\mathbf{a}\times(\mathbf{a}\times\mathbf{r})=(\mathbf{a}.\mathbf{r})\mathbf{a}-a^2\mathbf{r}.$$

Further application of operator $\mathbf{A}$ gives

$$\mathbf{A}^3\mathbf{r}=\mathbf{a}\times[(\mathbf{a}.\mathbf{r})\mathbf{a}-a^2\mathbf{r}]=-a^2(\mathbf{a}\times\mathbf{r})=-a^2\mathbf{A}\mathbf{r}$$
$$\Rightarrow \mathbf{A}^3\mathbf{r}+a^2\mathbf{A}\mathbf{r}=\mathbf{0}\ \forall\mathbf{r}\Rightarrow \mathbf{A}^3+a^2\mathbf{A}=\mathbf{0}.$$

Exercise 2

1. (c) $\mathbf{A}+\mathbf{B}=\begin{bmatrix} 7/3 & 8/15 & 48/5 & 37/7 \\ -21/8 & 1/3 & 2/3 & 53/4 \\ 1/11 & 5/7 & 9/11 & -1/2 \end{bmatrix}$;

$$\mathbf{A}-\mathbf{B}=\begin{bmatrix} -5/3 & -2/15 & -42/5 & -33/7 \\ 27/8 & 0 & 1 & -35/4 \\ 1/11 & -1/7 & -3/11 & 1 \end{bmatrix}.$$

2. (a) $\mathbf{AB}=\begin{bmatrix} -1 \\ -4 \\ -41 \end{bmatrix}$; (e) $\mathbf{AB}=\begin{bmatrix} -6 & 14 & 8 \\ 3 & -7 & -4 \\ -12 & 28 & 16 \end{bmatrix}$.

4. (a) $\mathbf{AB}=\begin{bmatrix} -8 & 21 \\ 16 & -6 \end{bmatrix}$, $\mathbf{BA}=\begin{bmatrix} -29 & -18 \\ 7 & 14 \end{bmatrix}$, $\mathbf{AB}\neq\mathbf{BA}$.

(d) $\mathbf{AB}=-46$, $\mathbf{BA}=\begin{bmatrix} 3 & -7 & 2 \\ 15 & -35 & 10 \\ -21 & 49 & -14 \end{bmatrix}$, $\mathbf{AB}\neq\mathbf{BA}$.

5. $\mathbf{A}^2=\begin{bmatrix} 2 & -9 & 28 \\ -12 & 22 & -8 \\ -2 & -3 & -4 \end{bmatrix}$; $\mathbf{A}^3=\begin{bmatrix} -44 & 42 & -16 \\ 40 & -124 & 64 \\ -4 & 6 & -32 \end{bmatrix}$;

$$\mathbf{A}^4=\begin{bmatrix} 56 & -300 & 16 \\ -272 & 616 & -480 \\ 40 & -36 & -16 \end{bmatrix}.$$

8. $(A^2)_{ij}=(A)_{ij}\Rightarrow\sum_{k=1}^{n}(A)_{ik}(A)_{kj}=(A)_{ij}\Rightarrow p_ip_j\sum_{k=1}^{n}p_k^2=p_ip_j\ \forall\, i,j\Rightarrow\sum_{k=1}^{n}p_k^2=1.$

13. Let $A_1, A_2, \ldots, A_n$ be such matrices that the matrix product of two successive matrices, i.e., A_iA_{i+1} for $1\leqslant i\leqslant n-1$, is defined. Assume that

$$(A_1A_2\ldots A_{n-1})^T=\widetilde{A}_{n-1}\ldots\widetilde{A}_2\widetilde{A}_1. \tag{A.1}$$

Then

$$(A_1A_2\ldots A_n)^T=[(A_1A_2\ldots A_{n-1})(A_n)]^T=\widetilde{A}_n\widetilde{A}_{n-1}\ldots\widetilde{A}_2\widetilde{A}_1, \tag{A.2}$$

by using Eq. (2.45) and Eq. (A.1). This is of the same form as Eq. (A.1) with $n-1$ replaced by n. Hence, by mathematical induction, Eq. (A.2) is the desired result. The result for hermitian conjugation can be obtained in a similar way.

16. $AB=\begin{bmatrix} 21+14i & 9+8i & -9+10i \\ -4-34i & -7i & 1+17i \\ 44-5i & 67+5i & -18+3i \end{bmatrix}.$

Exercise 3

1. If the given matrix is denoted by A, then

(a) $A=\begin{bmatrix} 3 & 1 & 5 \\ -2 & 6 & -1 \\ 4 & 2 & 6 \end{bmatrix}+\begin{bmatrix} 2i & -i & 0 \\ 7i & 6i & 4i \\ -3i & 2i & i \end{bmatrix};$

(b) $A=\frac{1}{2}\begin{bmatrix} 6+4i & -1+6i & 9-3i \\ -1+6i & 12i & 1+6i \\ 9-3i & 1+6i & 12+2i \end{bmatrix}+\frac{1}{2}\begin{bmatrix} 0 & 3-8i & 1+3i \\ -3+8i & 0 & -3+2i \\ -1-3i & 3-2i & 0 \end{bmatrix}.$

4. (b) Let A and B be two hermitian matrices, so that $A^\dagger=A$, $B^\dagger=B$. Hence

$$(AB)^\dagger=B^\dagger A^\dagger=BA. \tag{A.3}$$

Firstly, let A and B commute with each other. Then Eq. (A.3) reduces to $(AB)^\dagger=AB$, showing that AB is a hermitian matrix. Secondly, let AB be a hermitian matrix, so that $(AB)^\dagger=AB$. Comparing this with Eq. (A.3), we have $AB=BA$, showing that A and B commute with each other. The result is thus proved.

7. $a=-3/17$, $b=1/17$.

15. Let $\begin{bmatrix} a & b \\ c & d \end{bmatrix}$ be the matrix whose square equals the given matrix. Then

$$\begin{bmatrix} a & b \\ c & d \end{bmatrix}\begin{bmatrix} a & b \\ c & d \end{bmatrix}=\begin{bmatrix} a^2+bc & b(a+d) \\ c(a+d) & bc+d^2 \end{bmatrix}=\begin{bmatrix} 2 & 1 \\ 0 & 2 \end{bmatrix}.$$

This gives

$$a^2+bc=2,\ b(a+d)=1,\ c(a+d)=0,\ bc+d^2=2. \tag{A.4}$$

We have $c=0$ or $a+d=0$. But if $a+d=0$, b does not exist: hence $a\neq -d$ and $c=0$, giving $a=\pm\sqrt{2}$, $d=\pm\sqrt{2}$. This gives the two desired matrices

$$\begin{bmatrix} \pm\sqrt{2} & \pm 1/2\sqrt{2} \\ 0 & \pm\sqrt{2} \end{bmatrix},$$

where either all the upper signs or all the lower signs should be taken.

16. The most general antisymmetric matrix of order 3 is

$$\begin{bmatrix} 0 & a & b \\ -a & 0 & c \\ -b & -c & 0 \end{bmatrix}.$$

A set of three linearly independent antisymmetric matrices of order three is

$$\begin{bmatrix} 0 & 1 & 0 \\ -1 & 0 & 0 \\ 0 & 0 & 0 \end{bmatrix}, \begin{bmatrix} 0 & 0 & 1 \\ 0 & 0 & 0 \\ -1 & 0 & 0 \end{bmatrix}, \begin{bmatrix} 0 & 0 & 0 \\ 0 & 0 & 1 \\ 0 & -1 & 0 \end{bmatrix}.$$

Exercise 4

1. (a) $4a^2b^2c^2$; (c) $(a-b)(b-c)(c-a)(a+b+c)$.

6. $\mathbf{H}$ is hermitian $\Rightarrow \mathbf{H}^\dagger = \mathbf{H} \Rightarrow \det \mathbf{H}^\dagger = \det \mathbf{H}$. But $\det \mathbf{H}^\dagger = (\det \mathbf{H})^*$, so that $(\det \mathbf{H})^* = \det \mathbf{H} \Rightarrow \det \mathbf{H}$ is real.

Exercise 5

2. (b) $\dfrac{1}{12}\begin{bmatrix} 4 & 4 & -4 \\ -6 & -3 & 12 \\ 2 & -1 & 4 \end{bmatrix}$; (d) $\dfrac{1}{4}\begin{bmatrix} -16 & 4 & 4 & 0 \\ -31 & 8 & 6 & 1 \\ 34 & -8 & -8 & 2 \\ -14 & 4 & 4 & -2 \end{bmatrix}$.

4. If any one of the matrices $\mathbf{A}$ and $\mathbf{B}$ is nonsingular, its inverse exists, and multiplying suitably by the inverse (from left or right, as the case may be), it follows that the other must be a zero matrix. Hence the result follows.

8. $q \neq p$, $q \neq -2p$. When both the conditions are satisfied, the inverse is

$$\frac{1}{D}\begin{bmatrix} q+p & -p & -p \\ -p & q+p & -p \\ -p & -p & q+p \end{bmatrix}, \text{ with } D=(q-p)(q+2p).$$

The inverse matrix is of the same form as the given matrix, i.e., a matrix all the diagonal elements of which are equal to each other and all its nondiagonal elements are equal to each other.

9. Since two diagonal matrices commute with each other,

$$\mathbf{P}^{-1}\mathbf{A}\mathbf{P}\mathbf{P}^{-1}\mathbf{B}\mathbf{P} = \mathbf{P}^{-1}\mathbf{B}\mathbf{P}\mathbf{P}^{-1}\mathbf{A}\mathbf{P} \Rightarrow \mathbf{P}^{-1}\mathbf{A}\mathbf{B}\mathbf{P} = \mathbf{P}^{-1}\mathbf{B}\mathbf{A}\mathbf{P} \Rightarrow \mathbf{A}\mathbf{B} = \mathbf{B}\mathbf{A}.$$

11. Given $\mathbf{AB}=\mathbf{BA}$. Multiplying from the left and also from the right by $\mathbf{B}^{-1}$, we have $\mathbf{B}^{-1}\mathbf{A} = \mathbf{A}\mathbf{B}^{-1}$.

13. $2(a^2+b^2)=1$.

14. If $\mathbf{A} = \mathbf{I} - (2/\tilde{\mathbf{x}}\mathbf{x})\mathbf{x}\tilde{\mathbf{x}}$, $\tilde{\mathbf{A}} = \mathbf{I} - (2/\tilde{\mathbf{x}}\mathbf{x})[(\mathbf{x}\tilde{\mathbf{x}})^T = \mathbf{I} - (2/\tilde{\mathbf{x}}\mathbf{x})\mathbf{x}\tilde{\mathbf{x}} = \mathbf{A} \Rightarrow \mathbf{A}$ is symmetric. Further, $\mathbf{A}\tilde{\mathbf{A}} = \mathbf{A}^2 = [\mathbf{I} - (2/\tilde{\mathbf{x}}\mathbf{x})\mathbf{x}\tilde{\mathbf{x}}]^2 = \mathbf{I} - (4/\tilde{\mathbf{x}}\mathbf{x})\mathbf{x}\tilde{\mathbf{x}} + (2/\tilde{\mathbf{x}}\mathbf{x})^2\mathbf{x}\tilde{\mathbf{x}}\mathbf{x}\tilde{\mathbf{x}} = \mathbf{I} - (4/\tilde{\mathbf{x}}\mathbf{x})\mathbf{x}\tilde{\mathbf{x}} + (4/\tilde{\mathbf{x}}\mathbf{x})\mathbf{x}\tilde{\mathbf{x}} = \mathbf{I} \Rightarrow \mathbf{A}$ is orthogonal.

18. $U^\dagger=[(H+iI)^{-1}]^\dagger (H-iI)^\dagger=[(H+iI)^\dagger]^{-1}(H+iI)=(H-iI)^{-1}(H+iI)$;

$$U^\dagger U=(H-iI)^{-1}(H+iI)(H-iI)(H+iI)^{-1}$$
$$=(H-iI)^{-1}(H-iI)(H+iI)(H+iI)^{-1};$$

since $H-iI$ commutes with $H+iI$. This gives $U^\dagger U=I$.

Exercise 6

1. $$(A-BD^{-1}C)^{-1}=\begin{bmatrix} 0 & 0 & 1 \\ 1 & 1 & 0 \\ 2 & 1 & -1 \end{bmatrix},\ (D-CA^{-1}B)^{-1}=\begin{bmatrix} 0 & 1 \\ 1 & 1 \end{bmatrix}. \qquad (A.5)$$

The inverse of the given matrix $=\left[\begin{array}{ccc|cc} 0 & 0 & 1 & 0 & 0 \\ 1 & 1 & 0 & 0 & -1 \\ 2 & 1 & -1 & -1 & -1 \\ \hline -1 & 0 & 0 & 0 & 1 \\ -1 & -1 & 0 & 1 & 1 \end{array}\right].$

4. If the given matrix is denoted by P, then

$$P^*=\begin{bmatrix} A^* & B^* & C^* \\ D^* & E^* & F^* \end{bmatrix},\ \tilde{P}=\begin{bmatrix} \tilde{A} & \tilde{D} \\ \tilde{B} & \tilde{E} \\ \tilde{C} & \tilde{F} \end{bmatrix},\ P^\dagger=\begin{bmatrix} A^\dagger & D^\dagger \\ B^\dagger & E^\dagger \\ C^\dagger & F^\dagger \end{bmatrix}. \qquad (A.6)$$

6. (a) 2; (c) 3.

Exercise 7

1. $x=u+2v+5w,\ y=-v+2w,\ z=2u+4v+11w.$
2. (a) $x_1=x_2=x_3$, x_3 arbitrary; (c) $\mathbf{x}=0$.
3. (a) No solution; (b) $x=-14,\ y=14,\ z=2$.
5. The three equations for the three currents are

$$I_1-I_2-I_3=0,$$
$$I_3R_4-I_2R_2=-E_2, \qquad (A.7)$$
$$I_1(R_1+R_3)+I_3R_4=E_1.$$

The solution is

$$I_1=[E_1(R_2+R_4)+E_2R_4]/D,\ I_2=[E_1R_4+E_2(R_1+R_3+R_4)]/D,$$
$$I_3=[E_1R_2-E_2(R_1+R_3)]/D,\ D=R_2(R_1+R_3+R_4)+R_4(R_1+R_3).$$

With the given values, this becomes

$$I_1=7/29 \text{ amp.},\ I_2=61/145 \text{ amp.},\ I_3=-26/145 \text{ amp.} \qquad (A.8)$$

6. If we take $B=\begin{bmatrix} a & b \\ c & d \\ e & f \end{bmatrix}$, AB=I gives the four equations

$$a+2c+3e=1,\quad b+2d+3f=0,$$

$$4a+5c+6e=0,\ 4b+5d+6f=1. \tag{A.9}$$

This gives $\mathbf{B}=\begin{bmatrix} a & b \\ -2-2a & 1-2b \\ (5+3a)/3 & (3b-2)/3 \end{bmatrix}$, which contains two arbitrary parameters a and b.

Next, if we take $\mathbf{B}=\begin{bmatrix} x & r \\ y & s \\ z & t \end{bmatrix}$, $\mathbf{BA}=\mathbf{I}$ gives the nine equations

$$\begin{aligned} x+4r&=1, & y+4s&=0, & z+4t&=0, \\ 2x+5r&=0, & 2y+5s&=1, & 2z+5t&=0, \\ 3x+6r&=0, & 3y+6s&=0, & 3z+6t&=1, \end{aligned} \tag{A.10}$$

which have no solution.

Exercise 8

1. (a) (i) $\lambda=1$, $x=-4y$, y arbitrary; (ii) $\lambda=-2$, $x=-y$, y arbitrary.
2. (b) 4.
3. $u=5/2$, $v=-\frac{1}{2}$, $w=-\frac{1}{2}$, $x=-\frac{1}{2}$.
6. (a) $z=x+3y$, $t=-x-y$, x and y arbitrary;
   (b) $x_3=-2x_1-x_2$, $x_4=-x_2$, $x_5=x_1+x_2$, x_1 and x_2 arbitrary.

Exercise 9

1. If $\mathbf{A}\equiv[a_{ij}]$ is a triangular matrix of order n, its characteristic polynomial is $(a_{11}-\lambda)(a_{22}-\lambda)\ldots(a_{nn}-\lambda)$ whose roots are a_{ii}, $1\leqslant i\leqslant n$, proving the desired result.

4. Let $\mathbf{A}$ be a real orthogonal matrix and let

$$\mathbf{Ax}=\lambda\mathbf{x} \tag{A.11}$$

($\mathbf{x}$ and λ may be complex)$\Rightarrow\mathbf{x}^\dagger\,\widetilde{\mathbf{A}}=\lambda^*\mathbf{x}^\dagger\Rightarrow\mathbf{x}^\dagger\,\widetilde{\mathbf{A}}\mathbf{Ax}=|\lambda|^2\mathbf{x}^\dagger\,\mathbf{x}\Rightarrow\mathbf{x}^\dagger\,\mathbf{x}(1-|\lambda|^2)=0$ $\Rightarrow|\lambda|=1$. All the eigenvalues of an orthogonal matrix thus have unit magnitude. The eigenvalues of a complex orthogonal matrix have no constraint. Note that if we take the transpose of Eq. (A.11), we can obtain $\widetilde{\mathbf{x}}\widetilde{\mathbf{A}}\mathbf{A}\,\mathbf{x}=\lambda^2\widetilde{\mathbf{x}}\,\mathbf{x}\Rightarrow$ $\widetilde{\mathbf{x}}\,\mathbf{x}(1-\lambda^2)=0$. From this we cannot conclude that $\lambda=\pm1$ because $\widetilde{\mathbf{x}}\,\mathbf{x}$ may be zero even if $\mathbf{x}\neq0$; see Exercise 15(a) below.

7. The characteristic polynomial of a real matrix has real coefficients. Now it is well-known that the roots of a polynomial with real coefficients either must be all real or must appear in pairs of complex conjugates, from which the first part of the Exercise follows. The second part is now easy.

8. The characteristic polynomial of $\mathbf{B}$ is $\det(\mathbf{B}-\lambda\mathbf{I})=\det(\mathbf{P}^{-1}\mathbf{AP}-\lambda\mathbf{I})$ $=\det[\mathbf{P}^{-1}(\mathbf{A}-\lambda\mathbf{I})\mathbf{P}]=\det(\mathbf{A}-\lambda\mathbf{I})=$ the characteristic polynomial of $\mathbf{A}$.

13. Eigenvalues of $\mathbf{A}=1,\ 3\pm2\sqrt{2}$; $\mathbf{A}^{-1}=\begin{bmatrix} 3 & -2 & -12 \\ -4 & 3 & 17 \\ 0 & 0 & 1 \end{bmatrix}$

15. (a) $\lambda_1=e^{i\theta}$, $\mathbf{x}_1=\{1 \quad i\}$; $\lambda_2=e^{-i\theta}$, $\mathbf{x}_2=\{1 \quad -i\}$.

16. It is given that $\mathbf{x}^\dagger\mathbf{y}=0$. Let $\mathbf{H}\mathbf{x}=\lambda\mathbf{x}\Rightarrow\mathbf{x}^\dagger\mathbf{H}=\lambda^*\mathbf{x}^\dagger\Rightarrow\mathbf{x}^\dagger(\mathbf{H}\mathbf{y})=\lambda^*\mathbf{x}^\dagger\mathbf{y}=0$; which shows that $\mathbf{x}$ is orthogonal to $\mathbf{H}\mathbf{y}$.

20. General solution: $\mathbf{x}=a\{1 \quad -1\}e^{t}+b\{1 \quad 1\}e^{5t}$;
Solution with given boundary condition: $\mathbf{x}=\{1 \quad 1\}e^{5t}$.

23. Let A denote the given matrix. Tr $\mathbf{A}=nq=\Sigma\lambda_i$. We notice that if q is replaced by p, the rank of the resulting matrix is 1, that is, rank of $[\mathbf{A}-(q-p)\mathbf{I}]=1$, so that $q-p$ is an eigenvalue of A with multiplicity $n-1$. The remaining eigenvalue is $nq-(n-1)(q-p)=q+(n-1)p$. Eigenvector corresponding to $\lambda=q+(n-1)p$: $\mathbf{x}=\{1 \quad 1\ldots1\}$; $n-1$ eigenvectors corresponding to $\lambda=q-p$ are of the form $\mathbf{x}=\{x_1 \quad x_2\ldots x_n\}$ with $\Sigma x_i=0$.

Exercise 10

1. (b) Characteristic polynomial$=-\lambda^3+6\lambda^2-12\lambda+8=(2-\lambda)^3$; Minimal polynomial$=(2-\lambda)^3$; Since the minimal polynomial has a multiple root, the matrix is not diagonalizable; The only eigenvector is $\{1 \quad 0 \quad -2\}$; The matrix is similar to the triangular matrix $\mathbf{T}$ through the similarity transformation matrix $\mathbf{P}$, where

$$\mathbf{T}=\begin{bmatrix} 2 & 5 & 0.6 \\ 0 & 2 & 0.4 \\ 0 & 0 & 2 \end{bmatrix}, \quad \mathbf{P}=\begin{bmatrix} 1 & -4 & 0 \\ 0 & 5 & 1 \\ -2 & -2 & 0 \end{bmatrix}.$$

4. The eigenvalues of the matrix are found to be $\lambda=\frac{1}{2}(a+d)\pm\frac{1}{2}[(a-d)^2+4bc]^{\frac{1}{2}}$. The matrix is nondiagonalizable only if the eigenvalues are equal, that is, only if $(a-d)^2+4bc=0$. Now, if this condition is satisfied, the only distinct eigenvalue of the matrix is $\frac{1}{2}(a+d)$. Denoting the given matrix by A, we find that $[\mathbf{A}-\frac{1}{2}(a+d)\mathbf{I}]\neq\mathbf{0}$, but $[\mathbf{A}-\frac{1}{2}(a+d)\mathbf{I}]^2=\mathbf{0}$, so that the minimal polynomial has a double root and the matrix is nondiagonalizable.

5. $\mathbf{A}^3=3(p^2+pq+q^2)\mathbf{A}+(2p^3+3p^2q-3pq^2-2q^3)\mathbf{I}$;
$\mathbf{A}^{-1}=[\mathbf{A}-(2q+p)\mathbf{I}]/[(p-q)(2p+q)]$, $q\neq p$, $q\neq-2p$. Compare this with Exercise 5.8.

7. (a) Parts (i), (ii) and (iii) are proved by direct substitution of $\mathbf{P}_1$, $\mathbf{P}_2$ and $\mathbf{P}_3$. For part (iv), we note that $\mathbf{P}_i^2$ will be a polynomial in A of degree 4. This is reduced to a polynomial of degree 2 by using the result of the Cayley-Hamilton theorem:

$$(\mathbf{A}-\lambda_1\mathbf{I})(\mathbf{A}-\lambda_2\mathbf{I})(\mathbf{A}-\lambda_3\mathbf{I})=\mathbf{0}$$
$$\Rightarrow\mathbf{A}^3-(\lambda_1+\lambda_2+\lambda_3)\mathbf{A}^2+(\lambda_2\lambda_3+\lambda_3\lambda_1+\lambda_1\lambda_2)\mathbf{A}-\lambda_1\lambda_2\lambda_3\mathbf{I}=\mathbf{0}. \quad (A.12)$$

(b) $\mathbf{A}^3=\lambda_1^3\mathbf{P}_1+\lambda_2^3\mathbf{P}_2+\lambda_3^3\mathbf{P}_3$.

Exercise 11

2. The matrix A has elements $(\mathbf{A})_{ij}=a_ia_j$. If we take a_1 common from the first row, a_2 from the second row, etc., then

$$\det \mathbf{A}=a_1a_2\ldots a_n \begin{vmatrix} a_1 & a_2\ldots a_n \\ a_1 & a_2\ldots a_n \\ \vdots & \\ a_1 & a_2\ldots a_n \end{vmatrix},$$

from which it is evident that rank of **A**=1.

6. $\mathbf{x}^\dagger \mathbf{A}\mathbf{x}$ can be expressed as $\mathbf{y}^\dagger \mathbf{A}\mathbf{y}$ with $\mathbf{y}=\mathbf{P}^{-1}\mathbf{x}$. Now,

$$\mathbf{y}^\dagger \mathbf{A}\mathbf{y}=\sum_i \lambda_i\,|y_i|^2, \tag{A.13}$$

where λ_i are the eigenvalues of **A**. This is evidently positive if all λ_i are positive. On the other hand, it may be positive even if some of the λ_i are negative, so that the converse is not true.

9. The potential is $V(\mathbf{r})=\tilde{\mathbf{r}}\mathbf{A}\mathbf{r}$, where $\mathbf{r}=\{x \;\; y \;\; z\}$ and

$$\mathbf{A}=\begin{bmatrix} 2 & -1 & -1 \\ -1 & 2 & -1 \\ -1 & -1 & 2 \end{bmatrix}.$$

This can be expressed as $V=\tilde{\mathbf{r}}'\mathbf{A}\mathbf{r}'$, where **A** has the diagonal elements 3, 3, 0, and $\mathbf{r}'=\{x' \;\; y' \;\; z'\}=\mathbf{P}^{-1}\mathbf{r}$ with

$$\mathbf{P}=\begin{bmatrix} 1/\sqrt{3} & 1/\sqrt{2} & 1/\sqrt{6} \\ 1/\sqrt{3} & 0 & -2/\sqrt{6} \\ 1/\sqrt{3} & -1/\sqrt{2} & 1/\sqrt{6} \end{bmatrix}.$$

In terms of new coordinates (which are also orthogonal and cartesian), $V=3(x'^2+y'^2)$. It is evident that the potential is constant (and minimum) along the z' axis and V increases as the distance from the z' axis (that is, $(x'^2+y'^2)^{\frac{1}{2}}$) increases. We have $z'=(x+y+z)/\sqrt{3}$.

Exercise 12

2. If **A** has eigenvalues λ_i, $e^{\mathbf{A}}$ has eigenvalues $\exp \lambda_i$, which cannot be zero for any λ_i. Hence, from the result of Exercise 9.2, we see that $e^{\mathbf{A}}$ is nonsingular. Further, since **A** and $-\mathbf{A}$ commute with each other, $e^{\mathbf{A}}\,e^{-\mathbf{A}} = e^{0} = \mathbf{I}$.

5. The eigenvalues of **A** are $\frac{1}{2}$, $\frac{1}{3}$, $\frac{1}{4}$, and the diagonalizing matrix is

$$\mathbf{P}=\begin{bmatrix} 1 & 0 & -1 \\ 0 & 1 & 0 \\ 1 & -1 & 1 \end{bmatrix}.$$

From the identity $[(\mathbf{A}+\mathbf{I})-(\lambda+1)\mathbf{I}]=\mathbf{A}-\lambda\mathbf{I}$, it follows that the eigenvalues of $\mathbf{I}+\mathbf{A}$ are 3/2, 4/3, 5/4, and that **P** also diagonalizes $\mathbf{I}+\mathbf{A}$. Hence

$$\ln(\mathbf{I}+\mathbf{A})=\mathbf{P}\begin{bmatrix} \ln\frac{3}{2} & 0 & 0 \\ 0 & \ln\frac{4}{3} & 0 \\ 0 & 0 & \ln\frac{5}{4} \end{bmatrix}\mathbf{P}^{-1}=\tfrac{1}{2}\begin{bmatrix} \ln\frac{15}{8} & \ln\frac{6}{5} & \ln\frac{6}{5} \\ 0 & \ln\frac{16}{9} & 0 \\ \ln\frac{6}{5} & \ln\frac{135}{128} & \ln\frac{15}{8} \end{bmatrix}$$

$$=\begin{bmatrix} 0.3143 & 0.0912 & 0.0912 \\ 0 & 0.2877 & 0 \\ 0.0912 & 0.0267 & 0.3143 \end{bmatrix} \quad \text{(A.14)}$$

Further,

$$\mathbf{A}^2/2=(1/576)\begin{bmatrix} 45 & 27 & 27 \\ 0 & 32 & 0 \\ 27 & 13 & 45 \end{bmatrix}=\begin{bmatrix} 0.0781 & 0.0469 & 0.0469 \\ 0 & 0.0556 & 0 \\ 0.0469 & 0.0226 & 0.0781 \end{bmatrix},$$

$$\mathbf{A}^3/3=\begin{bmatrix} 0.0234 & 0.0182 & 0.0182 \\ 0 & 0.0123 & 0 \\ 0.0182 & 0.0111 & 0.0234 \end{bmatrix},$$

$$\mathbf{A}^4/4=\begin{bmatrix} 0.0083 & 0.0073 & 0.0073 \\ 0 & 0.0031 & 0 \\ 0.0073 & 0.0052 & 0.0083 \end{bmatrix},$$

$$\mathbf{A}-\frac{\mathbf{A}^2}{2}+\frac{\mathbf{A}^3}{3}-\frac{\mathbf{A}^4}{4}=\begin{bmatrix} 0.3120 & 0.0890 & 0.0890 \\ 0 & 0.2869 & 0 \\ 0.0890 & 0.0304 & 0.3120 \end{bmatrix}, \quad \text{(A.15)}$$

which compares well with the exact matrix $\ln(\mathbf{I}+\mathbf{A})$ of Eq. (A.14).

8. $\mathbf{A}^k=\begin{bmatrix} 1 & k & (k^2+k)/2 \\ 0 & 1 & k \\ 0 & 0 & 1 \end{bmatrix}$, $e^{\mathbf{A}}=\begin{bmatrix} e & e & 3e/2 \\ 0 & e & e \\ 0 & 0 & e \end{bmatrix}$,

$$\ln \mathbf{A}=\begin{bmatrix} 0 & 1 & \frac{1}{2} \\ 0 & 0 & 1 \\ 0 & 0 & 0 \end{bmatrix} \quad \text{(A.16)}$$

Let $\mathbf{S}=\sum_{k=0}^{\infty} \mathbf{A}^k/k!$. From the above result, we have

$$(\mathbf{S})_{11}=\sum_{k=0}^{\infty} [(\mathbf{A})^k]_{11}/k!=\sum_{k=0}^{\infty} 1/k!=e;$$

$$(\mathbf{S})_{12}=\sum_{k=0}^{\infty} [(\mathbf{A})^k]_{12}/k!=\sum_{k=0}^{\infty} k/k!=\sum_{k=1}^{\infty} 1/(k-1)!=\sum_{n=0}^{\infty} 1/n!=e;$$

$$(\mathbf{S})_{13}=\sum_{k=0}^{\infty} (k^2+k)/2k!=\tfrac{1}{2}\sum_{k=1}^{\infty} (k+1)/(k-1)!$$

$$=\tfrac{1}{2}\sum_{n=0}^{\infty} (n+2)/n!=3e/2. \quad \text{(A.17)}$$

It is then evident that this result agrees with $e^{\mathbf{A}}$ given above.

10. Expand $e^{\lambda\mathbf{A}}$ and $e^{-\lambda\mathbf{A}}$ in powers of λ and pick up the coefficients of the successive powers of λ.

Exercise 13

1. (b) $\begin{bmatrix} a^2 & ab & 0 & 0 \\ bc & c^2 & 0 & 0 \\ 0 & 0 & c-a & b^2 \\ 0 & 0 & b^2 & a-c \end{bmatrix}$.

4. (c) $\begin{bmatrix} a^2b^2 & a^2bd & ab^3 & ab^2d \\ a^2cd & a^2c^2 & abcd & abc^2 \\ b^3c & b^2cd & bc^2d & bc^3 \\ b^2c^2 & bc^2d & c^3d & c^4 \end{bmatrix}$.

Exercise 14

3. Transformation matrix: $\mathbf{P}=\begin{bmatrix} 1 & -1 \\ 1 & 1 \end{bmatrix}$;

$$\mathbf{P}^{-1}\,\sigma_x\mathbf{P}=\begin{bmatrix} 1 & 0 \\ 0 & -1 \end{bmatrix};\ \mathbf{P}^{-1}\,\sigma_y\mathbf{P}=\begin{bmatrix} 0 & -i \\ i & 0 \end{bmatrix};\ \mathbf{P}^{-1}\,\sigma_z\mathbf{P}=\begin{bmatrix} 0 & -1 \\ -1 & 0 \end{bmatrix}.$$

Exercise 15

4. $\delta^i{}_i=N$; $\delta^i{}_j\delta^j{}_k\delta^k{}_i=\delta^i{}_i=N$.
5. a_ρ and a_ϕ as in Eqs. (51); $a_z=\ddot{z}$.
6. Polar coordinates: $\partial f/\partial\rho$, $\partial f/\partial\phi$; Spherical polar coordinates: $\partial f/\partial r$, $\partial f/\partial\theta$, $\partial f/\partial\phi$, where f is the scalar field.
7. See the beginning of Sec. 20; Eq. (20.6).
9. $v^\rho=\rho^2(\cos^3\phi+\sin^3\phi)$; $v^\phi=\rho\sin\phi\cos\phi\,(\sin\phi-\cos\phi)$.

Exercise 16

2. The given tensor has four free indices (p, t, j, s) and three dummy indices (i, k, r). Hence the symbol $A^{ip}_j\,B^k_{ir}\,C^{rt}_{sk}$ represents N^4 different expressions. Each expression contains N^3 terms.
6. The proof is similar to that in the case of matrices: see Example 3.3.

Exercise 17

2. It can be shown that $B^{ij}E_{ik}=\delta^j{}_k$. Multiply Eq. (12) by E_{ks}, giving
$$A_{ij}\,B^{ik}E_{ks}=\delta^k{}_jE_{ks}\Rightarrow A_{ij}\,\delta^i{}_s=\delta^k{}_jE_{ks}\Rightarrow A_{sj}=E_{js}=E_{sj},$$
since E_{sj} is symmetric.
4. (a) $A^{ik}{}_j\,B^j{}_k+C^{ri}\,D_r=E^{st}{}_r\,F^{ri}{}_{st}$.

Exercise 18

3. $\partial f/\partial r$, $(1/r^2)\partial f/\partial\theta$, $(1/r^2 \sin^2\theta)\, \partial f/\partial\phi$.

4. We have $ds^2=(dz^1)^2+(dz^2)^2+2dz^1\, dz^2 \cos\alpha$, giving

$$[g_{ij}]=\begin{bmatrix} 1 & \cos\alpha \\ \cos\alpha & 1 \end{bmatrix},\ [g^{ij}]=\operatorname{cosec}^2\alpha \begin{bmatrix} 1 & -\cos\alpha \\ -\cos\alpha & 1 \end{bmatrix}$$

8. In fact, we have $\delta_{jk}=\delta^{i}_{.k}\, g_{ij}=g_{kj}=g_{jk}$.

Exercise 20

1. (c) $$A^{ijk}_{...l;\,q}=A^{ijk}_{...l,\,q}+A^{pjk}_{...l}\Gamma^{i}_{pq}+A^{ipk}_{...l}\Gamma^{j}_{pq}$$
$$+A^{ijp}_{...l}\Gamma^{k}_{pq}-A^{ijk}_{...p}\Gamma^{p}_{lq}.$$

3. We have

$$A_{i;j}=A_{i,j}-A_p\Gamma^p_{ij},\quad A_{j;i}=A_{j,i}-A_p\Gamma^p_{ij}$$

$\Rightarrow A_{i;j}-A_{j;i}=A_{i,j}-A_{j,i}$. Since the left hand side is a tensor, the right hand side also is.

4. We have

$$A_{ij;k}=A_{ij,k}-A_{pj}\Gamma^p_{ik}-A_{ip}\Gamma^p_{jk};$$

on cyclically permuting i, j, k twice and adding the three equations (noting that $A_{ij}=-A_{ji}$), we get

$$A_{ij;\,k}+A_{jk;\,i}+A_{ki;\,j}=A_{ij,\,k}+A_{jk,\,i}+A_{ki,\,j}.$$

Since the left hand side is a tensor, the right hand side also is. Moreover, denoting $B_{ijk}\equiv A_{ij,\,k}+A_{jk,\,i}+A_{ki,\,j}$, it can be easily seen that $B_{jik}=A_{ji,\,k}+A_{ik,\,j}+A_{kj,\,i}=-B_{ijk}$, etc.; hence the result.

10. The Christoffel symbols of the first kind are:

$$[11, 1]=\tfrac{1}{2}\,\partial f/\partial u,\ [22, 2]=\tfrac{1}{2}\,\partial h/\partial v,$$
$$[12, 1]=[21, 1]=-[11, 2]=\tfrac{1}{2}\,\partial f/dv,$$
$$[12, 2]=[21, 2]=-[22, 1]=\tfrac{1}{2}\,\partial h/\partial u.$$

11. Christoffel symbols of the first kind:

$$[11, 1]=\tfrac{1}{2}\,\partial R/\partial u,\ [22, 2]=\tfrac{1}{2}\,\partial T/\partial v,$$
$$[12, 1]=[21, 1]=\tfrac{1}{2}\,\partial R/\partial v,\ [12, 2]=[21, 2]=\tfrac{1}{2}\,\partial T/\partial u,$$
$$[11, 2]=\partial S/\partial u-\tfrac{1}{2}\partial R/\partial v,\quad [22, 2]=\partial S/\partial v-\tfrac{1}{2}\partial T/\partial u.$$

Christoffel symbols of the second kind (with $D=RT-S^2$):
$\Gamma^1_{11}=(T\partial R/\partial u-2S\partial S/\partial u+S\partial R/\partial v)/2D$, $\Gamma^2_{22}=(R\partial T/\partial v-2S\partial S/\partial v+S\partial T/\partial u)/2D$,
$\Gamma^1_{12}=\Gamma^1_{21}=(T\partial R/\partial v-S\partial T/\partial u)/2D$, $\Gamma^2_{12}=\Gamma^2_{21}=(R\partial T/\partial u-S\partial R/\partial v)/2D$,
$\Gamma^1_{22}=(2T\partial S/\partial v-T\partial T/\partial u-S\partial T/\partial v)/2D$, $\Gamma^2_{11}=(2R\partial S/\partial u-R\partial R/\partial v-S\partial R/\partial u)/2D$.

13. Use Eq. (58) and the fact that covariant differentiation satisfies the distributive law.

Exercise 21

4. $\dfrac{d^2u}{ds^2}+\dfrac{1}{f}\dfrac{\partial f}{\partial u}\left(\dfrac{du}{ds}\right)^2=0,\ \dfrac{d^2v}{ds^2}+\dfrac{1}{f}\dfrac{\partial f}{\partial v}\left(\dfrac{dv}{ds}\right)^2=0.$

Exercise 22

3. $$R_{1212}=-\frac{1}{2}\frac{\partial f}{\partial v}-\frac{1}{2}\frac{\partial u}{\partial h}$$
$$+\frac{1}{4f}\left[\left(\frac{\partial f}{\partial v}\right)^2+\frac{\partial f}{\partial u}\frac{\partial h}{\partial u}\right]+\frac{1}{4h}\left[\left(\frac{\partial h}{\partial u}\right)^2+\frac{\partial f}{\partial v}\frac{\partial h}{\partial v}\right].$$

5. $R_{1212}=\dfrac{\partial^2 S}{\partial u\partial v}-\dfrac{1}{S}\dfrac{\partial S}{\partial u}\dfrac{\partial S}{\partial v},\ R^{1}_{1\cdot 12}=R_{1212}/S.$

Index